Dependable Computing and Fault-Tolerant Systems

Edited by
A. Avižienis, H. Kopetz, J. C. Laprie

Volume 6

Springer-Verlag Wien New York

J. F. Meyer, R. D. Schlichting (eds.)

Dependable Computing for Critical Applications 2

Springer-Verlag Wien New York

Prof. John F. Meyer, The University of Michigan, Ann Arbor, MI 48104, USA
Prof. Richard D. Schlichting, The University of Arizona, Tucson, AZ 85721, USA

With 114 Figures

ISSN 0932-5581
ISBN-13:978-3-7091-9200-9   e-ISBN-13:978-3-7091-9198-9
DOI: 10.1007/978-3-7091-9198-9

FOREWORD

This volume contains the papers presented at the Second International Working Conference on *Dependable Computing for Critical Applications*, sponsored by IFIP Working Group 10.4 and held in Tucson, Arizona on February 18-20, 1991. In keeping with the first such conference on this topic, which took place at the University of California, Santa Barbara in 1989, this meeting was likewise concerned with an important basic question: Can we rely on Computers? In more precise terms, it addressed various aspects of computer system *dependability*, a broad concept defined as the trustworthiness of computer service such that reliance can justifiably be placed on this service. Given that this term includes attributes such as reliability, availability, safety, and security, it is our hope that these papers will contribute to further integration of these ideas in the context of critical applications.

The program consisted of 20 papers and three panel sessions. The papers were selected from a total of 61 submissions at a November 1990 meeting of the Program Committee in Ann Arbor, Michigan. We were very fortunate to have a broad spectrum of interests represented, with papers in the final program coming from seven different countries, representing work at universities, corporations, and government agencies. The process was greatly facilitated by the diligent work of the Program Committee and the quality of reviews provided by outside referees.

In addition to the paper presentations, there were three panel sessions organized to examine particular topics in detail. The first, entitled "Security and Dependability: Two Sides of a Coin or Different Currencies?," addressed the relationship between these two basic attributes. The second, entitled "Formal Verification for Security and Dependability: Differences and Similarities," contrasted the role of formal verification in these two areas, as well as discussing the techniques that are used in each. The final panel session was entitled "Static vs. Dynamic Mechanisms in Critical System Applications;" this panel debated the tradeoffs and limitations of static and dynamic mechanisms in critical applications.

Finally, we wish to acknowledge the sponsoring and cooperating organizations for their collective effort in guaranteeing the success of this forum. Help from many individuals is likewise gratefully acknowledged, including the generous assistance provided by WG 10.4 Chair Jean-Claude Laprie from the

LAAS-CNRS, Local Arrangements Chair Bill Sanders from the University of Arizona, and, for most of the clerical work associated with determining the technical program, Nancy Leach from the University of Michigan. We would also like to thank Vicraj Thomas for his excellent work in reformatting the papers for this volume.

Richard D. Schlichting	John F. Meyer
General Chair	Program Chair
Department of Computer Science	Department of EECS
The University of Arizona	The University of Michigan

Sponsors

IFIP Working Group 10.4 on *Dependable Computing and Fault Tolerance*

In cooperation with:

IFIP Technical Committee 11 on *Security and Protection in Information Processing Systems*

IEEE Computer Society Technical Committee on *Fault-Tolerant Computing*

EWICS Technical Committee 7 on *Systems Reliability, Safety and Security*

The University of Arizona

Conference Organization

General Chair:

R. D. Schlichting
The University of Arizona
Tucson, Arizona, USA

Vice-General Chair:

J. J. Quisquater
Philips Research
Brussels, Belgium

Ex Officio:

J.-C. Laprie
LAAS-CNRS
Toulouse, France

Program Chair:

J. F. Meyer
University of Michigan
Ann Arbor, Michigan, USA

Local Arrangements Chair:

W. H. Sanders
The University of Arizona
Tucson, Arizona, USA

Session Chairs

J. Abraham
University of Texas at Austin
Austin, Texas, USA

A. Costes
LAAS-CNRS
Tolouse, France

M.-C. Gaudel
Universite de Paris-Sud LRI
Paris, France

D. Gollmann
Royal Holloway Bedford New College
Surrey, England

H. Kopetz
Institut fur Technische Informatik
Vienna, Austria

J.-C. Laprie
LAAS-CNRS
Tolouse, France

S. Natkin
CEDRIC
Paris, France

U. Voges
Institut fur Datenverarbeitung
Karlsruhe, Germany

Program Committee

J. Abraham
University of Texas at Austin
Austin, Texas, USA

A. Costes
LAAS-CNRS
Toulouse, France

M.-C. Gaudel
Universite de Paris-Sud LRI
Paris, France

V. Gligor
University of Maryland
College Park, Maryland, USA

J. Goldberg
SRI International
Menlo Park, California, USA

D. Gollmann
Royal Holloway Bedford New College
Surrey, England

G. Hagelin
Erricsson Signal Systems AB
Stockholm, Sweden

H. Ihara
Hitachi, Ltd.
Yokohama, Japan

H. Kopetz
Institut fur Technische Informatik
Vienna, Austria

J. Lala
C. S. Draper Laboratory
Cambridge, Massachusetts, USA

C. Landwehr
Naval Research Laboratory
Washington, DC, USA

G. Le Lann
INRIA
Paris, France

J. McDermid
University of York
York, UK

M. Morganti
ITALTEL - Central Research Labs
Italy

J.-M. Rata
EDF-DER Service IMA
Clamart Cedix, France

D. Rennels
University of California
Los Angeles, California, USA

J. Rushby
SRI International
Menlo Park, California, USA

E. Schmitter
Siemens AG
Munich, Germany

S. Shrivastava
University of Newcastle upon Tyne
Newcastle upon Tyne, UK

D. Siewiorek
Carnegie-Mellon University
Pittsburgh, Pennsylvania, USA

L. Simoncini
IEI del CNR
Pisa, Italy

R. Turn
California State University
Northridge, California, USA

U. Voges
Institut fur Datenverarbeitung
Karlsruhe, Germany

Referees

J. Abraham

M. Ahamad

J. Arlat

K. Birman

E. Brickell

W. Carter

G. Chiola

E. Chow

M. Dal Cin

A. Costes

B. Courtois

F. Cristian

M. Diaz

B. Di Vito

J. Dugan

K. Echtle

W. Ehrenberger

M. Ercegovac

P. Ezhilchelvan

G. Florin

T. Frazier

J. Gannon

M.-C. Gaudel

R. Genser

G. Gilley

V. Gligor

J. Goldberg

D. Gollmann

M. Gordon

B. Grey

K. Grosspietsch

G. Hagelin

M. Harrison

H. Ihara

R. Jagganathan

J. Joyce

H. Kantz

R. Kemmerer

R. Kieckhafer

K. Kim

J. Knight

H. Kopetz

J. Lala

M. Lam

C. Landwehr

J.-C. Laprie

G. Le Lann

C. Liceaga

B. Littlewood

M. Malhotra

R. Manning

M. McDermid

P. Melliar-Smith

J. Meyer

C. Mitchell

M. Morganti

E. Nassor

J. Nehmer

V. Nicola

M. Nicolaidis

R. Oberparleitner

E. Pilaud

D. Powell

M. Pozzo

D. Pradhan

J. Quisquater

K. Ramamritham

J.-M. Rata

M. Raynal

D. Rennels

H. Riess

J. Rohr

J. Rushby

F. Saglietti

R. Schlichting

E. Schmitter

L. Sha

C. Shamris

K. Shin

S. Shrivastava

M. Sievers

D. Siewiorek

M. Sifakis

L. Simoncini

N. Speirs

M. Srivas

J. Stankovic

B. Sterner

Y. Tamir

P. Thevenod-Fosse

W. Tichy

K. Trivedi

R. Turn

U. Voges

Contents

Distributed Systems I

Distributed Systems I

Architectural Foundations, Concepts and Methods Behind ISACS – A Real-Time Intelligent System for Critical Applications

SANDRO BOLOGNA,* ØIVIND BERG, KJELL HAUGSET, JON KVALEM

Institute for Energy Technology

OECD Halden Reactor Project

N-1750 Halden – Norway

Abstract

In this paper we report about application requirements, design principles and methodologies used in the development of an integrated intelligent support system, intended to be used by the operator of a Nuclear Power Plant. The system, for its own nature, is an example of a class of applications normally referred to as critical applications. Dependability is a key requirement for the system.

Key Words: Operator support system, intelligent system, man-machine interface, incremental prototyping.

1 Introduction

ISACS (Integrated Surveillance And Control System) is a multicomputer system comprising several cooperating computer systems that together act as an advisor for the operator of a Nuclear Power Plant (NPP), by providing him/her with intelligent support during fault investigation, emergency containment, as well as during normal operation. In contrast to single operator support systems, for instance diagnosis, alarm filtering, procedure following, etc., ISACS considers the interaction between man and machine as a whole and integrates several specialized support systems by means of a knowledge-based, graphical

---

*On leave from ENEA, Italy

dialogue. ISACS tasks require both knowledge-based reasoning and interaction with dynamic entities in the environment, such as human beings, physical process, and other computer systems. It integrates information from the NPP internal sensors, produces plans in response to that information, presents the operator with high-level decision-oriented information, and interacts with the operator in the successful execution of those plans. To perform such tasks, the system must possess capabilities for: Perception – acquiring and interpreting sensed data to obtain knowledge of external entities; cognition – knowledge-based reasoning to assess situations, solve problems, and determine actions; and action – actuating effects to execute intended actions and influence dynamics, interacting with them imposes some real-time constraints, whose violations may preclude a successful result or may merely degrade the usefulness of the result.

2   Application Requirements

ISACS is an example of a broad class of applications characterized by several key requirements [1, 2, 3, 4]. These requirements cannot be addressed after building the system, but must be considered at all stages of the development.

Real-time.   ISACS has event-response requirements on the order of seconds, with a response generated by a sequence of tasks each running on the order of fractions of a second. A system like ISACS can afford to delay an operator-interaction by a few seconds without significantly degrading system performance. However, this does not give the application permission to spend an arbitrary amount of time computing a response.

Multiplicity of conditions.   It is not feasible to enumerate all the possible conditions that the system will encounter. For such a reason it is necessary to process information on an event basis. An event is defined as either a planned or unexpected plant transient. Planned transients are initiated by the operator and could be for example activation of a power reduction procedure, while unexpected transients are caused by failures in the process or control systems. The importance of events vary as a function of when they occur. This requires the possibility to consider the absolute or relative time of occurrence of events.

Prioritisation of events. Given that there may be several things the system can do at any one time, it is necessary to focus attention on which is to be done, and for how long. This problem is difficult because the answer depends upon what is happening at the time. For a given set of events to be considered at any one time, a priority must be established.

System dependability. We are concerned with an application where system failures will have strong safety implications and in which the human operator requires the system to function in a dependable way. That is, we seek predictability of the system response to event occurrences. An operator will use an automated aid only if he/she can acquire an accurate model of how it will perform and if, in a day-to-day use, the aid does not violate his/her expectations. On the other hand the operator may base his/her decisions on what to do, exclusively on the information provided by the system. This mission-critical nature stresses requirements such as rigorous engineering design, verification and validation, robustness and fault tolerance.

Multiple interacting functions. The system must perform many distinct functions simultaneously and must receive results from each function in accordance with a specific time limit requirement. Each of these single functions, as well as their interaction, must be clearly specified and documented.

Limited resources. The resources may be inadequate to perform all the functions the application would require to perform at every point in time, above all from the side of data transfer and data retrieval. Thus, significant compromises and prioritisations have to be done at the system design level in order to make the system work properly.

Incremental prototyping. This approach is necessitated by several common properties of the system:

- The functional requirements of the system change, either by request from the plant operators or because the developer's understanding of the problem changes.

- The ways the system will interact with other systems in its environment are difficult to specify completely in advance.

- The developer does not have a proven application architecture, previous to start the development phase, that assures the key performance objectives for the application.

- The inability to produce a system model during the requirement analysis, complete and accurate enough to be fully tested before finalizing the design.

Man-Machine Interface (MMI). An important feature of ISACS, as of many computing systems used for industrial process control, is that it both represents and participates in the external world. By participation we mean that it influences human decision making through the manipulation of symbols in a way that is controlled by humans. This may seem a very abstract definition, but it turns out that it leads to some interesting yet simple classification principles. We can distinguish between faults in the representation of the world, faults in manipulation of the representation, and faults in the interpretation back in the world. In the ISACS system, the third is one of major significance. This implies a need to identify in advance certain principles to be followed throughout the design of human-machine interaction.

Use of AI Techniques. Artificial Intelligence techniques are necessary to implement several of the ISACS's functions. Unfortunately, introducing AI techniques within the application aggravates an already difficult problem because of the lack of experience building dependable intelligent systems to work with real-time constraints [5].

3 System Goals and Functionalities

An overview of ISACS and its motivations is presented in this chapter. More details are found in [6, 7, 8].

3.1 Motivations

Even if todays nuclear power plants have a very good safety record, there is a continuous search for still improving safety. One direction of this effort address operational safety, trying to improve the handling of disturbances and accidents partly by further automation, partly by creating a better control room environment, providing the operator with intelligent support systems to help him/her in his/her decision making process.

Introduction of computerised operator support systems has proved to be an efficient way of improving the operators performance. A number of systems have been developed worldwide, assisting in tasks like process fault detection and diagnosis, selection and implementation of proper remedial actions. But, as the amount of computerised support increases, there is a need for coordinating the various support systems and presenting them in an integrated manner through a new man-machine interface. This is the basic motivation for the development of ISACS.

3.2　Problems in today's control rooms

Control rooms with conventional instrumentations or only limited use of computer-based support systems may face a number of weaknesses. Limited instrumentation, combined with the absence of signal processing, limits the amount and quality of process information made available to the operator. In other cases, extensive alarm systems without filtering or prioritisation of alarms may create an overflow of information, especially in disturbance situations. The standard "one sensor one instrument" technique, leaves it to the operator to integrate separate pieces of information. Also, information may be wrong or inconsistent, confusing the operator and possibly misleading him to perform wrong actions.

In the conventional control room, all information is presented in parallel, making it difficult to keep an overview and find the relevant information. Often, information is lacking when the operator wants to find the cause of a problem, or for planning counteractions. Also, mistakes are made in implementation of control actions, both non-procedural and procedural tasks where procedure steps may be omitted or wrong steps implemented.

By use of new information presentation techniques and introduction of Computerised Operator Support Systems (COSSs), between the process and the operator, most of the problems of conventional control rooms discussed above may be solved. Model-based techniques may identify disturbances at an early stage or predict plant behaviour. Knowledge- based systems may diagnose problems, and by use of fully graphic colour CRTs, information may be presented in a more clear way. The success of such systems is, however, dependent on a careful design taking into account human factors aspects. Normally, COSSs have to be evaluated in realistic environments before they are actually taken into use.

3.3   Improved operational safety through integration

A single COSS that has been validated and found useful to assist the operator in solving a specific task, may be added in an existing control room or included in the design of a new control room, and expected to function efficiently. On the other hand, if a large number of specific COSSs are to be installed, a number of problems are faced:

- Each new COSS adds more information in the control room, so there is an increased danger of information overflow.

- If the MMIs of the COSSs are not standardised, the operator will have problems switching from one to the other.

- The operator may become so involved in the use of a COSS that he may overlook more important tasks to be performed.

- Implementing a large number of COSSs that are not coordinated with respect to process coupling and computer application is inefficient and expensive.

The points given above point toward the need for an integrated approach when taking advanced computer technology into extensive use. If a large number of COSSs are designed and implemented as single systems without coordination, the improvement in operational safety could easily become much smaller than the sum of the effect of the single COSSs, or even negative. On the other hand, if careful integration of COSSs into the advanced control room is made, the hypothesis is put forward that the overall operational safety may be better than the sum of the effect of single COSSs. One reason for this is that new, important information can be generated based on input from several COSSs, as when high confidence in a diagnosis is reached when two independent systems confirm eachother. To reach this synergetic effect, the following features, among others, should be included in the integration concept:

- The integration concept should cover the total MMI.

- Seen from the operator, the MMI should function as a unified interface where information presented will change with plant conditions.

- An intelligent coordinator should keep an overview of the process and information available from the COSSs, as a basis for analysing the plant conditions and development.

- The coordinator should identify information of high importance and present that to the operator while less relevant information should not be directly available.

- While the intelligent coordinator is in control of which information to present in some fixed part of the MMI, the operator should be able to access any information he wants through other parts of the MMI.

- Even if the coordinator presents information such as probable diagnosis or recommended actions, the operator remains responsible and controls the process.

When classifying COSSs according to their function, they are seen to support the operator in one of the following tasks: plant state identification, action planning or action implementation. These are the three steps the operator takes in handling of any plant situation. In ISACS, the division of the operator's tasks into these three categories is reflected both in the structure of the ISACS intelligent coordinator and the layout of the man-machine interface.

4 System Structure

The multicomputer system which forms the basis for the implementation of the ISACS system, is a multi-vendor network-based system integrating conventional and knowledge processing capabilities. The previously existing computer system in the Halden Man-Machine Laboratory, mainly consisting of mini computers and LISP-machines, was used as a basis. However, the new UNIX-based workstation technology is heavily used in the ISACS configuration. Emphasis is placed on the distribution of tasks upon the different computers in the configuration. Since the number of tasks in ISACS are manyfold, it is very important to find an optimal distribution philosophy.

4.1 Hardware Configuration

The hardware configuration of the first version of ISACS, ISACS-1, is depicted in Figure 1. The basis for the communication is an Ethernet Local Area

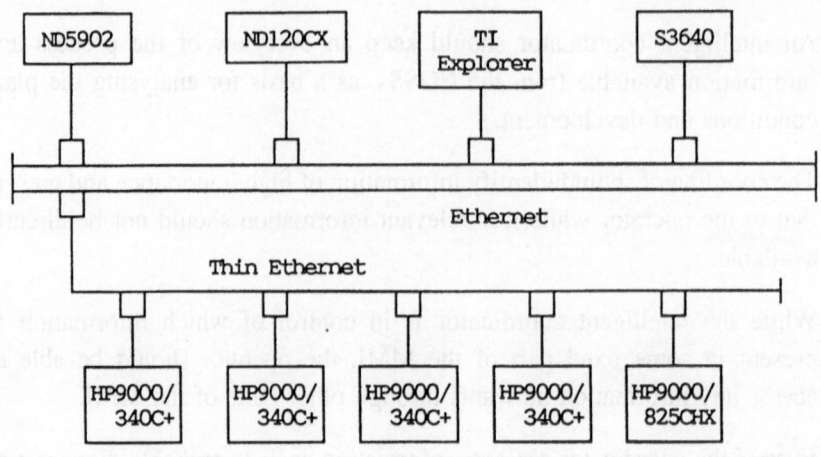

Figure 1: ISACS-1 Hardware Configuration

Network (LAN) and all computers are placed as nodes in this network.

As can be seen from the figure, there is a mixture of traditional mini computers from Norsk Data, LISP-machines from Texas Instruments and Symbolics and colour graphic workstations from Hewlett Packard. The mini computers are hosting applications requiring number crunching facilities, such as the process simulator models, the LISP-machines are hosting expert system applications, while the graphic workstations are hosting the MMI applications along with the ISACS coordinator and Data Base Management System.

4.2   Software Configuration

It is required that ISACS is developed in such a way that optimal flexibility and modularity are maintained. It is very important to keep the retrofitting possibilities alive all the way through development, which means that ISACS must be developed such that it is possible to

- remove parts of ISACS for integration into other environments than the existing in the Halden Man-Machine Laboratory.

- remove one or several COSSs from an ISACS environment and integrate with other systems.

- easily include new COSSs into ISACS.

- easily expand ISACS with new functions.

To meet the requirements specified above, it is essential that

- all communication between tasks in ISACS is message based.

- the interfaces between the different subsystems within ISACS are strictly defined.

The ISACS system, as a total, consists of several heterogeneous types of already developed systems, namely the COSSs, plus the intelligent coordinator and the MMI part. The different COSSs do logically belong to the ISACS system, but they do also in their own way act as complete stand- alone systems, and will live their own life independent of ISACS.

The COSSs are very different, not only regarding their functionality, but also with respect to software solution and implementation. Some of the COSSs are implemented using traditional programming techniques in the FORTRAN programming language, some are implemented using expert system techniques in LISP or PROLOG. This means that the different COSSs have different demands to computing power and environment, and that it is very important to distribute the different COSSs on suitable computers in the network.

4.3 Communication System

The communication facilities are a crucial part of ISACS. Since it was required that all communication should be message based, the flexibility and capability of the communication system was essential for developing ISACS-1. Since ISACS-1 itself and the COSSs within ISACS-1 reside physically on different computer nodes in a network, the use of standardised communication protocols was essential. Since the original network configuration was based on an Ethernet Local Area network, Ethernet and the TCP/IP protocol is also used for communication in ISACS-1. However, due to specific time constraints and the usage of old software packages, the internal communication on the ND computers was made using a proprietary communication protocol.

To ease the application programmer's work there was decided to make a simple and reliable Communication System SuperStructure, CSSS, to be implemented in the heterogeneous man-machine laboratory environment. The CSSS consists of an administration system, a supervisory system and a set of defined library functions.

The overall intention of the CSSS is that all tasks shall be able to communicate with any other task, regardless of the machine on which it is located and the protocol used to reach the remote system. The tasks shall also be able to communicate with several tasks simultaneously.

4.4 The Intelligent Coordinator

The intelligent coordinator in ISACS covers several functions:

Supervision and control of COSSs.   Continuously supervise appropriate COSS analyses for current situation, activate passive COSSs when necessary and summarise status for the operator.

Act on operator request.   When requested by the operator, the coordinator must interpret and convey requests to COSSs. Since there are several systems available, one must be able to switch easily between the various systems as the operator demands are changing.

Correlate information from different COSSs, the process and the operator, and draw conclusions.   Typically this will be important when several detection and diagnostic systems are active, coming up with conclusions that may disagree. One also must distinguish between parallel problems which may arise in plant-wide coupled systems, concluding which one has most safety significance and should be prioritised. Information and knowledge which logically belong to one problem must be collected in a suitable form for operators to get an overview of the particular problem considered.

The coordinator must interface with the complete MMI system defined for the operator and with every COSS defined in ISACS. The interfaces are realised with messages which are exchanged between the coordinator and the external systems (MMI, COSSs). Internally the coordinator has 5 functional blocks: one set of functions which handles the MMI, three set of functions to handle the different COSSs, for status identification, action planning and action implementation, and one central coordinator grouping one set of functions handling with events and operator actions. The various tasks identified for the coordinator are grouped logically according to the scheme in Figure 2. The first version comprise a total of 25 functions.

The MMI-Coordinator. The MMI-coordinator must convey information to each display and interpret requests from the operator. The overview picture will be continuously updated with plant data and synthesised information from the various COSSs. More detailed information is presented on operator workstations where the coordinator facilitates switching between different COSSs dependent on operator demands.

For example, the coordinator may report on a top-level display that a problem has been identified in one of the plant systems, and the cause of the problem is given. However, the operator may wish to access the detailed information behind this message and retrieve a procedure to remedy the failure. In such cases the coordinator must propose the relevant COSSs and displays for the operator. A major task will be to assist operators in navigating between different types of displays.

Further the coordinator must alert/remind the operator on his workstation when new important messages have been received from COSSs. This is of particular importance in cases when the operator is working on one task and a higher priority task should draw his attention.

The Central Coordinator. The major role of the central coordinator is to identify and follow events in the plant, as well as to interprete operator actions/request. Example of a planned event can be power reduction or load cycling. An unexpected event typically originates from problems with process equipment like stuck valves, pump failures or pipe breaks.

An event is created when one of the diagnostic systems concludes that a new problem has been identified. Then all relevant data associated with this event is collected like essential alarms, prognoses, possible countermeasures and history of actions taken. An event is also created when the operator initiates a planned transient such as a power reduction.

It is important for the coordinator to assign priorities to various planned and unplanned events to facilitate shifting operator attention to the most important tasks when new problems arise, while returning to the less important activities when the most critical problems have been solved. Priority is closely coupled to the safety impact of an event, with highest priority to those reducing safety most. Whenever the procedure started by the operator is not that one required by the most important event, the operator attention is drawn.

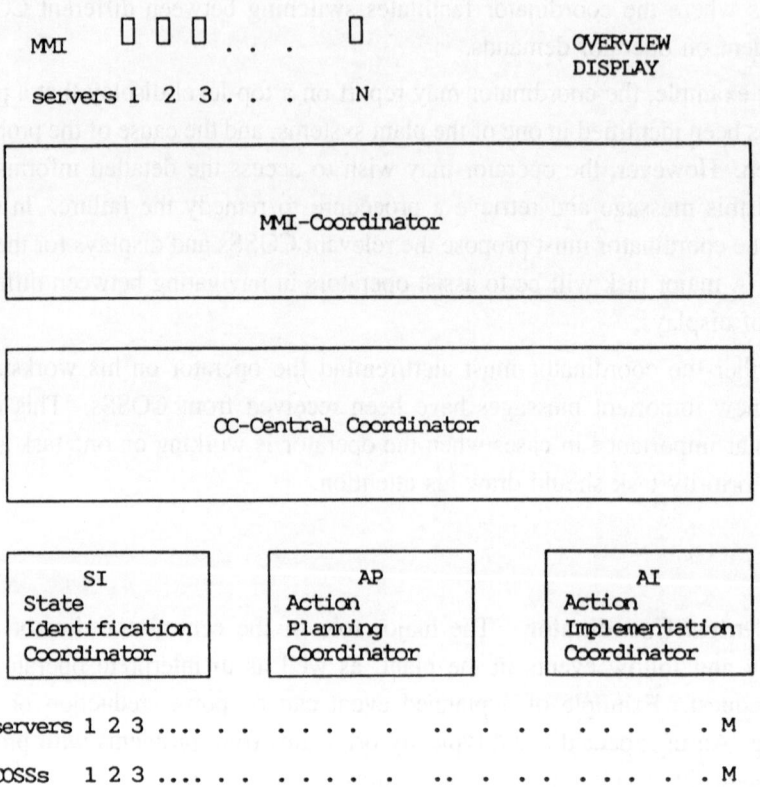

Figure 2: ISACS-1 Intelligent Coordinator Functional Blocks

The State Identification Coordinator. Many of the COSSs in ISACS deal with status identification, for instance various types of alarm and diagnosis systems. This means that this part of the coordinator has required most of the total effort.

The first task of this coordinator function is to determine the plant state (normal/disturbance/accident) based on the alarm status information. A more detailed assessment will be performed in case of disturbances.

When possible the coordinator will provide the cause of a problem as obtained from diagnosis systems. Synthesis of different information sources will either increase or decrease the certainty of a diagnosis.

With the diversity of COSSs available the coordinator will try to identify parallel problems which sometimes may be difficult for operators to do when just relying on conventional information systems.

The Action Planning Coordinator. The task of this part of the coordinator is to provide the operator with strategies and plans for how to solve identified problems or fulfill operational goals. If several plans exist they will be organised according to the corresponding events to avoid confusion about which problems the particular actions are supposed to solve. A prioritised list of reccomended actions will be made as well as the goals of the actions.

The Action Implementation Coordinator. In ISACS it will still be the operator which makes decisions and take actions. However, in addition to taking actions via the basic displays, the Computerised Procedures System (COPMA) will be available for guiding the operator through procedures.

The actions taken must be related to each specific event considered. The operator may switch between parallel events and the status of action implementation will be maintained for each event. A feedback on the success of actions and possible problems in carrying out actions are key tasks of this coordinator.

4.5 The Man-Machine Interface

ISACS is intended to act as a single, integrated interface for the operator for all operational situations. As such, all information from the process, and all commands to the process will be passed through ISACS. Therefore, the design of the ISACS MMI is critical, as the operator's ability to interpret and control the process is entirely dependent on the MMI and its underlaying software. In order to design the ISACS MMI, it was necessary to identify in advance certain principles to be followed throughout the design. The following general principles are being followed in the design of the ISACS man-machine interface:

- The operator has ultimate responsability for the safe operation of the process. ISACS simply acts to provide information to him at various levels of abstraction. Therefore, ISACS takes no direct actions on its own, but serves to present information to the operator, and performs analyses under the direction of the operator.

- Other than the requirement to actively aknowledge (e.g. by accessing the recommended display) high-level ISACS alarms, the operator will be free to use or ignore ISACS as he/she desires.

- ISACS will be able to present high level alarms to the operator, but cannot force additional information (e.g. COSS displays) on the operator. ISACS can recommend specific displays to look at to understand specific ISACS conclusions and recommend- ations, and can facilitate access to that display, i.e. by providing a mouse-sensitive area such that the display can be accessed by a single mouse click.

- Based on the above, ISACS as seen by the operator will consist of two different modules: the "autonomous ISACS" and the "operator-controlled ISACS". The autonomous ISACS will perform activities under its own control and report the summary on the overview display, as well as provide recommendations on displays to access or procedures to implement. In addition, autonomous ISACS will present high-level "IS alarms" to the operator. The operator- controlled ISACS will perform activities under the direction of the operator.

Figure 3 shows the functional layout for the ISACS MMI. It has been divided into three functional areas: the overview display (4 screens), operating level displays (8 screens) and alarms (1 screen plus audible and voice alarms).

The overview display consists of four screens: the alarm overview where groups of alarms are given in a mimic diagram, a Rankine Cycle display giving process status overview, a state identification display presenting plant safety state including a description and prioritisation of events, and finally an action planning and implementation overview giving recommendations for actions to

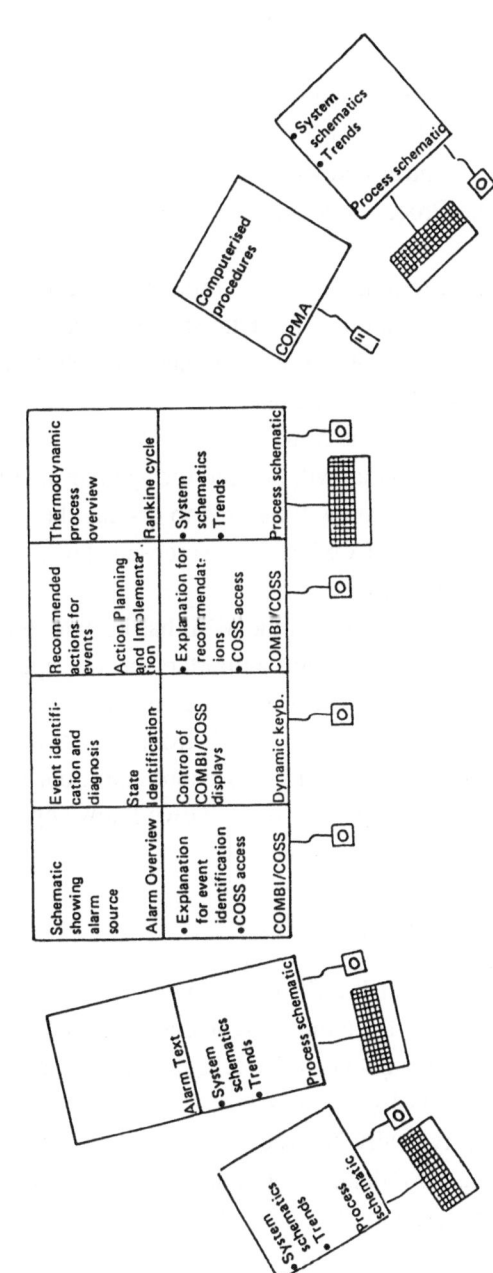

Figure 3: Functional Layout of ISACS-1 MMI

be taken. The information presented in the overview is controlled by ISACS. References are given to more detailed displays that can be fetched on the other screens.

On the COMBI/COSS screens, the operator may take up specific ISACS or COSS displays. The Process Schematic screens present detailed process diagrams. Process control takes place by using the keyboard and tracker ball associated with the Process Schematic screens by addressing components on the screen.

Further, a separate terminal for the computerised procedure system COPMA simplifies the implementation of predefined procedures.

Alarms are presented both as text, as audible alarms and, in the case of severe alarms, by use of voice output.

An important aspect of the Man-Machine Interface is the design of the dynamic multitask keyboard that will be used for operator interaction with ISACS. A separate screen/workstation will be used to display the various keyboards associated with the different COSSs. The screen-displayed keys are operated by use of the tracker-ball device.

A basic concern in the design of the MMI has been to find a balance between "forcing" information ISACS finds important on the operator (the overview) and allowing the operator to select information (the lower level screens). With a proper balance, the operator will benefit from the knowledge made available to him from the COSSs and the ISACS coordinator, but at the same time feel that he is in charge and inspired to use his own knowledge. Only experimental evaluations can show if this goal is met.

4.6 The Common Database

The Common Database in ISACS is the main repository of data to be shared by the different tasks. To avoid a tight coupling between tasks, a solution was chosen to let tasks deliver data to one central data pool, the Common Database, while other tasks are allowed to read the required data from this data pool. The data will typically be process related data, both static and dynamic, COSSs related data, MMI data, etc. In this way, the different COSSs transfer all relevant data to the Common Database, from which the Intelligent Coordinator can access the data which it finds interesting at all times. In addition, the COSSs data which are relevant for presentation in displays can be accessed by the User Interface System from the Common Database.

Modifications to data output from a COSS will not affect any other task receiving information, since only the content and structure of the Common Database are changed. As long as the original data is present, there is no need for doing modifications in the data accessing task.

The requirements to the implementation of the Common Database were quite rigorous, with respect to flexibility, real-time, easy data access, security, speed capacity, etc. These requirements have fixed severe restrictions to the Database Management System used for implementing the Common Database.

5 Development Methodology

A rigorous development methodology is of vital importance in the development of a complex computer based system for critical applications. The development methodology we have used is an amalgamation of ideas from design for reliability, software engineering and knowledge engineering. Here follows is a list of the key rules adopted:

Diversity of methods

Disturbance analysis may be addressed with different methods depending on the nature of the problem to be solved. In cases where a clear pattern of a disturbance is obtained from conventional alarms and a few plant parameters, cause-consequence analysis and symbolic reasoning is a viable method. However, when there is no clear signature of failures and it is hard to distinguish between the primary alarms and the consequence alarms, a reasoning scheme should be applied where the process behaviour is modelled. In many cases, both methods may be applied, opening for a comparison of results from the two methods.

For more severe transients, the cause of the problem may not be so important as to whether the safety systems are functioning properly. A functional approach to disturbance analysis, based on the availability of critical safety functions and success paths, should then be applied.

In ISACS all these paradigms are contained in the various COSSs supporting the coordinator with information during disturbances. In this way the coordinator can choose the most relevant COSS data in context with different types of disturbances. The diversity in methods applied guarantees a more roubust disturbance analysis compared to systems relying on one single basic methodology. This is important for potential dangerous processes, where all disturbances should be identified and treated properly.

Modular Approach

It is vitally important that the system can interact with a dynamically changing environment, both receiving input from the sensors and affecting the environment through actuators. Different modules have been built for the different functions. By providing different functions in separate modules, other modules, e.g. MMI, are relieved from the burden of having to know the exact details of how the process should be monitored and controlled. The functional modules are coupled by using a cooperating peers approach, that allows each module to request services of the other.

Concept of EVENT

An EVENT is defined as either a planned or unexpected plant transient. Planned transients are initiated by the operator and could be for example activation of a power reduction procedure, while unexpe transients are caused by failures in the process or control systems.

Information belonging to one event (alarms, diagnosis, relevant procedures and actions taken) are collected by the coordinator and stored in one event object. If parallel events exist the coordinator decides what information belongs to which event and puts it into the right event object. Information from the various diagnostic systems are basis for defining events and distinguishing between possible parallel transients.

Object Oriented Design and Implementation

The functional modules identified for the intelligent coordinator, have been implemented by objects. There are three main classes of objects:

Process objects. They are the main functional components of the application. A process is a persistent object triggered by the occurrence of one of a specific set of events. When a process completes its job, it suspends itself and waits for the next occurrence of one of its triggering events. The occurrence of an event will be notified by sending a message. Results produced by the processes are stored into process buffers and used by any other process that may need them.

Event objects. They provide a complete set of information around the occurrence of an event. Event objects allow for disjunction of events and the distribution of events to multiple processes. Event objects provide additional modularity between the occurrence of events and the triggering of processes based on those events.

Message objects. Communication among the external world and the coordinator functions is based on message objects. A new message object is issued every time any of the attributes, that compose the frame structure of the knowledge, change value. To keep a historical record of the different messages issued by the different COSSs, transient objects are created and temporarily stored in the knowledge base.

The object-oriented development strategy facilitates a step-wise implementation and testing of coordinator functions.

Hierarchical Knowledge Base Organisation

A special attention deserves the organisation of the Knowledge Base of the Intelligent Coordinator. To satisfy the requirements of incremental knowledge acquisition as well as easy to access and use, the concept of hierarchical structure has been adopted. The Knowledge Base consists of a number of workspaces which have a hierarchical structure. This means that there exists only one top-level workspace which is completely independent of any object, while other workspaces are defined as subworkspaces of certain objects and cannot exist without those objects. The advantages of using such a structure are:

- easy to save and/or to retrieve part of the Knowledge Base

- possible to activate/deactivate distinct workspaces

An example of the content of a typical function subworkspace is shown in Figure 4. More details are found in [9].

The GSI [10] software package is used to communicate with user programs and database system. Communicate means both getting data and sending data. This data communication can be implemented within G2 world through certain objects that are referred as GSI sensors. Sensor values have to be polled from G2 side, either by rules, formulas, procedures or other data seeking mechanism, because user programs residing outside G2 cannot set any sensor values by themselves.

version REG 09.10.1990

whenever the message-identifier ID of any
message-from-copma-object M receives a
value
then conclude that the procedure-event-
safety-state of analyse-copma - (if the
content-type of M is procedure-
deselected then the symbol normal)

whenever the active-procedure P of analyse-
copma receives a value and when P is
not none
then invoke procedure-event-safety-state
rules

whenever the message-identifier ID of any
message-from-copma-object M receives a
value.
then conclude that the active-procedure of
analyse-copma - (if the content-type C
of M is procedure-deselected or C is
procedure-selected then the symbol none
else if C is component-manipulation then
the procedure-name of M)

whenever the safety-state S of any event E
receives a value and when the active-
procedure P of analyse-copma has a
current value and P is not none
then invoke procedure-event-safety-state
rules

Category : procedure-event-safety-state

when there exists an event E such that (the
ai-procedure of E - the active-procedure
of analyse-copma)
then conclude that the procedure-event-
safety-state of analyse-copma - the
safety-state of E

24-ANALYSE-COPMA

Scan interval - 5 sec

If on-line-flag-for-coss
then start analyse-copma-procedure ()

ANALYSE-COPMA-PROCEDURE

| the count of each message-from-copma- | ==== |
| object | |

OLD-MESSAGE-IDENTIFIER-OF-COPMA

| old-message-identifier-of-copma | ==== |

Figure 4: An Example of a Function Subworkspace

Incremental Prototyping

The primary need is in the difficulty of specifying completely the problem and the solution until at least one trial system has been built and tested. That is particularly true for the Intelligent Coordinator.

The difficulty of specifying the function. It is extremely difficult to specify a full set of detailed requirements for the different functions of the intelligent coordinator until experimental ideas have been tried out using prototypes.

The difficulty of specifying the solution. It is extremely difficult to specify all the knowledge required to achieve the desired level of system performance until at least one trial system has been built and tested.

Incremental prototyping approach has been made possible by the development tool adopted for the system.

Distinct Development Activities

Development is divided into four distinct activities: requirement definition, conceptual modeling, implementation and evaluation, carried out in a cyclical manner.

Requirement definition. Since the primary aim of the ISACS system is to provide the operator with an integrated decision support system, a clear statement was needed as to what should be provided. Basically this specification document describe the motivation behind ISACS and the operator needs for the safety operation of the plant.

Conceptual modeling. It consists of a modular breakdown of the system, and an abstract description of the entities that will constitute the system with their relations, represented at a level which is independent of how these components will be implemented. The conceptual model, represented in a mixture of structured English, pseudo-code and data and control flow diagrams, provided a detailed "blueprint" for implementing the system.

System Implementation. The system has been implemented by the use of G2, a development environment supporting object oriented programming. It was particularly easy to associate a specific object to each of the function identified in the conceptual model.

System evaluation.   Evaluation is conducted at many different levels. From component level, using the different simulation capabilities offered by the development environment in use, to system level, using plant operators with the system linked to a full scale simulator. A careful planning and reporting of the all evaluation activities is a must due to the nature of the system.

Good Engineering Practices

ISACS is intended for critical applications but not highly safety-critical. This allow to relax the requirements that normally apply to the development of highly safety-critical systems and requires only good engineering practices to be applied over the entire life-cycle, from concept to final evaluation. Some of the basic principles adopted are as follows:

Resources and Organization.   All activities are properly planned. Methods, tools and human resources used to carry on the activities are clearly identified before start. An ad hoc steering committee meets each week to discuss progresses and problems. A progress report is issued weekly.

Documentation.   A complete set of documentation is produced for the project, in parallel with the development activites. All documents produced inside the project are univocally identified.

Documentation Reviews.   All documents produced by the project are submitted to review before to be delivered for use in the next phase.

Different development tools for different purposes

For the development of a system of the size and complexity of ISACS, it is extremely important to identify the right tools to be used for the different purposes. Three different kinds of main blocks cooperating through a network have been identified:

computation: such as inference engines and algorithms calculation

data storage: such as data base managers

environment interaction: such as man-machine interface.

Different tools have been identified and used for the different purposes, and precisely:

- G2 for development of the Knowledge Base [10]

- SYBASE for development of the Common Database [11]

- PICASSO for development of the User Interface [12].

6 Conclusions, Results and Status

ISACS belongs to a class of computer based systems with very little proven experience from the point of view of how to assure the level of dependability required by the application. This is for at least two reasons: the major significance of the man-machine interface and the inherently complex nature and difficulty of testing of knowledge-based systems. This makes it quite difficult to refer to any previous experience. Nearly all the theory, techniques and tools available for specifying, designing, implementing and validating computer based systems are concerned with how to prevent or to discover faults in the representation of the external world and/or in the manipulation of the representation, when using conventional software techniques. Very little is available on how to deal with the new problems introduced by the use of AI techniques and very little attention has been given to the problem of how to avoid faults in the interpretation back to the world of the information provided by the system.

To gain experience on these two aspects, we have planned several versions of ISACS. The first version, ISACS-1, has been released for total system testing and evaluation at the end of 1990. System testing and evaluation is planned to start in the early 1991. The purpose with the first version is to check the feasibility of the concept, including design of the MMI and functions of the intelligent coordinator, as well as the applicability of the methodology we have adopted. Also, we will gain experience with the use of current technology in the areas of real-time data bases, computer communication and expert system tools for solving the very complex tasks of ISACS.

References

[1] J. Lark, L. Erman, S. Forrest, K. Gostelow, F. Hayes-Roth, and D. Smith, "Concepts, methods, and languages for building timely intelligent systems," *The Journal of Real-Time Systems*, vol. 2, May 1990.

[2] B. Hayes-Roth, "Architectural foundations for real-time performance in intelligent agents," *The Journal of Real-Time Systems*, vol. 2, May 1990.

[3] T. Laffey, P. Cox, J. Schmidt, S. Kao, and J. Read, "Real-time knowledge-based systems," *AI Magazine*, Spring 1988.

[4] K. Arzen, "An architecture for expert system based feedback control," *Automatica*, vol. 25, no. 6, 1989.

[5] S. Bologna, E. Ness, and T. Sivertsen, "Dependable knowledge-based systems development and verification: What we can learn from software engineering and what we need," in *IEEE International Conference on Tools for Artificial Intelligence 90*, November 1990.

[6] K. Haugset *et al.*, "ISACS-1, the prototype of an advanced control room," in *IAEA International Symposium on Balancing Automation and Human Action in Nuclear Power Plants*, July 1990.

[7] K. Haugset *et al.*, "ISACS-1 motivation, general description," Tech. Rep. HWR-265, OECD Halden Reactor Project, 1990.

[8] J. Kvalem *et al.*, "ISACS-1 hardware/software tools and environment," Tech. Rep. HWR-266, OECD Halden Reactor Project, 1990.

[9] R. Grini *et al.*, "Intelligent coordinator for ISACS-1, description and implementation," tech. rep., OECD Halden Reactor Project, 1991. To be published.

[10] Gensym Corporation, USA, *G2 Users' Manuals*, 1989.

[11] *SYBASE Users' Manuals*.

[12] *PICASSO-2 User Guide*, 1990.

Implementing Forward Recovery Using Checkpoints in Distributed Systems *

JUNSHENG LONG, W. KENT FUCHS

Center for Reliable and High-Performance Computing

Coordinated Science Laboratory

University of Illinois

Urbana, IL 61801, USA

JACOB A. ABRAHAM

Computer Engineering Research Center

Department of Electrical and Computer Engineering

University of Texas

Austin, TX 78712, USA

Abstract

This paper describes the implementation of a forward recovery strategy in a Sun NFS environment. The implementation is based on the concept of lookahead execution with rollback validation. It uses replicated tasks executing on different processors for forward recovery and checkpoint comparison for error detection. In the experiment described, the recovery strategy has nearly error-free execution time and an average redundancy lower than TMR.

1 Introduction

Rollback recovery using checkpoints is a classic method of backward recovery widely used in fault-tolerant systems. Checkpoint-based backward recovery has two drawbacks: an execution time penalty due to rollback and the problem

*This research is supported in part by the National Aeronautics and Space Administration (NASA) under Contract NAG 1-613 and in part by the Department of the Navy and managed by the Office of the Chief of Naval Research under Contract N00014-91-J-1283.

of determining if a checkpoint is error free. Several studies have investigated reducing the performance degradation due to checkpointing and rollback by selecting the optimal checkpoint placement [1-4]. Also several approaches have been developed to solve the checkpoint validation problem. One approach is to validate the system state before a checkpoint is taken via concurrent error detection or system diagnosis [5-6]. Another is to simply keep a series of consecutive checkpoints and perform multiple rollbacks when necessary [7]. In contrast to backward recovery, forward recovery attempts to reduce the time required for recovery by manipulating some portion of the current state to produce an error-free new state. However, forward recovery generally depends on accurate damage assessment, a correction mechanism, and sometimes massive redundancy [8-10].

In this paper, we describe an implementation of a forward recovery scheme which is intended to exploit parallelism in multiprocessor and distributed computer systems [11]. Compared with the standard NMR approach to forward recovery, this strategy uses, on average, fewer processors and can provide an execution time close to that of error-free execution. If one is willing to replicate processes over the network, our approach can be an attractive alternative to achieving forward recovery in a distributed environment. Our implementation is on a Sun NFS (Network File System) distributed network. The objective of the implementation is to investigate the feasibility for distributed systems, and to measure the performance overhead of our recovery technique in an actual implementation.

The following section gives an overview of our forward recovery strategy. The subsequent two sections describe the details of our implementation and discuss experimental results from four application programs.

2   Forward Recovery with Checkpointing and Replicated Tasks

2.1   Computation and System Model

The system considered consists of multiple homogeneous computer systems connected to each other through a network. A task is an independent computation and it can be a group of related sub-tasks. A task is divided into a series of sequential sub-computations by checkpoints. Each checkpoint provides a complete recovery line for a task [3,12]. Checkpoints are stored in a reliable

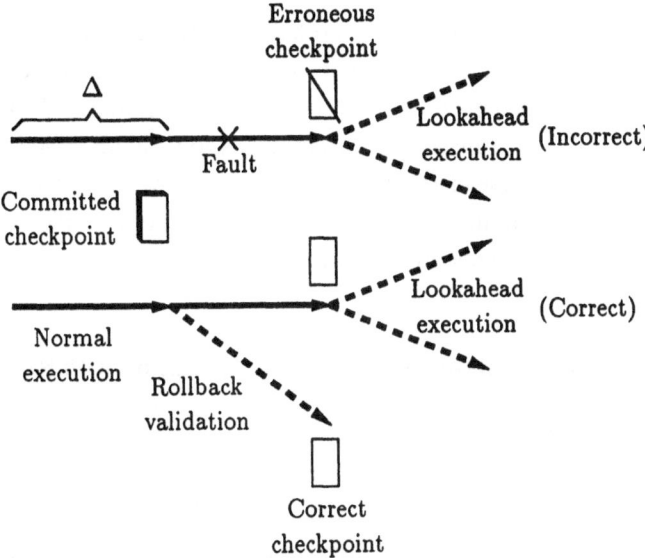

Figure 1: Lookahead Execution and Validation

network file system.

The scheme is able to detect and recover from faults in individual processing nodes, which result in a single erroneous (corrupted) checkpoint. Faults in the software design, network, voter task, or secondary storage may not be detectable nor recoverable.

2.2 Forward Recovery through Lookahead Execution with Rollback Validation

The essence of our forward recovery scheme is lookahead execution with rollback (checkpoint) validation. Checkpoint comparison is used for checkpoint validation, and lookahead execution for forward recovery. This concept is illustrated in Figure 1.

A task is replicated and executed concurrently on two different processors. After each checkpoint interval, Δ, two checkpoints are produced by the replicated task pair. The checkpoints are compared by a voter task. If there is a match in the checkpoints, the corresponding checkpoint is committed and the task advances to the next Δ.

If two checkpoints disagree, then, instead of rolling back, two identical tasks are scheduled to continue from both of the uncommitted checkpoints (lookahead execution) on two separate processors. Meanwhile, another task (the validation task) is restarted from the last committed checkpoint on a fifth processor. After Δ, a new checkpoint is produced by the validation task. This checkpoint is compared to the two previously (uncommitted) disagreeing checkpoints. If there is a match, a correct checkpoint is identified. The task pair that was executed ahead from the disagreeing erroneous checkpoint and the validation task are terminated. The correct checkpoint is then committed and the incorrect checkpoints are removed.

2.3   Recovery Schemes Based on Lookahead Execution with Rollback Validation

In our recovery strategy, the potential for forward recovery lies in the fact that there should be at least one correct execution (thus, one valid checkpoint) during the normal run, since the lookahead execution from this valid checkpoint advances the computation without rollback. Rollback may not be avoided when the validation checkpoint does not agree with the two disagreeing normal task checkpoints, since all lookahead runs may be incorrect.

If the number of validation re-tries is unlimited (the rollback of a validation task when its checkpoint disagrees with the two checkpoints of the normal tasks), the tasks executed ahead can spawn their children of lookahead and validation tasks, as the number of validation re-tries increases. This recursive scheme [13-14] can maximally utilize the forward recovery capability of the lookahead execution, and rollback can potentially reach its lower-bound. If the probability of multiple failures during a checkpoint period is low, then it is unlikely that recursive validation and process spawning will be required.

In this paper, we report an implementation based on a non-recursive design. We select two tasks for the normal execution and one rollback validation task. We call this scheme DMR-F-1 for **D**ual **M**odule **R**edundancy with one try of **ONE** validation rollback task. The reduced cost argument for DMR-F-1 is that it requires fewer processors on average than TMR (for most error rates and distributions of checkpoints) and two more processors momentarily during lookahead than TMR. This makes DMR-F-1 suitable for distributed systems since some lightly-loaded processing elements (workstations) may be used temporally for recovery. The temporal use of these workstations for recovery does not exclude their use for other purposes, The close to error-free execution time is particularly suitable for long-life CPU-bound (e.g., numerical) applications.

3   Implementation

3.1   Host Environment

Our implementation environment consists of a Sun 3/280 server and a pool of 12 Sun 3/50 diskless workstations. The server provides a Sun NFS transparent access to remote file systems under SunOS 4.0. A voter task for the checkpoint comparison and recovery initiation is also run on this server. All checkpoints are kept in the server. The Sun 3/50 workstations are used as the processing units. Our implementations are entirely user level with no kernel modifications required.

3.2   Basic Problems

In order for our recovery strategy to work, two basic problems have to be resolved: the (remote) restartability of a checkpoint and the capability of comparing checkpoints. That is, a task must be able to be restarted from a checkpoint produced on other nodes and a checkpoint produced on a node must be identical to any checkpoint from any other nodes if both are correct and from the same checkpoint interval. The former is required for lookahead execution and the latter for checkpoint comparison (validation).

Due to different workloads at each node, the processing speed may vary on each node. In our recovery scheme, the task execution time is determined by the slowest process in the replicated task pair. The mismatch in processing speed (or workload) prolongs the completion of the task computation. It also causes the problem of uncommitted checkpoints accumulating in the file system.

3.3   Checkpointing

A checkpoint is a snapshot of a process state at the time of checkpointing. There has been considerable research concerning checkpoint construction in UNIX. Smith implemented a mechanism for checkpoint construction in UNIX for the purpose of process migration [15]. His checkpoint is an executable file generated by a checkpoint operation. It contains the text segment, the data segment, as well as the stack segment of the process state. The stack segment is treated as a part of the data segment. The processor state (e.g., registers) is saved by a *setjmp()* system call. The restart of the checkpointed process is simply the re-execution of this executable file on another processor. Li and

Fuchs developed a checkpointing scheme in their research on compiler-assisted checkpoint insertion [16]. Their checkpoint is a data file that contains the data segment and partial stack segment of the checkpointed process. The checkpoint is intended for use in the same shell process on the same machine.

Our checkpoint structure is a superset of that of Li and Fuchs. In addition to complete stack and data segments, our checkpoint also contains a segment for the file I/O output data during that checkpoint interval. The inclusion of the file output as a part of a checkpoint makes checkpoint comparison effective for error detection (described later). The process registers are saved as a part of the stack. The omission of the text segment is possible because the original executable file is already available through NFS. There is no need to transfer the executable file in order to perform a remote restart.

Checkpoint Operations

The checkpoint/restart operations include three routines: *_checkpoint()*, *_restart()*, and *_terminate()*. They are all user level and can be inserted into user applications either automatically by a compiler [16] or manually. The routines are described as follows:

- *_checkpoint()* is placed in user application programs. When executed, it saves the processor state on the stack, stores both data segment and stack segment in a data file and signals to the voter that a new checkpoint has been generated.

- *_restart()* is inserted in *main()* as the first instruction to execute. It checks with the voter if there is a checkpoint from which to start. If there is a checkpoint, it reads in and restores the data and stack segments, and resumes from the checkpoint; otherwise, it does a normal return.

- *_terminate()* is inserted before every *exit()* by manual placement and inside *exit()* by the compiler. Its execution signals to the voter that the task has terminated.

Restartability and Comparability

The virtual and uniform memory layout of UNIX in homogeneous machines provides the basis for the restart of a checkpointed process on a remote node.

In UNIX, however, some state attributes are kernel-dependent. They cannot be saved and carried across kernels (i.e., nodes) in a sensible fashion [15]. Examples are *process group*, *signal* received, the value of real time clock, children the process may have spawned with *fork()*, and file descriptors.

The kernel-dependent attributes cause problems for checkpoint comparability. The valid checkpoints from the same execution on different nodes may not be the same if the program has these attributes stored in its variables. Furthermore, some kernel-dependent attributes (e.g., the internal file table maintained by the operating system) are usually not a part of a checkpoint but are related to some variables (e.g., _iobuf) in the checkpoint. This can make a checkpoint unrestartable across kernels (or network).

In the current implementation, we focused our effort on the comparability and restartability of a checkpoint with respect to file I/O. We included information stored in the operating system as a part of our checkpoint. A set of library routines was developed which provide the semantics of the *standard I/O* library. The library keeps extra data as a part of the checkpoint, such as file name, access mode, and file position, associated with the opened files. During checkpointing, all file buffers are flushed for opened files, and the file positions are updated and stored in the checkpoint. During a restart, those files are re-opened and re-positioned according to the previously saved information in the checkpoint. In this manner, the attributes of file I/O can be saved and restored easily across the network. These file I/O routines together with _checkpoint(), _restart(), _terminate() comprise the checkpoint library(ckp.a).

For all other kernel-dependent attributes, we enforce the following restrictions in order to make the checkpoint comparable and restartable. For restartability, a program may not use or depend on these kernel-dependent attributes that have partial information internal to the operating system, except file I/O. For comparability, we either eliminate the use of variables to store such kernel-specific attributes, carefully place them in local variables (on the stack) whose scope does not includes a checkpoint routine, or clear these variables before checkpointing. Fortunately, most numeric applications seldom use kernel-dependent values but file I/O, and thus meet the restrictions we put on checkpoint restartability and comparability.

3.4 Checkpoint Comparison and Recovery Initiation

In our DMR-F-1 implementation, the checkpoint comparison and recovery initiation are managed by a voter process running on the Sun NFS server. Our current implementation assumes the voter is reliable. During a checkpoint operation, the communication between the voter and the tasks being executed uses the Internet *socket*. On each processing node, there is a simple RPC-based (Remote Procedure Call) daemon process that schedules or kills a task in that processing node on the behalf of the voter. Originally, a **rsh** call from the voter was used for scheduling and killing a task on a remote node. However, we found the performance overhead of a **rsh** call to be unacceptable. Therefore, the voter was designed to be capable of scheduling and terminating a remote task, comparing checkpoints, and initiating recovery across a network.

The voter is invoked with the name of the task program to be executed and its arguments. The voter then schedules two replicated processes from this program and waits for messages from the scheduled tasks. When a task is initiated, the call to *_restart()* sends a *register_msg* to the voter. Upon receiving it, the voter sends back a checkpoint file name when recovery is needed. Otherwise, the voter replies with no checkpoint and the task does a normal start. When a task creates a checkpoint with *_checkpoint()*, the voter gets a *checkpoint_msg*. The voter either advances to the next checkpoint interval or does a lookahead/validation operation, depending on whether the checkpoint comparison fails or not. When a task is terminated, the voter receives an *exit_msg*. If all replicated tasks have exited, the voter terminates.

In a distributed environment, the processing speed of processing nodes may vary due to differences in hardware and workload. This mismatch in processing speed causes the replicated tasks to lag behind one another. Therefore, the task completion time is prolonged since it is a function of the slowest of the replicated pair. In addition, the uncommitted checkpoints produced by the faster task can accumulate in the file system. In a distributed environment, a checkpoint may be a natural place for migrating processes and redistributing workload. We added a simple mechanism in the voter algorithm to adjust the performance of the replicated task executions. If the voter detects the growth in the checkpoint count by two for a task, the two replicated tasks are switched to other nodes.

4   Experimental Results

4.1   Experiment Programs

Four scientific programs were used in our experiments. They are described as follows:

rkf uses the Runge-Kutta-Fehlberg method for solving the ordinary differential equation $y' = x + y$, $y(0) = 2$ with step size 0.25 and error tolerance $5 \ 10^{-7}$. A table of function values was generated for $x = 0$ to 1.5 every 0.0001. This is a computation-intensive program with a small data set.

convlv uses the FFT algorithm to find the convolution of 1024 signals with one response. The length of each signal was 256 bytes. The length of the response was 99 bytes. The size of the entire data set was over 1 Mbyte but the size of the memory-resident data set was only a few Kbytes.

rsimp uses the revised Simplex method to solve the linear optimization problem for the *BRANDY* set from the Argonne National Laboratory. One characteristic of this program is its large memory-resident data set (about 1 Mbyte). **rsimp** has no file output during execution.

ludcmp uses an LU decomposition algorithm to decompose 100 randomly generated matrices with size uniformly distributed from 50 to 60. Although it has a larger data set (2.4 Mbytes) than that of **rsimp**, this program occupies less main memory because memory is reused, i.e., a memory block is allocated before a new matrix is read in, and is deallocated after the result is written out.

4.2   Checkpoint Placement

_checkpoint() is inserted manually with roughly constant checkpoint interval time. We simply place *_checkpoint()* in the main control loop in the following fashion:

```
main()
{
        int leverage = certain value;
        int count = 0;
        _restart();

        while (expr) {
        /* major work */
        if ((count = (count+1) % leverage) == 0)
                _checkpoint();
        }
        _terminate();
}
```

This gives an adjustable checkpoint interval by choosing proper leverage with a small variance (see Table 1). Research concerning the optimal placement of checkpoints has shown that the optimal checkpoint interval is typically quite large (in tens seconds) and insensitive to small displacement [1-4].

4.3   Error Injection

In our experiments, errors are injected into checkpoints in order to study the recovery behavior of the programs. An error is injected into a checkpoint by randomly flipping bits in the data or stack segment of the checkpoint. This type of error is intended to model the changes in variables due to possible errors in memory and in data manipulation (ALU).

The probability of a node failure during a checkpoint interval in our experiments, p_f, is independent and constant. The range of p_f is selected as $[0, 0.1]$. Each program is run five times for each failure probability.

4.4   Experiment Results

Program Characteristics

The overhead measures we considered consist of the checkpoint size (ckp_size), checkpointing time (ckp_time), and checkpoint comparison time (cmp_time). Other measures of interest are the checkpoint interval (ckp_int) and the error free execution time with and without checkpointing. The restart time of a task is about the same order as checkpointing time in magnitude and was not listed explicitly. Table 1 summarizes these overhead measures for each program. ckp_size consists of three parts: data segment, stack segment and the file output during the checkpoint interval. $rsimp$ gives an example of a large checkpoint. Most applications we examined have checkpoints of size (64-128K bytes). The stack size is small in all four programs. This is not unexpected for scientific applications in which the calling depth is rather limited. The file output size can be large in some applications (e.g., **convlv**).

In Table 1, both ckp_time and cmp_time do not include the processing time for the file output portion of the checkpoint. For ckp_time, the file output portion is already written to disk during execution; thus, it is not necessary to re-write this portion again to the checkpoint. Three variables in a checkpoint are enough to locate this file output portion (file name, starting position and length for each output file). The small standard deviation in checkpoint interval indicates that the manual insertion of _checkpoint( ) has produced a nearly constant checkpoint placement.

Table 1: Overhead Measurements

Programs Name	# ckp (per run)	ckp_size (data/stack/file) (in bytes)	ckp_time (std. dev.) (in sec.)	cmp_time (std. dev.) (in sec.)	ckp_interval (std. dev.) (in sec.)	exec_time (w/o. ckp) (in sec.)
rkf	88	51777 (46972/1734/3071)	0.1477 (2.563e-2)	0.1492 (7.2498e-3)	29.7202 (1.0840)	2638.58 (2625.58)
convlv	128	75950 (66196/1554/8200)	0.2172 (0.3411)	0.1608 (8.6302e-3)	13.917 (0.90787)	1809.22 (1781.42)
ludcmp	50	121510 (71708/1550/48252)	0.2408 (3.428e-2)	0.2030 (1.8224e-2)	20.626 (2.1092)	1043.38 (1031.34)
rsimp	59	995314 (991676/3638/0)	2.411 (0.3767)	3.8286 (0.21893)	42.8063 (8.6359)	2713.04 (2568.38)

Detectability of Checkpoint Comparison

Checkpoint comparison for error detection is the key to checkpoint validation in our recovery strategy. The effectiveness of checkpoint comparison is studied for the four selected programs. In order to avoid the interference of run-time error injection with checkpoint comparability, a random bit or word error is

Table 2: Detectability Experiments

Program	# Errors Detected				#
	data	stack	file	abort	Missed
bit error					
rkf	78	1	22	0	0
convlv	68	3	30	0	0
ludcmp	43	0	58	0	0
rsimp	99	0	-	2	0
word error					
rkf	76	3	22	0	0
convlv	71	0	30	0	0
ludcmp	37	3	59	2	0
rsimp	98	0	-	2	1

injected in the previous checkpoint to model a transient error occurrence during its subsequent checkpoint interval. Then one task is started from this erroneous checkpoint and another task from the error free checkpoint. The checkpoints produced by the two tasks after one checkpoint interval are compared. A mismatch indicates a detected error. Table 2 summarizes the results for 101 injected random errors. The number of errors detected is categorized by where the error is detected: the data , stack and the file output segments of the checkpoints. The abortion of the task due to an error in the checkpoint can be treated as a special case of error detection by either a time-out mechanism implicitly or an abortion signal to the voter explicitly.

The errors detected by checkpoint comparison account for the majority of injected errors that occurred (about 98 percent) for all four programs. If the file output during the checkpoint interval is not included in the checkpoint structure, 22 to 59 percent of the errors would not be detected (**rkf, convlv** and **ludcmp**). There is one error missed in our experiments. In this case, we have a valid file output during execution and a valid checkpoint at the end; this missed error is actually masked off and causes no problems with respect to correct executions. This case occurs when the error is in a dead variable and this variable is re-initialized later. In sum, the checkpoint structure provided an effective error detection tool for the programs we studied.

Performance Results

The performance indices we considered are relative execution time, number of processors used, number of checkpoints, and voter processing overhead. The relative execution time is defined as the execution time over the error-free execution time. The execution time in our experiments is actually the program response time. It includes system time, user time and the blocking time. The voter processing overhead is the time spent in the voter program minus the checkpoint comparison time and divided by the checkpoint interval. The analytical predictions for relative execution time, number of processors and number of checkpoints are also included to compare against our experimental results. The analytical predictions are derived from the assumptions of constant failure rate and small overhead (checkpointing time and comparison time) [11].

The data were collected for two workload conditions: day time (10 AM to 6 PM) and night time (10 PM to 8 AM). During the day, the workload among our workstations was uneven and the NFS server was busy. During the night, our workstations and the NFS server were lightly loaded.

In Figures 2 and 3, the relative execution time is plotted for each of the four programs under two different workload conditions. The relative execution time for the programs with a moderate checkpoint size (**rkf, convlv** and **lud-cmp**) is close to the analytical prediction and nearly one. That is, the expected execution time with error injected is close to the error free execution time. However, the relative execution time for the program with large checkpoints (**rsimp**) is larger than the analytical prediction, especially with a high p_f. This can be explained by the fact that $rsimp$ is likely to be blocked due to its large file I/O operations during checkpointing and comparison. In fact, the limited speed of the NFS file handling and our use of the file server for managing checkpoints centrally resulted in a performance bottle neck. The relative execution time increases significantly for high error rates due to file server activity during checkpointing and comparison of checkpoints. This suggests that either a reduction in checkpoint size, an increase in file system speed, or other non-centralized server implementations may improve the relative execution over that of our current implementation.

The relative execution time fluctuates more for the day time condition than that for the night. The execution time is longer in the day runs than in the night (Figure 4). This reflects the fact that the workload is heavier and more likely to change during the day than during the night.

For the failure range we considered, the number of processors used, N_p,

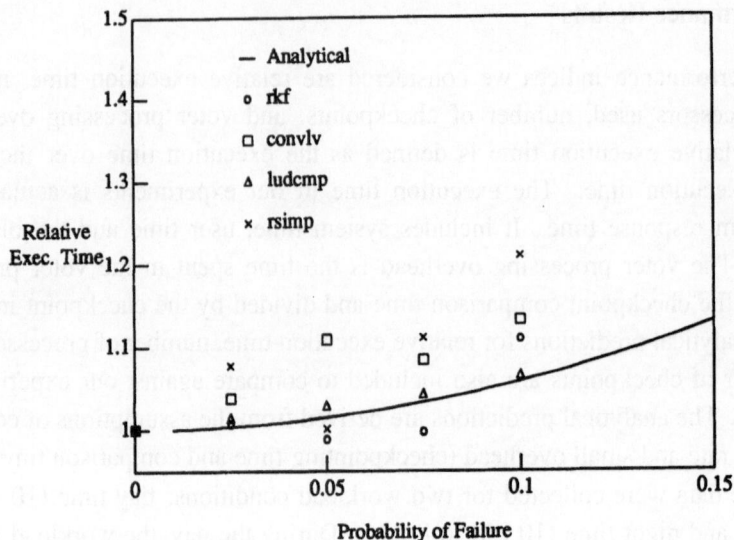

Figure 2: Relative Execution Time During the Day

is less than the three that TMR requires, although DMR-F-1 uses two more processors momentarily during lookahead/validation operations. Interestingly, N_p is quite insensitive to the workload distribution conditions and checkpoint size (Figures 6 and 5).

The number of checkpoints, N_c, is highly sensitive to the workload distributions (Figures 8 and 7). Without switching the task executions on the nodes that have different processing speed, N_c averaged 6.95 in five runs for **convlv**. In one of these runs, N_c reached 18.45. With the switch mechanism mentioned in the previous section, N_c is limited to about 2 or 3. The simple switch rule we use in the voter limits N_c by redistributing workload.

In Figure 9, the average processing overhead for the voter is plotted. The overhead is small compared to the checkpoint interval, and increases as the failure probability increases. The large checkpoint size also increased this overhead due to the waiting time for file I/O during checkpoint comparison.

5 Conclusion

We have described an implementation of a forward recovery strategy using checkpointing and replicated tasks. This recovery scheme exploits distributed

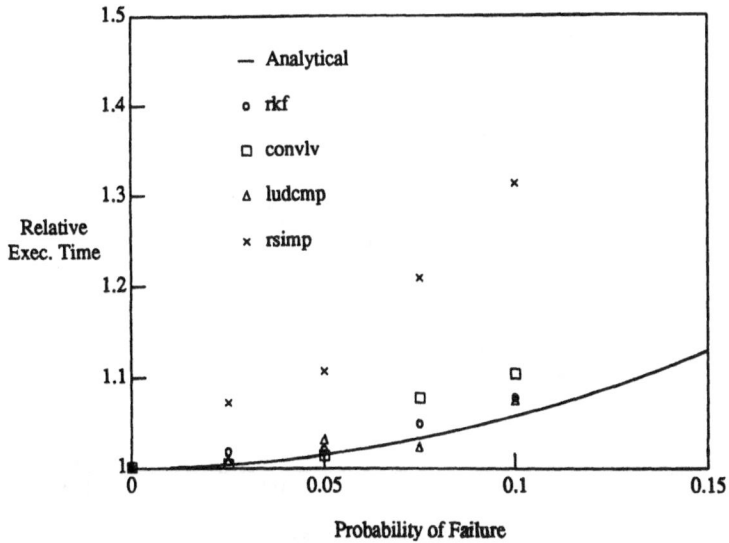

Figure 3: Relative Execution Time During the Night

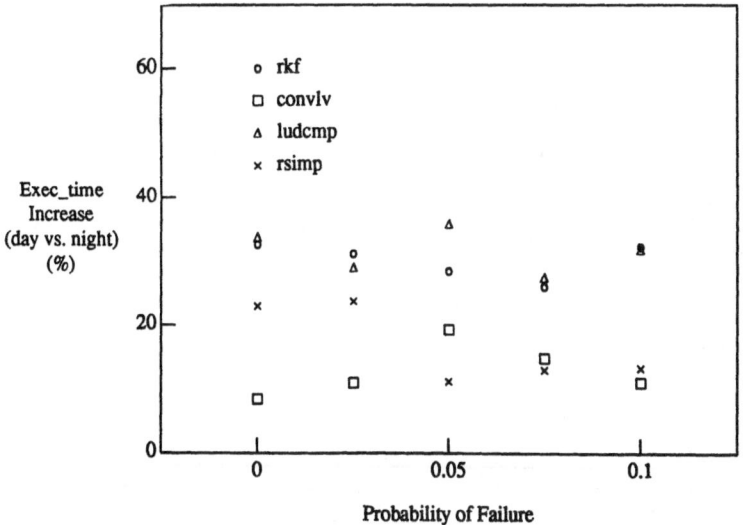

Figure 4: Difference in Execution Time

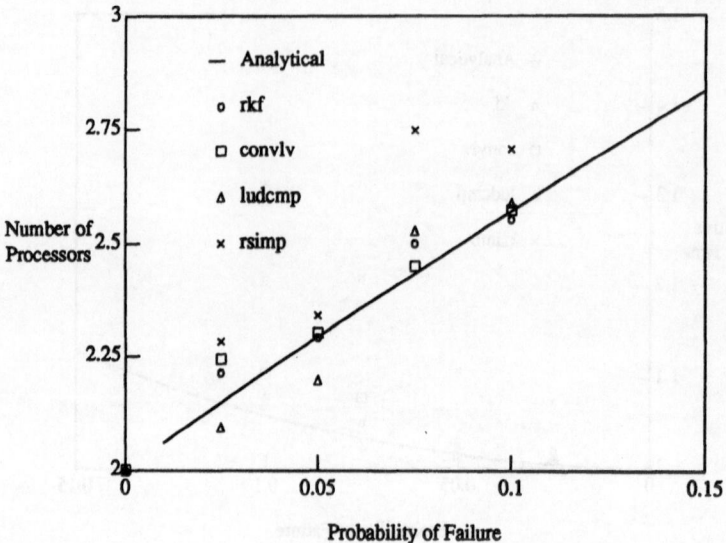

Figure 5: Number of Processors During the Day

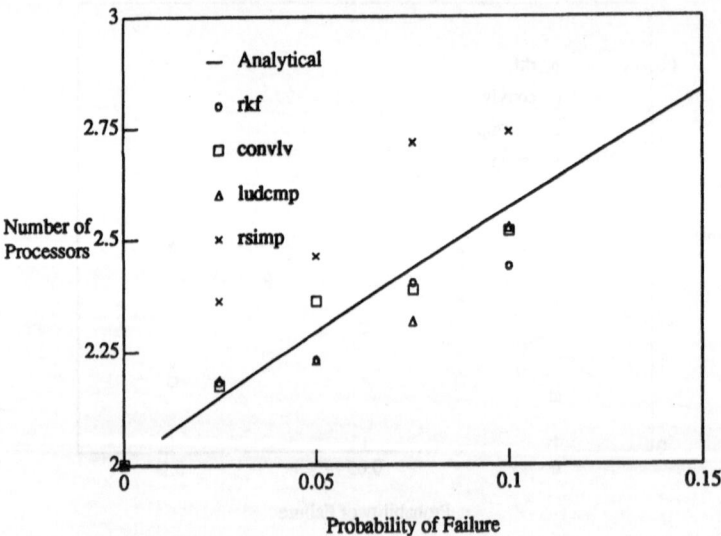

Figure 6: Number of Processors During the Night

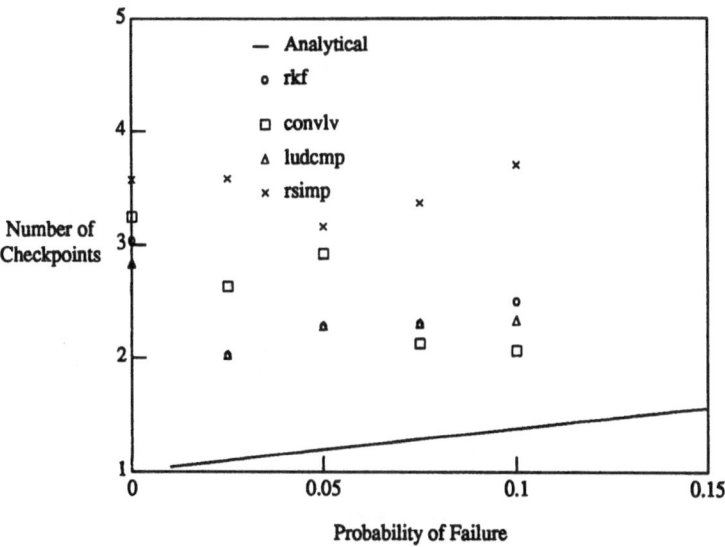

Figure 7: Number of Checkpoints During the Day

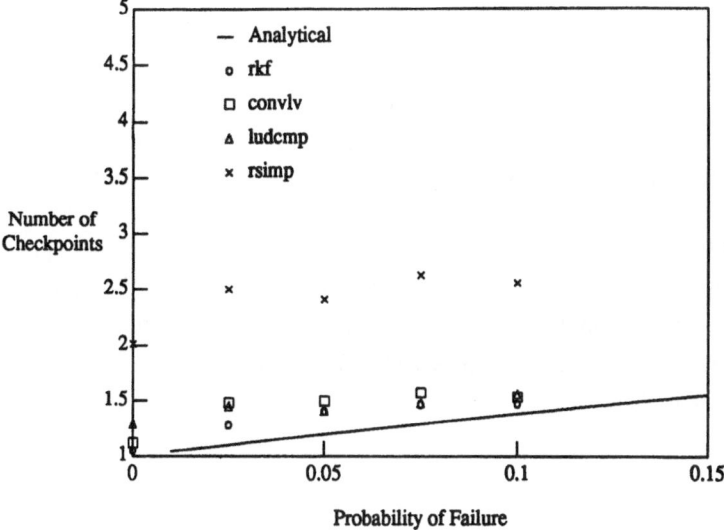

Figure 8: Number of Checkpoints During the Night

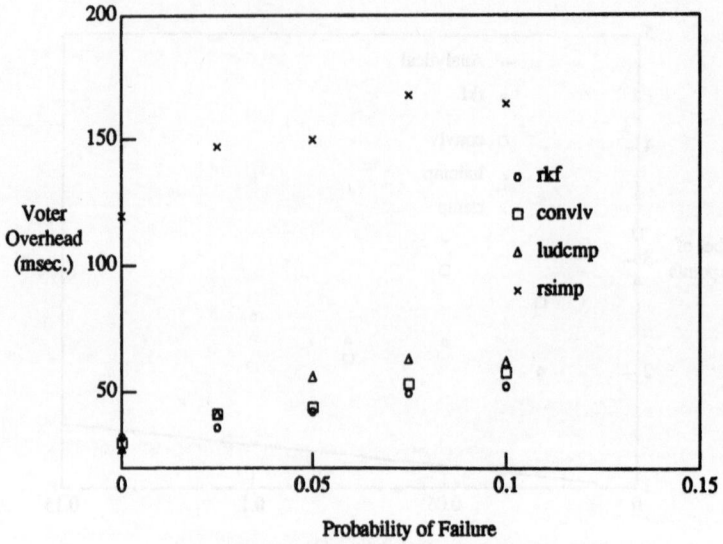

Figure 9: Voter Processing Overhead

computing to achieve forward recovery through lookahead execution and check-point validation through comparison.

Four scientific programs were evaluated. Checkpoint comparison was found to be an effective means of error detection in our experiments. For moderate checkpoint sizes (64-128K), checkpointing time and comparison time were reasonably small. The execution time was reasonably close to the error free execution time. For large checkpoint sizes, this overhead had an important impact on performance due to the centralized file server used in our implementation. The workload also had an important impact on the number of checkpoints and the execution time. The former resulted from the uneven distribution of the workload and the latter from both workload distribution and workload level. Finally, the average number of processors used in our forward recovery scheme was less than TMR.

References

[1] A. Duda, "The effects of checkpointing on program execution time," *Information Processing Letters*, vol. 16, pp. 221–229, 1983.

[2] E. Gelenbe and D. Derochette, "Performance of rollback recovery systems under intermittent failures," *CACM*, vol. 21, no. 6, pp. 493–499, 1978.

[3] C. M. Krishna, G. S. Kang, and Y.-H. Lee, "Optimization criteria for checkpoint placement," *CACM*, vol. 27, no. 6, pp. 1008–1012, Oct. 1984.

[4] J. W. Young, "A first order approximation to the optimal checkpoint interval," *CACM*, vol. 17, no. 9, pp. 530–531, Sept. 1974.

[5] A. Tantawi and M. Ruschitzka, "Performance analysis of checkpointing strategies," *ACM Trans. on Computer Systems*, vol. 2, no. 2, pp. 123–144, May 1984.

[6] S. Thanwastien, R. S. Pamula, and Y. L. Varol, "Evaluation of global rollback strategies for error recovery in concurrent processing systems," *Proc. 16th Int'l. Symp. on Fault-Tolerant Computing Systems*, pp. 246–251, 1986.

[7] Y.-H. Lee and G. S. Kang, "Design and evaluation of a fault-tolerant multiprocessor using hardware recovery blocks," *IEEE Trans. on Computers*, vol. 33, no. 2, pp. 113–124, 1984.

[8] D. J. Taylor and C.-J. H. Seger, "Robust storage structures for crash recovery," *IEEE Trans. on Computers*, vol. 35, no. 4, pp. 288–295, 1986.

[9] C.-C. J. Li, P. P. Chen, and W. K. Fuchs, "Local concurrent error detection and correction in data structure using virtual backpointers," *IEEE Trans. on Computers*, vol. 38, no 11, pp. 1481–1492, 1989.

[10] T. Anderson and P. Lee, *Fault Tolerance: Principles and Practice.* Englewood Cliffs, N.J.: Prentice-Hall, 1981.

[11] J. Long, W. K. Fuchs, and J. A. Abraham, "A forward recovery strategy using checkpointing in parallel systems," *Proc. Int'l. Conf. on Parallel Processing*, vol. 1, pp. 272–275, 1990.

[12] K. Tsuruoka, A. Kaneko, and Y. Nishihara, "Dynamic recovery schemes for distributed processes," *IEEE 2nd Symp. on Reliability in Distributed Software and DataBase Systems*, pp. 124–130, 1981.

[13] P. Agrawal, "Raft: A recursive algorithm for fault-tolerance," *Proc. Int'l. Conf. on Parallel Processing*, pp. 814–821, 1985.

[14] P. Agrawal and R. Agrawal, "Software implementation of a recursive fault-tolerance algorithm on a network of computers," *Proc. of the 13th Annual Symposium on Computer Architecture*, pp. 65–72, 1986.

[15] J. M. Smith, "Implementing remote fork() with checkpoint/restart," *Technical Committee on Operating Systems Newsletter*, vol. 3, no. 1, pp. 15–19, Winter, 1989.

[16] C. C. Li and W. K. Fuchs, "Catch: Compiler-assisted techniques for checkpointing," *Proc. 20th Int'l. Symp. on Fault-Tolerant Computing Systems*, pp. 74–81, 1990.

Replication

Replication

Evaluating the Impact of Network Partitions on Replicated Data Availability

JEHAN-FRANÇOIS PÂRIS

Department of Computer Science
University of Houston
Houston, TX 77204-3475, USA

Abstract

Many distributed systems maintain multiple replicas of their critical data to protect these data against equipment failures. When this is the case, a *replication control protocol* must be chosen to insure that a consistent view of the data is always presented.

In this paper, we present a simple aggregation technique leading to closed form estimates of the availability of replicated objects whose replicas reside on networks subject to communication failures. We illustrate our technique by comparing the availabilities of replicated objects with three replicas managed by majority consensus voting (MCV), and dynamic-linear voting (DLV), under three different network configurations.

Key Words: Fault-tolerance, replicated systems, redundancy, voting.

1 Introduction

Many applications depend on critical data that must remain available in the presence of equipment malfunctions. Recent advances in networking technology have made the replication of these data on several sites of a local area network a cost-effective proposition. First, having multiple replicas of the same data virtually eliminates the risk of permanent data loss. Second, distributing the replicas among distinct sites of a network increases the probability that the data will remain available in the presence of hardware faults. Managing replicated data presents however a special challenge as site failures and network malfunctions are likely to result in inconsistent replica updates. Special *replication control protocols* have been devised to avoid this occurrence and insure that a consistent view of the replicated data is always presented.

Various replication control protocols have been presented in the literature. These protocols vary greatly in their complexity, their communication overhead, the protection they provide or do not provide against communication failures, and the number of replicas they require to guarantee full access to the replicated data in the presence of a given number of site failures. As a result, the evaluation of the performance of replication control protocols has become an area of great practical interest. An important measure of this performance is the *availability* of the replicated data object managed by the protocol. By definition the availability of a replicated data object represents the steady-state probability that the object is available at any given moment.

Several techniques have been used to evaluate the availability of replicated data. Combinatorial models are very simple to use [1, 2] but cannot represent complex recovery modes as these found in available copies and dynamic voting protocols. Simulation models can be very accurate if all the parameters of the modeled system are known. They have two major disadvantages; the first is that they are computationally intensive and the second is that they provide only numerical results. As a result, stochastic models have become the method of choice for evaluating the availability of replicated data managed by protocols with complex recovery modes [3, 4, 5, 6]. These however suffer from two important limitations: First, stochastic models become quickly intractable unless all failure and repair processes have exponential distributions. Second, stochastic processes do not handle well communication failures as the number of distinct states in a model increases exponentially with the number of failure modes being considered. As a result, all recent studies of the availability of replicated data have either relied on simulation models or have totally neglected communication failures. This neglect has resulted in over-optimistic evaluations of the availability of the replicated data objects under study.

We present in this paper a simple aggregation technique leading to closed form estimates of the availability of replicated objects whose replicas reside on networks subject to communication failures. We illustrate our technique by comparing the availabilities of replicated objects with three replicas managed by majority consensus voting (MCV) and dynamic-linear voting (DLV) under three different network configurations. We show that communication failures have a very different impact on the availability of three replicas managed by MCV and DLV with DLV being the least affected.

The remainder of this paper is organized as follows: Section two reviews voting protocols; Section three introduces our aggregation technique; Section four illustrates our method on an example. Our conclusions appear in Section five.

2 Voting Protocols

Voting protocols [7] probably constitute the best known class of replication protocols. Voting protocols ensure the consistency of replicated data objects by disallowing all read and write requests that cannot collect an appropriate quorum of replicas. Different quorums for read and write operations can be defined, and different weights, including none, assigned to every replica [8]. Consistency is guaranteed as long as the write quorum W is high enough to disallow parallel writes on two disjoint subsets of replicas, and the read quorum R is high enough to ensure that read and write quorums always intersect.

These conditions are simple to verify, which accounts for the conceptual simplicity and the robustness of voting schemes. Voting has however some disadvantages. It requires a minimum number of three replicas to be of any practical use. Even then, quorum requirements tend to disallow a relatively high number of access requests.

Several solutions have been proposed to overcome these limitations. *Dynamic Voting* (DV) [9] and *dynamic-linear voting* (DLV) [4] adjust quorums to reflect changes in replica availability and network topology. Both protocols greatly improve the availability and reliability of replicated objects with more than three replicas. *Voting with witnesses* [3], *voting with ghosts* (VWG) [2] and *voting with bystanders* [10] share the common thread of introducing auxiliary entities that are used by the protocol to increase the availability of replicated data objects.

3 The State Aggregation Technique

The inability of stochastic models to model replicated data objects with multiple failure modes and complex recovery procedures is probably their important limitation as it severely restricts our ability to evaluate the availability of replicated data in the presence of network partitions. This inability is a direct consequence of the fact that the number of distinct states in the model increases exponentially with the number of failure modes being considered. Dugan and Ciardo

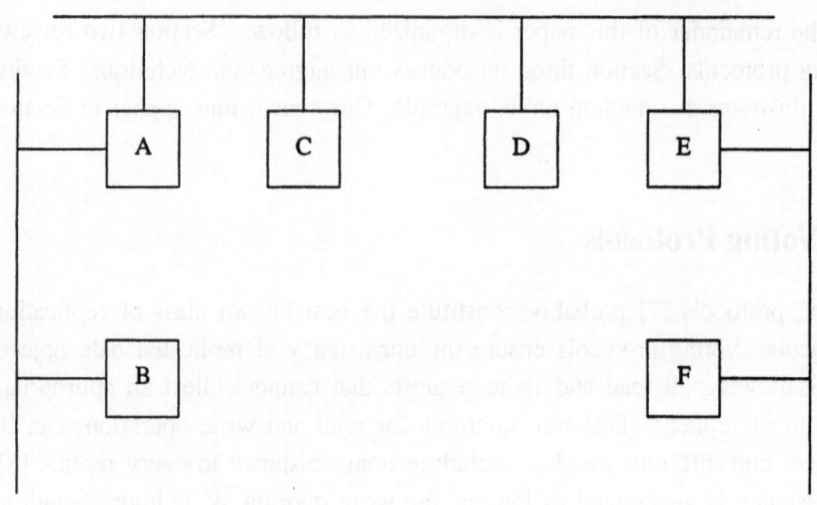

Figure 1: A LAN with Six Sites on Three Segments

have proposed to use Petri nets to generate stochastic models of replicated data object with witnesses managed by the MCV protocol [11].

Another solution consists of reducing the complexity of the model itself by identifying parts of the system that can be studied in isolation and replaced by simpler equivalent components [12, 13]. This technique has been widely used in computer systems performance evaluation to solve Markov models too complex to be directly tractable. We will show how it can be applied to the evaluation of the availability of replicated data objects.

Many local-area networks consist of several carrier-sense segments or token rings linked by selective repeaters or gateway hosts. Figure 1 shows one example of such networks: it contains three CSMA segments AB, $ACDE$ and EF. A is the gateway between AB and $ACDE$ while E is the gateway between $ACDE$ and EF. Since repeaters and gateways can fail without causing a total network failure, such networks can be partitioned. The key difference with conventional point-to-point networks is that sites that are on the same carrier-sense network or token ring will never be separated by a partition. We will refer to these entities as *LAN segments* [2].

Consider now the replicated object X represented on figure 2. It consists of two replicas A and B located on the same LAN segment and a third replica C on a second segment. Let us assume that the two LAN segments are linked by a gateway G. Under MCV, replica C will only be able to participate in elections

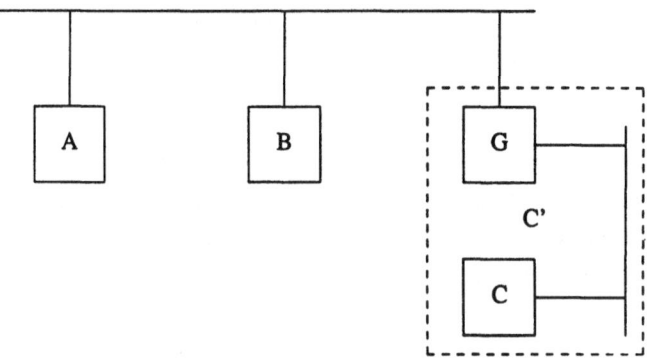

Figure 2: Three Replicas on Two LAN Segments

when the gateway G is operational. For all practical purposes, a failure of G will have the same effect as a failure of C. We propose therefore to replace the subsystem consisting of site C and its gateway G by an *aggregate site C'* that will remain operational as long as *both* C and G are operational. The replicated object consisting of sites A, B and the aggregate site C' will have the same availability as the replicated object X but will be much easier to investigate since we do not have to consider gateway failures.

This aggregation technique can be trivially extended to replicated objects consisting of an arbitrary number of replicas located on a network consisting of LAN segments linked by gateways provided that the following conditions are met:

(a) There is at most one LAN segment that contains more than one replica. (We will refer to that segment as the *backbone segment*.)

(b) If there is a backbone segment, all sites that are not on the backbone segment communicate with the sites on the backbone segment through their own gateways or sequences of gateways.

(c) If there is no backbone segment, there is at least one LAN segment such that all sites that are not on the segment communicate with the sites on the segment through their own gateways or sequences of gateways.

Condition (a) is necessary to ensure that the replicated object can be reduced to an equivalent object with all its aggregate sites on the same LAN segment. Conditions (b) and (c) are necessary to ensure that the aggregate sites do not include common gateways as common gateways would introduce non-independent failures of aggregate sites.

These conditions clearly restrict the number of replicated objects that can be analyzed through our aggregation method. Fortunately they are generally met by replicated objects with two or three replicas and these replicated objects are the most likely to be encountered in practice.

Another limitation of the method is its implicit assumption that sites that become part of an aggregate site can never become a single site majority. While this assumption is correct for all voting protocols that never allow single site majorities, it is not true for weighted voting and dynamic-linear voting protocols. Consider for instance the replicated data object represented in figure 2 and assume that its three replicas are managed by a weighted voting protocol assigning one vote to replica A, one vote to replica B, and three votes to replica C. Since replica C holds a majority of the votes, the replicated data object will remain available as long as C remains available. Failures of the gateway G will affect the accessibility of the object from sites A and B but not its overall availability. This is not true for the simplified "equivalent" model obtained by merging sites C and G into a single aggregate state C' since any failure of G results in a failure of C'.

This situation could be dismissed as an oddity since assigning a majority of the votes to a single site negates most of the benefits of replication. Single site majorities are however a feature of dynamic-linear voting protocols. Let us go back to the replicated object represented in figure 2 and assume that it is now managed by a dynamic-linear voting protocol with $C > B > A$. As long as A, B and C are operational, the majority partition will consist of these three sites: A, B, C. A failure of site B would result in the *exclusion* of B from the majority partition, which would now become A, C. Should site A fail while B is still unavailable, site C would become a single site majority partition C. Here again there would be a discrepancy between the original model and the equivalent model obtained by aggregating C and G into C'. The problem can be avoided by reordering the sites in such a way that C becomes the lowest site. This is not possible for a replicated object consisting of three replicas located on three distinct LAN segments as the one represented on figure 3.

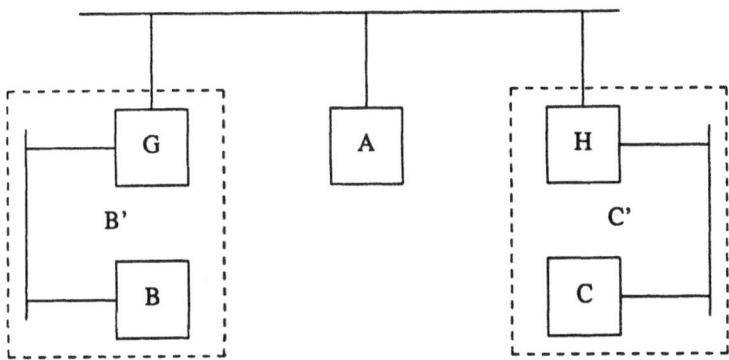

Figure 3: Three Replicas on Three LAN Segments

4 An Example

In this section we illustrate our aggregation technique by comparing the availabilities of replicated objects with three replicas managed by majority consensus voting (MCV) and dynamic-linear voting (DLV) under three different network configurations. Our model consists of a set of sites with independent failure modes that are connected via a network composed of LAN segments linked by gateways. When a site fails, a repair process is immediately initiated at that site. Should several sites fail, the repair process will be performed in parallel on those failed sites. We assume that failures are exponentially distributed with mean failure rate λ, and that repairs are exponentially distributed with mean repair rate μ. The system is assumed to exist in statistical equilibrium and to be characterized by a discrete-state Markov process.

The three configurations investigated are: (a) three replicas on the same LAN segment (1LS), (b) three replicas on two LAN segments linked by one gateway (2LS), and (c) three replicas on three LAN segments linked by two gateways (3LS). Configuration (a) is the only configuration that is immune to network partitions as the three replicas are on the same LAN segment. 3. Configuration (b) is represented on figure 2: it has one backbone segment containing replicas A and B and one LAN segment containing a single replica C. Its only aggregate site is site C', which results from the merge of gateway G with site C. Configuration (c) is represented on figure 3. Since replicas B and C communicate with replica A through distinct gateways, we will have the two aggregate sites B' and C' respectively consisting of G and B and H and C.

Observing that all aggregate sites consist of *one* gateway and *one* site holding a replica, we now derive the failure and repair rates of an aggregate site. Figure 4 (a) contains the state transition diagram for a subsystem consisting of a gateway and a site holding a replica. The diagram has four states. State $\langle 11 \rangle$ represents the state of the subsystem when the site and its gateway are both operational. States $\langle 01 \rangle$ and $\langle 10 \rangle$ represent states when either the site or its gateway have failed while site $\langle 00 \rangle$ corresponds to a failure of both entities.

As seen on figure 4 (b), the state transition diagram for the aggregate site has only two states. State 1 represents the state of the subsystem when the aggregate site is operational and can participate in elections. It corresponds to the state $\langle 11 \rangle$ of the subsystem and has one outbound transition whose rate 2λ is the sum of the rates of the two transitions leaving state $\langle 11 \rangle$. State 0 is a failure state that corresponds to the three other states of the subsystem. Its outbound transition has a rate μ' given by

$$\mu' p_0 = \mu(p_{01} + p_{10})$$

where p_0 is the probability of the aggregate site being in state 0 while p_{01} and p_{10} respectively represent the probabilities that the subsystem is in state $\langle 01 \rangle$ or $\langle 10 \rangle$.

Observing that

$$p_0 = 1 - p_1 = 1 - \frac{\mu^2}{(\mu + \lambda)^2}$$

and

$$p_{01} = p_{10} = \frac{\lambda\mu}{(\mu + \lambda)^2},$$

we have

$$\mu' = \frac{\mu}{(1+\lambda / 2\mu)}.$$

In the absence of network failures, the availability of a replicated data object with three replicas managed by MCV A_{MCV} is equal to the probability that at least two of the three sites holding replicas are operational. We have therefore

$$A_{MCV} = A_1 A_2 A_3 + (1 - A_1)A_2 A_3 + A_1(1 - A_2)A_3 + A_1 A_2(1 - A_3)$$

where A_1, A_2 and A_3 are the respective availabilities of the three replicas. Since the availability of a single replica is given by

$$A = \frac{\mu}{(\lambda + \mu)}$$

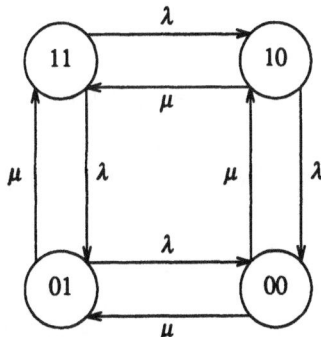

(a) State Transition Diagram for a Site and its Gateway

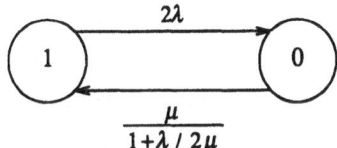

(b) State Transition Diagram for the Equivalent Aggregate Site

Figure 4: Aggregating a Site with its Gateway

and the availability of an aggregate site is given by

$$A' = p_1 = \frac{\mu^2}{(\mu + \lambda)^2},$$

we have

$$A_{MCV}(1LS) = A^3 + 3A^2(1 - A) = \frac{\mu^3 + 3\lambda\mu}{(\mu + \lambda)^3},$$

$$A_{MCV}(2LS) = A^2 A' + 2AA'(1 - A) + A^2(1 - A') = \frac{\mu^4 + 4\lambda\mu^3 + \lambda^2\mu^2}{(\mu + \lambda)^4},$$

and

$$A_{MCV}(3LS) = AA'^2 + 2AA'(1 - A') + A'^2(1 - A) = \frac{\mu^5 + 5\lambda\mu^4 + 2\lambda^2\mu^3}{(\mu + \lambda)^5}.$$

The graph in figure 5 represents the compared availabilities of the three configurations under study for values of $\rho = \lambda / mu$ between 0 and 0.2. The first value corresponds to perfectly reliable sites and the latter to sites that are repaired five times faster than they fail and have an individual availability of 0.833. The dotted curve at the bottom represents the availability of an unreplicated data object and was added to provide an element of comparison. As one can see, the availability of replicated objects managed by MCV is strongly affected by the possibility of network partitions. When $\rho > 0.1$, the availability of a replicated object with three replicas on three distinct LAN segments barely exceeds that of an unreplicated object and is even worse for $\rho > 0.2$.

The same approach can be followed for deriving expressions for the availabilities of the three configurations under dynamic linear voting (DLV). The derivations are somewhat more complex because DLV is a more sophisticated protocol.

Figure 6 contains the state transition rate diagram for three identical replicas managed by DLV in the absence of communication failures. Note that left-to-right and top-to-bottom transitions represent site failures while right-to-left and bottom-to-top transitions indicate site repairs. State 3 represents the state of the replicated object when all its three replicas are operational. The majority partition then comprises these three replicas. A failure of one of these three replicas brings the replicated object in state 2. The failure does not affect the availability of the replicated object since a majority of the replicas in the previous majority partition remain operational. The DLV protocol does however

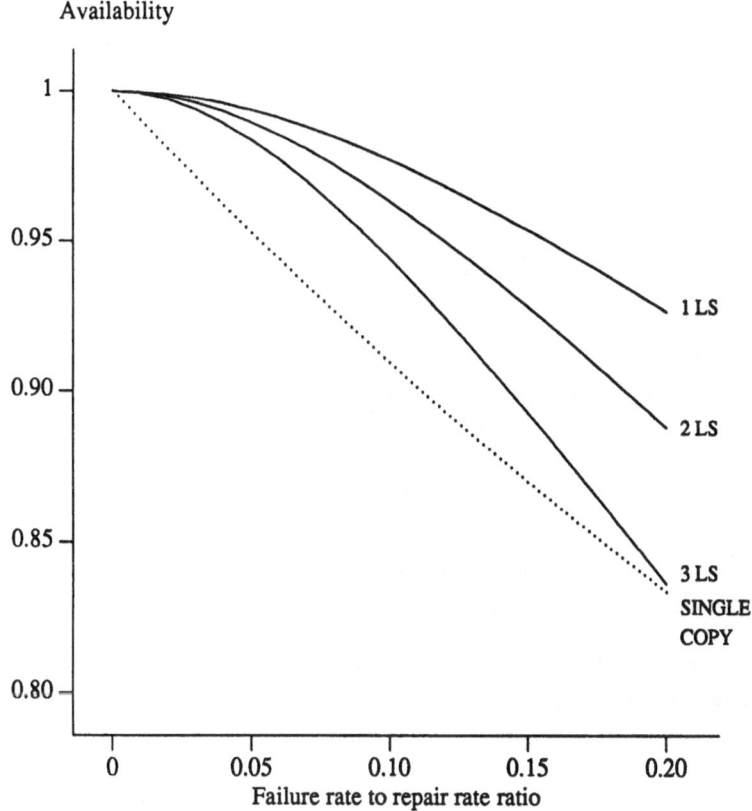

Figure 5: Compared Availabilities for Three Replicas (Majority Consensus Voting)

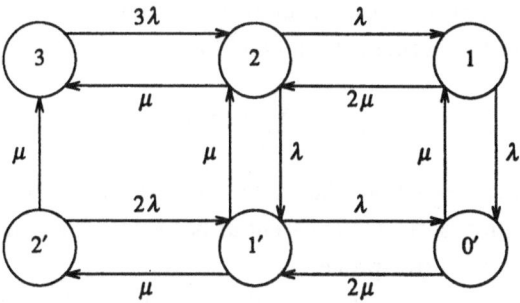

Figure 6: State Transition Diagram for DLV (Three Replicas on the Same LAN Segment)

update the majority partition which loses the replica that failed and is now comprised of the two replicas that remain operational.

A failure of one of the two replicas that are available when the replicated object is in state 2 would result in a tie because the remaining replica would constitute exactly one half of the current majority partition. The DLV protocol breaks such ties by using the linear ordering of the sites holding the replicas. Two cases need therefore to be considered:

(1) If the site holding the replica that failed is lower ranked than the site holding the replica that remains operational, that last operational replica becomes the new majority partition and the replicated object remains available. This corresponds to a transition from state 2 to state 1 on the diagram.

(2) If the site holding the replica that failed is higher ranked than the site holding the replica that remains operational, the replicated object becomes unavailable and the majority partition is not updated. This corresponds to a transition from state 2 to state $1'$ on the diagram.

State $0'$ represents the state of the replicated object after its three replicas have failed. Recovering from state 0 would bring the replicated object into state 1 if the site that recovers is the higher ranked of the two sites in the last majority partition or into state $1'$ if this is not the case. Finally state $2'$ represents the state of the replicated object when one of its two operational replicas does not belong to the current majority partition and the other one resides on the lower ranked of the two sites in the current majority partition. It is therefore an unavailable state.

Let p_i represent the probability that the system is in an available state i and q_j the probability of being in an unavailable state j'. The state transition diagram, along with the normalization condition

$$\sum_{i=1}^{3} p_i + \sum_{j=0}^{2} q_j = 1,$$

yield a system of linear equations that can be solved using standard techniques. Symbolic manipulation software is essential because, although the process is simple, it is tedious and error-prone.

The availability is given by the sum of probabilities of being in one of the three available states:

$$A_{DLV}(1LS) = \sum_{i=1}^{3} p_i = \frac{\rho^3 + 3\rho^2 + 4\rho + 1}{(\rho + 1)^4}$$

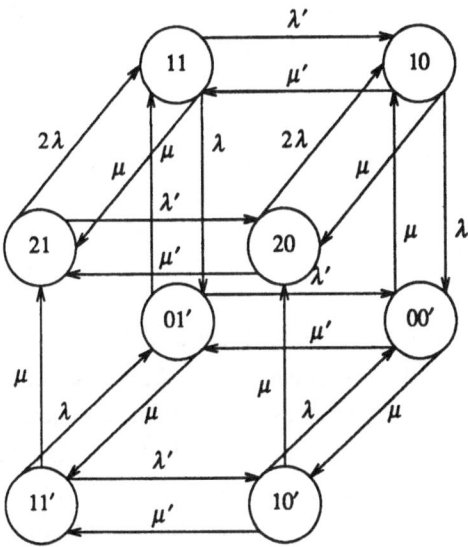

Figure 7: State Transition Diagram for DLV (Three Replicas on Two LAN Segments Separated by a Gateway)

with $\rho = \lambda/\mu$.

Figure 7 shows the state transition diagram for DLV when the three replicas are on two LAN segments. States are now identified by pairs of numbers $\langle mp \rangle$ where m is the number of operational replicas on the first LAN segment $(0 \leq m \leq 2)$ and n is the state of the aggregate site formed by the third replica and the gateway to its LAN segment. Hence state $\langle 21 \rangle$ represents the state of the replicated objects when its three replicas and the gateway linking them are all operational. States where the replicated object is unavailable are identified by a prime mark $(')$.

Transitions between states are similar to those observed on the diagram of figure 6 with one major exception: while the three replicas had previously the same failure and repair rates, the aggregate site has now a failure rate $\lambda' = 2\lambda$ and a repair rate

$$\mu' = \frac{mu}{(1 + \lambda/2\mu)}.$$

As before, failure transition reduce the number of operational sites while recovery transitions have the opposite effect. Some recovery transitions leave the replicated object in an unavailable state because the replica that failed last is still unavailable. For instance, state $\langle 11' \rangle$ is an unavailable state although two of the three replicas are operational because the replica that failed last has not yet recovered.

The availability of the replicated object is then given by the sum of the probabilities of being in one of the four available states:

$$A_{DLV}(2LS) = p_{21} + p_{20} + p_{11} + p_{10} = \frac{\rho^4 + 4\rho^3 + 6\rho^2 + 5\rho + 1}{(\rho + 1)^5}$$

where p_{ij} is the probability of being in state $\langle ij \rangle$.

The case where the three replicas are on three distinct LAN segments was easy to solve: As figure 3 indicates, the replicated object is then represented by one replica on the backbone segment and two aggregate sites. The state transition diagram for DLV with three replicas can therefore be derived from that for three replicas on two LAN segments by replacing all instances of λ and μ by λ' and μ' and vice versa. The availability of the replicated object is then given by:

$$
\begin{aligned}
A_{DLV}(3LS) &= p_{21} + p_{20} + p_{11} + p_{10} \\
&= \frac{3\rho^6 + 19\rho^5 + 49\rho^4 + 67\rho^3 + 54\rho^2 + 27\rho + 4}{(\rho + 1)^6(3\rho + 4)}
\end{aligned}
$$

The graph in figure 8 represents the compared availabilities of the three configurations under study for the same values of ρ as in figure 5. The availabilities afforded by DLV and MCV for three replicas in the absence of network partitions were known to be practically equal [14]. We were therefore very surprised to observe than DLV with three replicas on three distinct LAN segments performed almost as well as MCV with the same number of replicas on two segments. This result is even more impressive when one recalls that our aggregation technique tends to underestimate the availability afforded by protocols allowing single site majorities and that DLV belongs to that class of protocols.

Previous studies of the DLV protocol had concluded that it needed at least four replicas to outperform MCV in any significant fashion [14]. We have shown that this conclusion does not hold when communication failures are taken into account as MCV with three replicas tends to behave poorly when the failure-rate-to-repair-rate ratio exceeds 0.1 while DLV continue to provide a good availability.

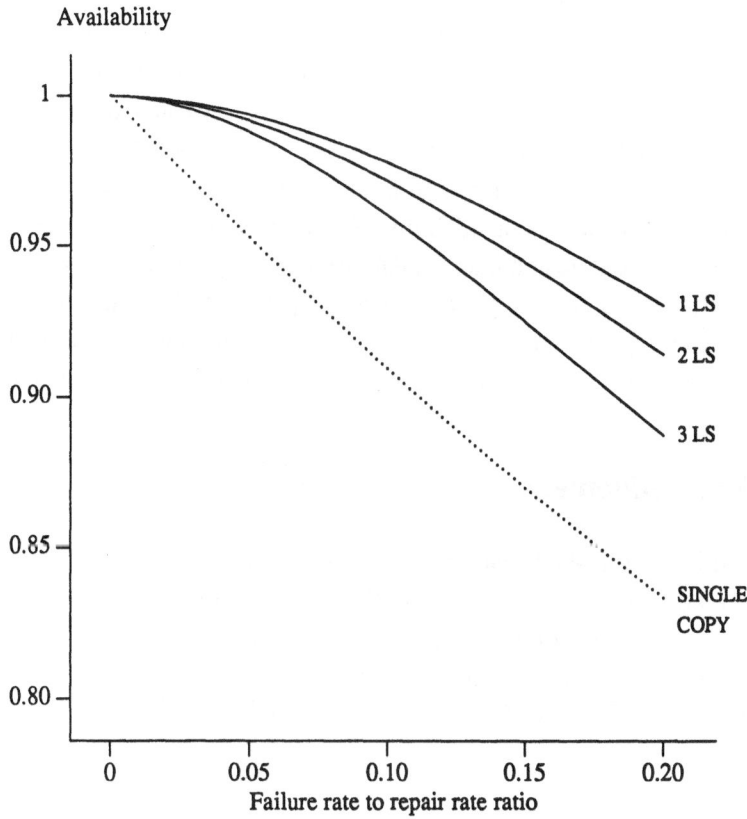

Figure 8: Compared Availabilities for Three Replicas (Dynamic-Linear Voting)

5 Conclusions

We have presented in this paper a simple aggregation technique leading to closed form estimates of the availability of replicated objects whose replicas reside on networks subject to communication failures. To illustrate our technique, we have compared the availabilities of replicated objects with three replicas managed by majority consensus voting (MCV) and dynamic-linear voting (DLV) under three different network configurations.

Two conclusions can be reached from our study. First, communication failures can severely reduce the availability of replicated data. Second, the effect of communication failures on data availability is not equally distributed among all protocols. Hence some replication control protocols (among which DLV and MCV) may appear equivalent when reliable communication is assumed and behave quite differently when communication failures are considered.

6 Acknowledgements

We wish to thank Elizabeth Pâris for her editorial comments.

The Markov analysis of the availability of the protocols under study has been done with the aid of MACSYMA, a large symbolic manipulation program developed at the Massachusetts Institute of Technology Laboratory for Computer Science. MACSYMA is a trademark of Symbolics, Inc.

References

[1] C. Pu, J. Noe, and A. Proudfoot, "Regeneration of replicated objects: A technique and its eden implementation," *IEEE Transactions on Software Engineering*, vol. SE-14, no. 7, pp. 936–945, 1988.

[2] R. van Renesse and A. Tanenbaum, "Voting with ghosts," in *Proceedings of the 8th International Conference on Distributed Computing Systems*, pp. 456–462, 1988.

[3] J.-F. Pâris, "Voting with witnesses: A consistency scheme for replicated files," in *Proceedings of the 6th International Conference on Distributed Computing Systems*, pp. 606–612, 1986.

[4] S. Jajodia and D. Mutchler, "Enhancements to the voting algorithm," in *Proceedings of the 13th VLDB Conference*, pp. 399–405, 1987.

[5] M. Ahamad and M. H. Ammar, "Performance characterization of quorum-consensus algorithms for replicated data," *IEEE Transactions on Software Engineering*, vol. SE-15, no. 4, pp. 492–496, 1989.

[6] J.-F. Pâris and D. Long, "On the performance of available copy protocols," *Performance Evaluation*, vol. 11, pp. 9–30, 1990.

[7] C. Ellis, "Consistency and correctness of duplicate database systems," *Operating Systems Review*, vol. 11, 1977.

[8] D. Gifford, "Weighted voting for replicated data," in *Proceedings of the 7th ACM Symposium on Operating System Principles*, pp. 150–161, 1979.

[9] D. Davcev and W. A. Burkhard, "Consistency and recovery control for replicated files," in *Proceedings of the 10th ACM Symposium on Operating System Principles*, pp. 87–96, 1985.

[10] J.-F. Pâris, "Voting with bystanders," in *Proceedings of the 9th International Conference on Distributed Computing Systems*, pp. 394–401, 1989.

[11] J. B. Dugan and G. Ciardo, "Stochastic petri net analysis of a replicated file system," *IEEE Transactions on Software Engineering*, vol. SE-15, no. 4, pp. 394–401, 1989.

[12] P. Courtois, *Decomposability: Queuing and Computer System Applications*. New York: Academic Press, 1977.

[13] D. Ferrari, G. Serazzi, and A. Zeigner, *Measurement and Tuning of Computer Systems*. Englewood Cliffs, NJ: Prentice-Hall, 1983.

[14] D. D. E. Long and J.-F. Pâris, "A realistic evaluation of optimistic dynamic voting," in *Proceedings of the 7th Symposium on Reliable Distributed Systems*, 1988.

A Distributed Systems Architecture Supporting High Availability and Reliability

PAUL D. EZHILCHELVAN, SANTOSH K. SHRIVASTAVA

Computing Laboratory
University of Newcastle upon Tyne
NE1 7RU, England, UK

Abstract

A reliable distributed systems architecture composed of fail-silent nodes connected by redundant networks is developed. A fail-silent node is constructed by replicating the computations on two distinct and dedicated processors which check each other's performance to form a self-checking processor pair. Given that no more than one processor in a fail-silent node fails, the node is guaranteed either to function correctly or to effectively stop functioning. High availability of distributed system services can be obtained by replicating application level processes on fail-silent nodes. Managing distributed computations in general, and replicated processes in particular, require group communication (multicast communication) services. The paper presents an integrated suit of protocols for performing reliable multicasts, and illustrates how the dual processor redundancy in a fail-silent node can be exploited for implementing these protocols.

Key Words: Distributed systems, reliability, replication, availability, fault tolerance, network protocols, multicast communication.

1 Introduction

Designing and implementing distributed systems which continue to provide specified services in the presence of a bounded number of processor and communication failures is a difficult task. One of the problems is that a failed processor might perform arbitrary state transitions (exhibit uncontrolled behaviour), the effects of which must be prevented from manifesting in the results delivered by an application program. Ideally one would like to program distributed systems with perfect (failure free) processor, storage and communications hardware. Unfortunately, since all hardware components must eventually fail, such perfection is unattainable under normal circumstances. A sensible approach is then to assume that the underlying components — such as processors and communication links — are *fail-controlled* (do not possess arbitrary failure modes but behave in a "disciplined" manner under failures) and then to develop programming systems capable of tolerating the assumed faulty behaviour. Intuitively it should be clear that the task of building such a programming system is considerably simplified if the underlying hardware components are considered to function, upon failure, in a "disciplined" manner. Of course, there is still the task of building hardware components, most notably processors, that do indeed behave only as assumed. Fortunately, by judicious utilisation of redundancy in the form of redundant processors and communication channels with independent failure modes, highly reliable and predictably dependable hardware components with fail-controlled behaviour can be built. Given such a hardware base, one can then replicate application level entities (processes, objects) on distinct fail-controlled processors to construct *highly available services*. Such services themselves require replica consistency protocols to ensure that replicated services appear functionally equivalent to their non-replicated counterparts. In general, at the heart of any distributed system capable of providing high availability of system services will lie a number of redundancy management protocols: those necessary for managing redundant networks, redundant processors and application level process replicas. The task of designing such protocols can get hopelessly complicated in the absence of an overall architectural framework. In this paper we develop one such framework which is summarised below:

(i) distributed computations are assumed to be structured as processes communicating via messages;

(ii) processes execute on fail-controlled nodes which satisfy the abstraction of *fail-silence:* a node either functions according to the specification or stops functioning;

(iii) each node is composed of two processors which together execute clock synchronisation and order protocols to provide the fail-silent abstraction;

(iv) nodes communicate with each other through redundant communication channels.

(v) processes can be replicated on distinct nodes for increased availability; the degree of replication (if any) for a process will be determined by the application level availability requirements.

Such a system architecture can support *K-resilient processes:* a $K+1$ replicated process ($K > 0$) can tolerate a maximum of K replica failures before it becomes unavailable. When computations are replicated, "one to one" process communication patterns common in non-replicated computations have to be replaced by "many to many" communication patterns. We identify such "group communication" services and present the main contribution of the paper: a reliable and fast group communication protocol ideally matched to the functionality of the two processor fail-silent nodes. If non-faulty processors of nodes use digital signature based message authentication, then the architecture presented here can provide a practical basis for building scalable and responsive distributed systems capable of tolerating large classes of processor failures (Byzantine failures in the limit). Indeed, the DELTA-4/XPA extra performance architecture [1], intended for real time applications, embodies the main features outlined above: application level processes use a "leader-follower" replication technique for achieving k-resiliency and communicate using a reliable group communication service; the underlying hardware is assumed to be fail-silent. The group communication protocol to be described here, which has been specially designed for two processor fail-silent nodes, provides the functionality required by this replication technique.

It should be noted that there are two distinct levels of process replication in our architecture: (i) *at node level* an application process could be replicated on K+1 distinct nodes for K-resiliency; and (ii) *within each node* where a process of a node is really made up of two processes each running on a distinct processor. Fig. 1. clarifies these replication levels. Here, each node N_i is composed of two dedicated processors P_1 and P_2; a replicated, 1-resilient process G is composed of node level processes P and Q; and process P (Q) itself is composed of processes p_1 and p_2 (q_1 and q_2) running on processors P_1 and P_2 respectively.

We next state the *fault assumptions*. Since a node is constructed to meet the fail-silent abstraction, the node level processes running on it will be assumed to possess the fail-silent property. The processors of a node, on the other hand, are assumed to be *fail-uncontrolled* in that a failed processor (and hence its processes) can perform arbitrary state transitions. At most one processor in a two-processor node is assumed to fail. Redundancy in the communication

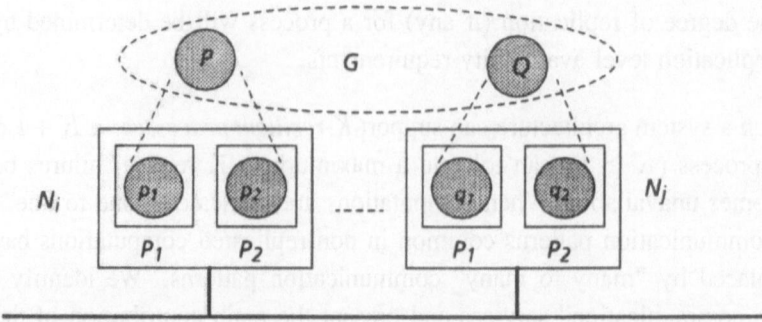

Figure 1: Replication Levels

subsystem is assumed to provide at least one fault-free communication path between non-faulty processors of any two nodes in the system, enabling processes on such processors to communicate with each other in a known and bounded time. It will also be assumed that the originator of a message can be *authenticated* by a non-faulty processor. Standard techniques employing checksums and message retransmissions are available to ensure that messages are delivered uncorrupted (with high probability); digital signatures implement message authentication (with high probability).

Replication plays different roles in the two levels: within a node, processor (process) replication is for maintaining the abstraction of a fail-silent node (a node level fail-silent process), while at the node level, process replication is for availability (since, at a node level, functioning processes do not perform incorrect state transitions). Note that we are not coping with *design faults*, since replication of faulty components cannot be expected to tolerate the consequences arising from those faults. The architecture nevertheless provides scope for incorporating *design diversity*; we will return to this subject very briefly towards the end of this paper.

We assume that distributed computations have been structured as a number of processes that interact only via messages. A process has an *input port* through which it receives all the messages directed to it and an *output port* through which it can send messages to other processes. This simple model is sufficiently general in that other models, such as clients and servers interacting through remote procedure calls, or objects communicating by messages can be seen as special cases (certain enhancements to the model are however possible and will be discussed in the last section). We assume that computations performed by processes are *deterministic* so that if all the correctly functioning replicas of a process have identical initial states then they will continue to produce identical responses to incoming messages provided the messages are processed in an identical order. The determinacy requirement is essential for achieving *active replication*, where each and every replica of a process carries out processing. In our architecture, active replication is employed within a node, so this requirement is fundamental to the preservation of the abstraction of a fail-silent node and node level processes.

The rest of the paper is structured as follows: we begin in section 2 by identifying group communication requirements for application level replicated computations, and then describe in section 3, a simple reliable multicast protocol. In both these sections we will be assuming that *each underlying node is a single entity* that has the fail-silent property, so, each node level process will also be regarded as a single fail-silent process. In section 4 we will describe how the abstraction of a fail-silent node can be implemented. In section 5, the protocol presented in section 3 will be adapted to exploit the particular features fail-silent nodes. Section 6 concludes this paper.

2   Replicating Application Level Processing

2.1   Process groups

The task of managing a distributed computation containing replicated processes can be best formulated in terms of the management of process groups (where each group will represent a replicated process) which are interacting via messages. To avoid any consistency problems, it is necessary to ensure that a group appears to behave like a single entity in the presence of concurrently arriving messages and failures. If not managed properly, such messages could be serviced in different order by the processes of a group, with the consequence that the states of replicas could diverge from each other. Group membership changes (caused by events such as replica failures and insertion of new replicas) can also cause problems if these events are observed in differing order by the users of the group. For example, consider the following scenario (see fig. 2), where process group G_A (replicas A_1, A_2) sends a message to a single process B, and is expecting a reply from B, but B fails during delivery of the reply to G_A. Suppose that the reply message is received by A_1 but not by A_2, in which case the subsequent actions taken by A_1 and A_2 can diverge. The

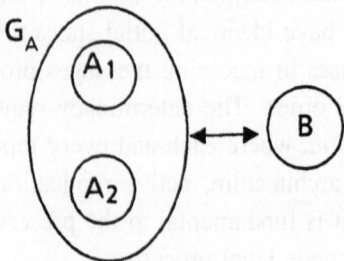

Figure 2: Process Groups

problem is caused by the fact that the failure of B has been "seen" by A_2 and not by A_1. To avoid these problems, communication between process groups is typically performed using *ordered reliable multicasts:* reliability ensures that all functioning members of a group receive messages intended for that group and ordering ensures that these messages are received in an identical order at each of the members. ISIS [2], *x*-kernel/Psync [3] and DELTA-4 [4] are examples of distributed systems relying on such multicasts for group management. Let us briefly examine the varieties of multicast communications.

2.2   Varieties of Multicast Services

The reliability and ordering attributes of multicast protocols provide a convenient means of classifying increasingly more sophisticated services. To start with, assume a sender, S_1, multicasting to a group $G = \{P_1 \dots P_n\}$.

(i) *Unordered Unreliable:* A multicast from S_1 is received by zero or more of the functioning $P_i \in G$; successive multicasts from S_1 are received in an unspecified order at the destinations. One case is depicted in fig.3, where the two multicasts from S_1 (m_1, m_2) are received at P_2 and P_j in differing order; further m_2 has not been received by P_1.

(ii) *Fifo Reliable (non atomic):* Provided the sender does not fail during a multicast, the message is received by all the functioning receivers. Further, all the multicasts from the same sender are received at the functioning destinations in the order the multicasts were made.

(iii) *Fifo Multicast:* As in (ii) above, except that if the multicast by the sender is received by a functioning receiver, then all the other functioning receivers

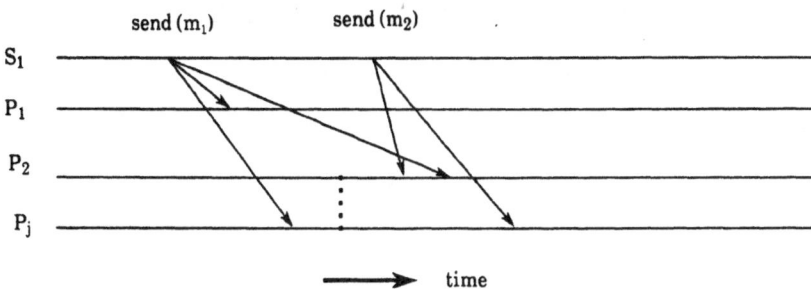

Figure 3: Unordered and Unreliable Multicasts

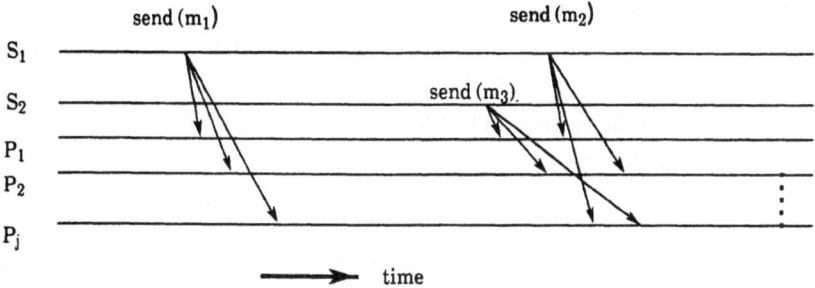

Figure 4: Multicasts with Fifo Ordering

will also receive it, even if the sender fails during the multicast. Fig. 4 shows a specific case where two senders (S_1, S_2) are multicasting to G. While m_1 and m_2 are received in *fifo* order at G, m_2 and m_3 are received in opposite order at P_2 and P_j (but this is permitted).

(iv) *Causal Multicast:* This multicast extends the ordering property of the previous service to causally related sends of different senders while meeting the reliability guarantee of (iii) (the concept of causally related events in a distributed system has been defined precisely by Lamport [5], here we will rely on the intuitive explanation given with reference to fig. 5). Fig. 5 depicts the following scenario: S_1 is multicasting to groups G_1 and G_2, P_1 is multicasting to group G_1, where $G_1 = \{P_2, P_3\}$ and $G_2 = \{P_1, P_4\}$. In this case, there is a potential flow of information from events send (m_1) to send (m_2), and from send(m_2) to send(m_3) (the potential information flow is shown by dotted arrows in the figure). Thus, the sending of message m_3 is (potentially) causally related to that of m_1 and, hence, the causal

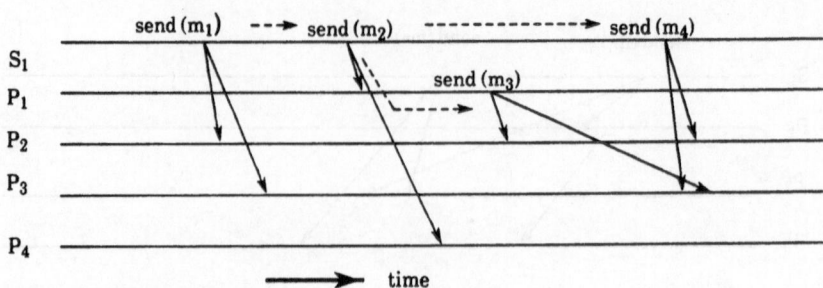

Figure 5: Causal Multicasts

multicast protocol ensures that all the functioning receivers of G_1 receive m_1 before m_3. However, events send (m_3) and send (m_4) are not causally related (as there is no potential information flow path between the send events of m_3 and m_4) so, these messages need not be received in identical order at the receivers (P_2, P_3).

(v) *Totally Ordered Multicast:* The causal order of the previous service is extended to *total order*, such that causal as well as non-causal (concurrent) messages are received in an identical order (which preserves the causality) at the receivers; the reliability guarantee of (iii) is maintained.

Although this is not a complete classification of all possible multicasting services (one could, for example, imagine a total order preserving unreliable multicast), it is sufficiently broad for our purposes (most published protocols that we know of provide services which fall into one of the classes described here).

Order preserving (causal or total) multicasts provide a convenient means of handling any consistency problems when dealing with resource allocation and related problems for distributed computations. There are several applications where it is preferable (for performance reasons) to impose an application specific ordering. For example, imagine a group representing a replicated file (each member is a process looking after a file copy) which can be used by remote client processes. To ensure that file copies remain mutually consistent, it is necessary to ensure that file operations (read and write) are applied in an identical order at all the functioning replicas. The use of a total order preserving multicast protocol by clients for sending operation requests to the group will ensure data consistency. However we can relax ordering requirement (and possibly gain some performance improvements) if the knowledge about the semantics of the operations is exploited. Thus, if the file is opened for read only operations, then the operations coming from different clients need not be ordered. Replicated object management in the DELTA-4/XPA extra performance architecture [1] and in the Arjuna distributed system [6] are two examples where total ordering at a group is imposed at a higher level in an application specific manner, rather than at the multicast protocol level. In particular, DELTA-4/XPA uses a clever "leader-follower" replication technique in which the message processing order is chosen by the "leader" of the group who forces the order on to the "followers". Such systems require only a fifo multicast service from the underlying communications system. Thus, in systems where application specific ordering is preferable and possible, only a fifo multicast service becomes the necessary facility required from the underlying communications system. The rel/REL_{fifo} protocol developed by us [7] and described in the next section provides such a service. It can also be extended easily to provide the causal multicast service.

3 A Reliable Multicast Protocol

We will now briefly describe the rel/REL_{fifo} protocol for reliable multicasts (as stated before, the fifo ordering is for a given sender; a sender's multicasts are received at the destinations in the order they were made). The development of rel/REL_{fifo} protocol assumes the existence of a fifo reliable (*non-atomic*) multicasting service, called *rel*, where by a *functioning* sender can reliably deliver a message to a set of functioning receivers in a known and bounded time. One way to realise *rel* would be for the sender to individually send the message to the members of the group and receive acknowledgements, and selectively retry this a finite number of times till acknowledgements are received from functioning members; another way, if the underlying network supports broadcasts, like in Ethernet, would be to broadcast the message and follow this up with selective retransmissions till acknowledgements are received. (Given that the redundancy in the communication subsystem assumed in our architecture is intended to ensure the existence of at least one fault free communication path between non-faulty processors in the system, realisation of *rel* with a known and bounded message delivery time is possible and perhaps straightforward). Note that we do not require *rel* to be *failure atomic:* a failure of the sender during the execution of *rel* can result in an incomplete multicast. Given the existence of *rel*, the sender uses a procedure *REL* for multicasting a message m in two rounds: *REL* transmits m using *rel* twice in a row. So every functioning receiver which receives the second multicast "knows" that the first one completed successfully, in which case it "forgets" about this multicast — in the sense that no actions are required for completing the multicast. However, if a receiver, after receiving the first one does not receive the second one within a "reasonable" timeout period, then it suspects a failure of the sender — in which case there could be a functioning receiver who has not received the message. The receiver therefore completes the multicast by (recursively) using *REL* to multicast m.

The protocol is discussed in detail elsewhere [7] where several optimizations are discussed and the protocol compared with other reliable multicast (broadcast) algorithms. In the absence of any failures, the above protocol can consume twice as many messages as its "non atomic" counterpart. If this price is not considered excessive — we expect this would be the case in the next generation of high bandwidth networks — then rel/REL_{fifo} offers an attractive alternative to existing protocols that work very hard to reduce the number of messages under no failure situations but at the expense of quite a sluggish performance when failures do occur. In addition, rel/REL_{fifo} does not require the maintenance of complex state information (such as message histories recorded in other protocols) or need the services of any underlying protocols for node failure detection, election of a new sender and so forth. The above protocol has the attractive property that a received message can be delivered to local processes straight away; monitoring and completion of the multicast can be carried out concurrently.

Let t_{rel} be the time taken for multicasting a given message to a given group by the protocol implementing *rel*, and t_d be the duration of the timeout interval at a receiver. Then the latency of the rel/REL_{fifo} protocol under a variety of situations can be stated as:

(i) no failures: t_{rel}

(ii) sender failure after first round: t_{rel}

(iii) sender failure in the first round: $t_{rel} + t_d + t_{rel}$

(iv) worst case f failures (sender and $(f - 1)$ receivers fail during their first rounds): $t_{rel} + f(t_d + t_{rel})$.

So far we have ignored any internal architectural features of fail-silent nodes. The protocol adapts very well when such features are taken into account for the case of two processor fail-silent nodes: *REL(m)* can be executed jointly by the two processors, each executing *rel(m)* once.

4   Fail-silent Nodes

We now describe how fail-silent nodes can be constructed. The basic idea is to replicate processing on two processors which can check each other's performance to form a self-checking processor pair; if one processor suspects a failure in the performance of the other processor, then one of the two processors in the node must have failed and therefore the suspecting processor stops functioning in an attempt to make the node fail silent. Fail-silent nodes have been used widely, for example in commercial transaction processing systems (eg. [8]). Such nodes have been designed with the assistance of specialized comparator hardware and clock circuits. A common (reliable) clock source is used for driving a pair of processors which execute in lock-step, with the outputs compared by a (reliable) comparator; no output is produced, once a disagreement is detected by the comparator. We term a node designed this way to be a *hard-fail-silent node*. A node constructed according to the design to be described below will be termed a *soft-fail-silent node*, since no special clock or comparator circuits are employed; rather processors use clock synchronisation and order protocols "to keep in step" [9]. Such nodes provide an attractive alternative to hard-fail-silent nodes since no special hardware assistance is required and further, as the principles behind the protocols do not change, the protocol software can be easily ported to any pair of processors (including the ones expected to be available in future). Since only two processors are used within a node to check on each other, the fail-silent characteristics of a node can be guaranteed only if no more than one processor within a node is faulty. A more detailed discussion of this and other fail-controlled processor architectures can be found in [9]. In the discussion to follow, we will be referring to processes running on the processors of a node (such as p_1, p_2, fig. 1) .

The overall node architecture is shown in fig. 6. Each of the two processors (P_1, P_2) have network interfaces (n_1, n_2) for inter-node communication over redundant networks; in addition, the processors are internally connected via a communication link, l, for intra-node communication. This link is considered to be reliable in the sense that a link failure will be considered in terms of a

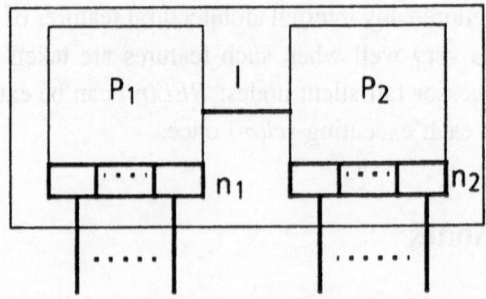

Figure 6: A Fail-Silent Node

processor failure. Each non-faulty processor in a node is assumed to be able to *sign* a message it sends by affixing the message with its (the processor's) unforgeable signature; it is also assumed to be able to *authenticate* any received message, thereby detect any attempts to corrupt the message. Digital signature based techniques (eg. [10]) can be relied upon to provide such functionality.

It is necessary that the replicas of computational processes (such as p_1 and p_2, fig. 1) on processors within a node select identical messages for processing, to ensure that they produce identical outputs. Identical message selection can be guaranteed by maintaining identical ordering of messages at input ports and ensuring that processes pick up messages at the head of their respective input ports. An *order protocol* is then required to ensure identical ordering if both processors are non-faulty.

An implementation of this order protocol will require that the clocks of both the processors of a node are synchronised such that the measurable difference between readings of clocks at any instant is bounded by a known constant ε. Algorithms for achieving this abstraction exist (see [11, 12]). Communication for clock synchronization and ordering takes place via the internal link. Given that the clocks of both the processors are synchronised, the order protocol can be implemented using a version of the signed message algorithm for Byzantine agreement [13].

As there are only two processors in a node, the clock synchronisation and message ordering protocols become particularly simple, since they are expected to work only in the absence of any failures. For example, an *order process* of a processor (which executes the order protocol) stamps a message to be ordered with its local clock reading; a copy of the timestamped message is signed and sent over the link to the *order process* of the other processor in the node. If T is the timestamp of the message received from or sent to the order process of the other processor, then the message becomes *stable* at local clock time $T + d + \varepsilon$, where d is the maximum transmission time taken for a timestamped message to travel from one order process to another order process over the link. A message with timestamp T will be said to become stable at a given instant of time, if no message with timestamp $T_1 < T$ will ever have to be accepted by an order process after that time. Stable messages are queued at the relevant input ports in the increasing timestamp order (with care being taken not to queue a stable message, if its replica has already been queued). If the order process receives, during the execution of the order protocol, a corrupt or untimely message, then one of the processors in the node must have failed ; the order process can subsequently insruct the local processor to stop functioning. Thus the clock synchronisation and message ordering protocols are not expected to work by masking a failure. This feature makes these protocols particularly simple to implement.

Each non-faulty processor of a node has five processes:

(i) *Sender Process:* this process takes the messages produced by the computational processes of that processor, signs them and sends them via the link to the neighbour processor of the node for comparison.

(ii) *Comparator Process:* this process authenticates all incoming messages from the neighbouring processor; an authenticated message (m_i) is compared with its counterpart produced locally. If the comparison succeeds, the authenticated message m_i is counter signed (by considering the first signature as a part of the message) and this double signed message is handed over to a *transmitter process* for network delivery to destination nodes. A locally produced message that cannot be compared because its counterpart does not arrive within a timeout period or a comparison which detects a disagreement indicates a failure. Once a failure is indicated, the comparator process stops, which results in no further double signed message being produced from that node.

(iii) *Transmitter Process:* this process is responsible for reliably multicasting the doubly signed messages to destination nodes using a version of the protocol discussed in the previous section (additional details will be presented in the next section).

(iv) *Receiver Process:* this process accepts authentic messages for processing from the network and considers only those authentic messages whose first two signatures are from distinct processors of the same node as valid messages; such valid messages are sent to the local *order process*. The receiver process is also responsible for completing an incomplete multicast, in a manner suggested earlier (again, additional details will be given in the next section).

(v) *Order Process:* this process executes the order protocol (mentioned earlier) with its counterpart in the other processor and attempts to construct identical queues of valid messages for processing by the computational processes.

When the comparator process of a non-faulty processor in a node detects a failure and therefore stops, no new double signed messages can be emitted by the node; any messages coming from this node that are not doubly signed will be found to be unauthentic at the receiving nodes. Thus a two-processor node exhibits fail-silent behaviour in the following sense: it either emits only valid messages or only *detectably invalid* messages; this behaviour is guaranteed so long as at most one processor in the node fails .

5 Reliable Multicasts over Two Processor Fail-silent Nodes

5.1 Fifo Multicasts

The protocol presented in section 3 proves particularly attractive in exploiting the features of two processor fail-silent nodes connected by redundant networks. Figure 7 shows one such configuration for duplicated networks (for the sake of illustration, a dual bus structure is shown; a variety of other schemes are possible, such as point to point links and so forth). The protocol retains its essential characteristics, except that the two sequential multicasts made by a sender using *REL* can now be replaced by two concurrent multicasts, each one made by the processors of the node over the redundant networks. As before, a receiving node must complete a multicast which is suspected to be incomplete. Throughout our discussion below, we will treat unicasting as a degenerate case of multicasting.

As stated in the previous section, every processor has a *transmitter process* which receives doubly signed messages for transmission from the local com-

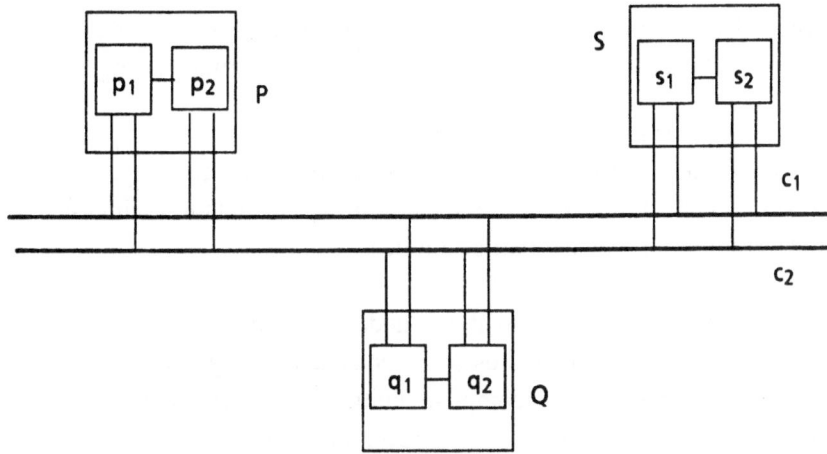

Figure 7: A Distributed System

parator process. We assume that the transmitter maintains an input message queue in which messages to be transmitted can be deposited. The algorithm for this process is shown below; the process uses *rel* for multicasting a message over n redundant networks (we assume that a message contains a list of destination addresses):

transmitter process:
   cycle
   get(m);/* get a message from the input queue */
   i := 1;
   do   /* loop for making a multicast over n networks */
        i ≤ n → rel(m,i); /* multicast m on network i */
               i := i + 1
   od
   end cycle

If both the processors of a transmitting node (say S in fig. 7) are non-faulty, then each of them will independently complete multicasting a message m. If one of them, say s_1 in S, is faulty, then the following mutually exclusive situations are possible:

(i) comparison for m succeeds at s_1 and s_2 but s_1 fails to complete the multicast; in this case only s_2 will be able to complete multicasting m.

(ii) comparison for m succeeds at s_1, but s_1 fails to send the locally produced version of m to s_2 for comparison; so s_2, being unable to compare, decides to stop (after a timeout); in the mean time s_1 attempts to multicast m but fails resulting in a single partial multicast from S.

Since at most one processor within a node is assumed to fail, these are the only two types of failure scenarios to be considered; in particular we do not have to deal with the case of incomplete multicasts by *both* the processors of a sending node. Thus, if a processor of a receiving node receives two messages — one from each of the processors in the sending node — then it is safe for the processor to assume that the multicast will be "seen" in every other receiving node. Thus a processor need only take measures for completing a multicast if only one of the two messages is received by it.

Every processor also has a *receiver process* (mentioned in the previous section) which is responsible for picking up authentic messages coming from the redundant network and, as we shall see, the internal link. Transmitter and receiver processes use the services of *rel* for message transmission and reception over the redundant networks. The main functions of the receiver process are: (i) to authenticate a received message; (ii) deliver a received message to the local order process for ordering (before eventual processing by the computational processes); and (iii) if necessary, complete a multicast.

The communication paths between the receiver and order processes are shown in fig. 8 (for a node Q with processors q_1 and q_2). We note that the order processes exchange messages with each other (for the order protocol) via the receiver processes, rather than directly. This arrangement is necessary to ensure that a receiver process of a non-faulty processor "sees" every message that is being ordered; thus a receiver is always in a position to complete, if necessary, a multicast for a message which has been (or is being) ordered. Consider the following scenario, which will emphasize the observation just made. Assume that node S is the sender of a message m, with P and Q as receivers (see fig. 7), and a processor within S fails during a multicast, resulting in only q_1 receiving m, but not q_2. If the order processes communicate directly with each other, then m can get ordered at q_2, without the receiver at q_2 being "aware" of it. Thus we have a situation where q_2 will consume m, without being prepared to complete a multicast for m. The arrangement of fig. 8 avoids

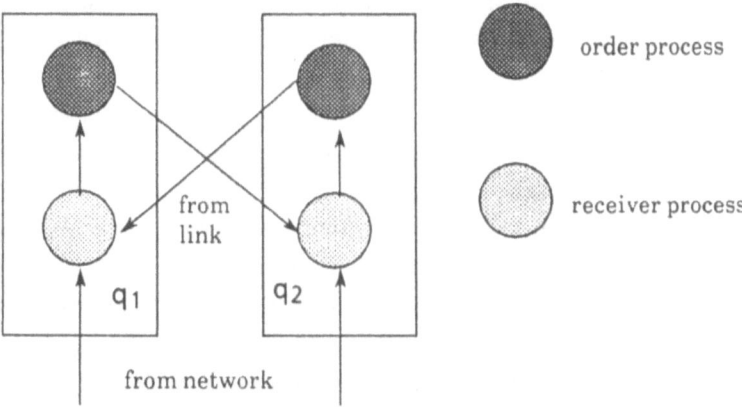

Figure 8: Order and Receiver Processes

this, as the receiver at q_2 does receive m, indirectly from the order process of q_1. A processor completing a multicast for a message signs the message before transmission; so a receiver process can expect *three types* of messages (as an example we will consider node S multicasting m to P and Q; we will use the notation $(msg)s_1$ to represent the message m from s_1; the notation shows the *signature-stripped* message *msg* signed by processor s_1 of node S):

(i) $type_1$: a double signed message from the network (the receiver at q_1 can get $((msg)s_1)s_2$ and $((msg)s_2)s_1$, if both s_1 and s_2 completed the multicast, or just one of the two, if either s_1 or s_2 did not complete the multicast);

(ii) $type_2$: a message for ordering, over the link, from the order process of the neighbour; this message is triple signed and timestamped (the receiver at q_1 can get $((((msg)s_1)s_2)t)q_2$ from q_2 for ordering, where, t is the timestamp put by the order process of q_2);

(iii) $type_3$: triple signed message from the network (for example, the receiver at q_1 can get $(((msg)s_1)s_2)p_1$, which is a message from p_1, if p_1 tries to complete the multicast initiated by S; a similar message from p_2, namely, $(((msg)s_1)s_2)p_2$ may also be received).

The algorithm for a receiver process is shown below, where it is assumed that the receive primitive performs authentication and accepts only those autentic messages whose first two signatures are from different processors of the same node; it is also assumed to discard any duplicate messages from a given sending processor.

The receiver process delivers authentic messages to the local order process (filtering out any replicas) so that these messages can be ordered identically in both the processors. At the same time it arranges, for a multicast message, the monitoring of the multicast. The algorithm for the procedure MONITOR(m) is given below. This procedure starts a thread (a lightweight process) to monitor the progress of the multicast; thus message ordering and monitoring the completion of a multicast can proceed concurrently.

Recall that two messages — one from each processor of the sending node — must be received for the multicast to be considered complete. The thread, upon being set up by the MONITOR process, looks for *dual messages* in its queue within a timeout period t_d. We define dual messages for a multicast as follows: m and m/ will be said to be dual messages if their signature stripped versions are identical and the last signatures in m and m/ are from different processors of the same fail-silent node. For example, if msg is the signature stripped version of m and m/, then for a processor in Q, $((msg)s_1)s_2$ and $((msg)s_2)s_1$ will be duals; similarly, $(((msg)s_1)s_2)p_1$ will be dual of either $(((msg)s_1)s_2)p_2$ or $(((msg)s_2)s_1)p_2$. If dual messages are received then this means that the multicast can be considered to be complete, so the thread is destroyed. If the timeout occurs then this indicates that the sender node most probably carried out a single partial multicast, so the thread arranges the multicast to be completed by depositing a triple signed message in the queue of the local transmitter. The multicasting of the triple signed message will be superfluous and harmless if appropriate messages arrive later to form dual messages. Indeed, it is possible that the timeout expires for one thread but not for its counterpart in the other processor. This, however, does not cause any inconsistency among the two processors. The processors of a node are expected to multicast a message sufficiently close to each other in time, so that the timeout period t_d should be largely influenced by the difference between maximum and minimum transmission delays for messages.

5.2 Correctness reasoning

Suppose that a node S carries out a multicast to a group G of nodes. Since each node is a two processor fail-silent node, there are only three mutually exclusive cases to consider for such a multicast: (i) the multicast succeeds,

receiver process:
  cycle
  receive(m) /* receive an authentic message from the network or link */
  if m is a multicast message →
    if m is $type_1$ →
      if a replica already received from the neighbour order process →
          skip
      □
      replica not received → send a copy to the local order process;
          MONITOR (m)
        /* a new message, needs ordering and monitoring */
      fi

    □
    m is $type_2$ → send a copy to the local order process
      /* message to be ordered */
    strip the timestamp and the third signature from m;
      /* m is now double signed */
    if a replica of m received from the network → skip
    □
    no replica received from the network → MONITOR (m)
      /* a new message, needs monitoring */
    fi

    □
    m is $type_3$ → MONITOR (m);
      strip the third signature from m /* m is now double signed */
      if a replica of m already received → skip
      □
      a replica not received → send a copy to the local order process;
      fi /* a new message, needs ordering */
    fi
  □
  m is not a multicast message → send to the local order process
      /* no further action necessary */
  fi
  endcycle

procedure MONITOR (m)
    { msg := m with all the signatures stripped
    if multicast for msg complete → skip
    □

    multicast for msg not complete →
        if thread for msg created → deposit m in input queue of thread
        □

        thread for msg not created → create a thread for msg;
                deposit m in input queue of the thread
                      /* start thread to monitor progress of the multicast */
        fi
    fi
    }
thread:
    { m := head of input queue
        /* message at the head of the queue copied into m */
    msg := m with all the signatures stripped
    start-timer(td)
        /* now look for the dual of m in the queue with a timeout */
    do
    look for dual messages in the input queue →
        record completion of msg multicast; die
        /* dual messages are received, so... the thread is killed */
    ||
    timeout →
        if m is $type_3$ → strip third signature /* m is now double signed */
        □

        m is $type_1$ → skip
        □

        m is $type_2$ → raise a failure exception
            /* error: incorrect output by receicer process */
        fi
    sign m; deposit m in the queue of the transmitter process;
    record completion of msg multicast; die
        /* timeout ... so suspect an incomplete multicast ;
        arrange for m to be multicast; no more work for msg needed */
    od
    }

with both the processors in S multicasting; (ii) a processor in S fails such that only a single multicast is made; and (iii) a processor in S fails such that only a single partial multicast is made. We show that the protocol described before meets the property required from fifo multicasts.

case 1: Each and every non-faulty processor of functioning nodes in G will "see" the multicast as complete.

case 2: Each and every processor of functioning nodes in G will "see" the multicast, detect it as incomplete and attempt to complete it individually. In both the cases, all the receiver processes in the functioning nodes in G will get the message, which will be passed on to local destination processes via local order processes.

case 3: Let the message be available for processing at a non-faulty processor, p_1. Since the message must pass through the receiver process of p_1, that receiver process will eventually attempt to complete the multicast. Thus all functioning nodes in G will "see" the multicast. If the neighbouring processor, p_2, is also non-faulty, then p_2 will also attempt to complete the multicast; so all functioning nodes in G can "see" the multicast as complete.

A more careful development of the correctness reasoning touching upon the failure atomicity and termination properties of the protocol will be presented in a subsequent publication.

5.3 Causal multicasts

The protocol described above can be made to respect causality quite easily. (Causal multicasts were discussed in section 2.2). First, let us see how causality can be broken. Consider the following case: node S multicasts m_1 to nodes P, Q and R; P upon consuming m_1, multicasts m_2 to Q and R; causality requires that Q and R receive m_1 first and then m_2. The following is possible: S fails such that only P gets m_1; this message is delivered to P who then multicasts m_2, and then P completes the multicast of m_1; however m_2 will now be received before m_1 at Q and R. This situation will not occur if the completing multicast for m_1 at P precedes the multicast for m_2. In our protocol, this can be guaranteed if the condition $t_d < \pi$ can be made to hold at a non-faulty processor, where π is the minimum time taken for an ordered message to be processed, the output compared with the neighbour, and a double signed message deposited in the queue of the transmitter process. Let t be the time a message (m_1) is received at a processor; so a causally related message (m_2) cannot be produced and queued in the queue of the transmitter before time $t+?$ and if the above condition holds, any completing multicast message (m_1) will get queued before m_2 in the queue of the transmitter, thereby ensuring that causality is preserved, since the transmitter will now multicast m_1 before m_2.

6   Concluding Remarks

We restate the main features of the architecture:

(i) distributed computations are assumed to be structured as processes communicating via messages;

(ii) processes execute on fail-controlled nodes with the *fail-silent* property: a node either functions according to the specification or stops functioning;

(iii) each node is composed of two processors which together execute clock synchronisation and order protocols to provide the abstraction of a fail-silent node;

(iv) nodes communicate with each other through redundant communication channels.

(v) processes can be replicated on distinct nodes for increased availability; the degree of replication (if any) for a process will be determined by the application level availability requirements.

Reliable multicast protocols are required for supporting application level distributed computations; several such multicasting services were identified. We noted that many systems (eg, [1, 6]) require fifo reliable multicasts. We have presented an *integrated* suit of protocols for supporting reliable fifo and causal multicasts over two processor fail-silent nodes. We have shown how dual redundancy within a node can be exploited for making multicasts using a version of the rel/REL protocol presented in section 3. Reliable multicast (broadcast) protocols have been studied extensively under the assumption that the nodes are fail-silent (eg, [2, 3, 14]); protocols under the fail-uncontrolled assumption, requiring Byzantine agreement, have also been studied (eg, [15]).

The protocols developed here differ from these in that the specific architecture that we have proposed permits the use of ordinary fail-uncontrolled processors within nodes, obviating at the same time any need for network wide Byzantine agreement and order protocols. In this respect the architecture presented here can provide a practical basis for building scalable and responsive distributed systems capable of tolerating large classes of processor failures (Byzantine failures in the limit).

We have assumed a simple processing model for active replication within a node: a process has a single input and output port and messages are picked up in fifo order from the input port. Many systems require additional functionality in the form of multiple ports, timeouts, alarm messages and preemptable processing. These features are potential sources for introducing divergence in the states of replicas, and require more sophisticated ordering techniques within a fail-silent node; such techniques are discussed in [16].

The architecture we have presented is in the spirit of the fault tolerant hardware-software architectures developed by other researchers (eg, [17, 18]). Interestingly, we note that by employing two different types of processors within a node, a measure of tolerance against design faults in processors can be obtained. We also note that, since the processors of a node are loosely synchronised, common mode transient failures are unlikely to affect the computations on the processor pairs in an identical fashion; thus a measure of tolerance against transient failures can be achieved. In this respect, soft-fail-silent nodes appear to have an advantage over hard-fail-silent nodes with their lock step execution. Systems such as Mars [19] have adopted a different approach than ours. Here, each node is assumed to be fail-silent; however a node is a single processor with self checking capability, and time redundancy at application level is used for coping with transient failures. We are currently in the process of building fail-silent (and failure masking, TMR) nodes; order and synchronization protocols are currently being implemented on transputer based nodes. It is our intension to evaluate the performance of systems built using such nodes in realistic settings.

Acknowledgements

This work has been supported in part by grants from the UK Science and Engineering Research Council and ESPRIT project 2252 (DELTA-4). Conversations with John Warne and colleagues from the DELTA-4 project, in particular, David

Drackley, David Powell, Douglas Seaton, Neil Speirs and Paulo Verissimo have been helpful.

References

[1] P. A. Barrett *et al.*, "The Delta-4 extra performance architecture (XPA)," in *Digest of papers, FTCS-20*, pp. 481–488, June 1990.

[2] K. Birman and T. Joseph, "Reliable communication in the presence of failures," *ACM TOCS*, vol. 5, pp. 47–76, Feb 1987.

[3] L. Peterson, N. Buchholz and R. Schlichting, "Preserving and using context information in interprocess communication," *ACM TOCS*, vol. 7, pp. 217–246, August 1989.

[4] D. Powell *et al.*, "The Delta-4 approach to dependability in open distributed computing systems," in *Digest of Papers, FTCS-18*, pp. 246–251, June 1988.

[5] L. Lamport, "Time, clocks, and the ordering of events in a distributed system," *Communications of the ACM*, vol. 21, pp. 558–565, July 1978.

[6] M.Little and S. Shrivastava, "Replicated k-resilient objects in Arjuna," in *Proceedings of the IEEE Workshop on the Management of Replicated Data*, pp. 53–58, November 1990.

[7] S. Shrivastava and P. Ezhilchelvan, "Rel/REL: A family of reliable multicast protocols for high speed networks," tech. rep., Computing Laboratory, University of Newcastle upon Tyne, UK, 1990.

[8] P. A. Bernstein, "Sequoia: A fault-tolerant tightly coupled multiprocessor for transaction processing," *IEEE Computer*, pp. 37–45, February 1988.

[9] S. Shrivastava *et al.*, "Fail-controlled processor architectures for distributed systems," tech. rep., Computing Laboratory, University of Newcastle upon Tyne, UK, 1990.

[10] R. Rivest, A.Shamir, and L. Adleman, "A method of obtaining digital signatures and public-key cryptosystems," *Communications of the ACM*, pp. 120–126, February 1978.

[11] J.Y.Halpern, B.Simons, H.R.Strong, and D.Dolev, "Fault tolerant clock synchronisation," in *Proceedings of Third ACM Symposium on PODC*, pp. 89–102, August 1984.

[12] L.Lamport and P.M.Melliar-Smith, "Synchronising clocks in the presence of faults," *Journal of the ACM*, vol. 32, pp. 52–78, January 1985.

[13] L.Lamport, R.Shostak, and M.Pease, "The byzantine generals problem," *ACM Transactions on Programming Languages and Systems*, vol. 4, no. 3, pp. 382–401, 1982.

[14] J. Chang and N. Maxemchuck, "Reliable broadcast protocols," *ACM TOCS*, vol. 2, pp. 251–273, August 1984.

[15] H. F.Cristian, H.Aghili and D.Dolev, "Atomic broadcast: from simple message diffusion to byzantine agreement," in *Digest of papers, FTCS-15*, pp. 200–206, June 1985.

[16] A.Tully and S. Shrivastava, "Preventing state divergence in replicated distributed programs," in *Ninth IEEE Symposium on Reliable Distributed Systems*, pp. 104–113, Oct 1990.

[17] J. -C. Laprie *et al.*, "Definition and analysis of hardware and software fault tolerant architectures," *IEEE Computer*, pp. 39–51, July 1990.

[18] R. Harper, J. Lala, and J. Deyst, "Fault tolerant parallel processor architecture overview," in *Digest of papers, FTCS-18*, July 1988.

[19] H. Kopetz *et al.*, "Tolerating transient faults in Mars," in *Digest of papers, FTCS-20*, pp. 466–473, June 1990.

Evaluation

A Numerical Technique for the Hierarchical Evaluation of Large, Closed Fault-Tolerant Systems

DON LEE

Aerospace Corporation

Mail Stop M1-046

P.O. Box 92957

Los Angeles, CA 90009, USA

JACOB ABRAHAM

Computer Engineering Research Center

University of Texas at Austin

2201 Donley Dr., Ste. 395

Austin, TX 78758, USA

DAVID RENNELS

Computer Science Department

University of California at Los Angeles

Boelter Hall 3732

Los Angeles, CA 90024, USA

GEORGE GILLEY

Aerospace Corporation

Mail Stop M1-046

P.O. Box 92957

Los Angeles, CA 90009, USA

Abstract

This paper describes a novel approach for evaluating the reliability of large fault-tolerant systems. The design hierarchy of the system is preserved during the evaluation, allowing large systems to be analyzed. Semi-Markov models are used at each level in the hierarchy, and a numerical technique is used to combine models from a given level for use at the next level. Different values of parameters, such as coverage, can then be used appropriately at any level, resulting in a much more accurate prediction of reliability. The proposed technique has been validated through comparison with analytical calculations, results from existing tools and Monte-Carlo simulation.

1 Introduction

When designing complex fault-tolerant systems with high mission reliability requirements, reliability evaluation aids designers in comparing computer design alternatives and the impact of various subsystem parameters and dependencies on system reliability, as well as in estimating the values of design parameters (i.e., coverages and the number of spares per subsystem) that will satisfy the reliability requirements of the design. To predict a design's reliability, a model of the design must be created, so that the necessary mathematical manipulations can be performed conveniently to make predictions and investigations into the design's reliability. However, reliability prediction is only as accurate as the models and assumptions (such as that the occurrence of faults follows a Poisson process) on which it is based. The models used must approximate as closely as possible the system being designed or evaluated. Otherwise, the system as designed will be inappropriate, or the system chosen after evaluation will not be the best choice [1].

Markov Models

Markov models are widely used to model complex systems, and the reliability of fault-tolerant systems can be evaluated by solving the models [2, 3]. Unlike other methods (e.g., fault trees and reliability block diagrams), Markov models are capable of including the details of the design that handles a fault and the resulting errors. In particular, Markov models can include coverages which are extremely important parameters in a fault-tolerant design. Design reliability is extremely sensitive to the values of coverage [4]. Additionally, Markov models can represent system dependencies; transient, intermittent, and near-coincident faults; and standby systems with "warm" spares. Fault trees and reliability block diagrams are unable to model the impact of these design features and fault types on a system's reliability with the same ease as the Markov models.

However, the use of Markov models for the reliability prediction of large, closed fault-tolerant systems introduces several problems. A major problem is dealing with large state spaces. Markov representations of designs that use many different components, (e.g., CPUs, buses, and memories), high redundancy levels, and complex mechanisms to detect and recover from errors result in an explosion in the number of states that represent the design. For example, if we have a system with n components, it is possible to have 2^n states in the design model. Such large design models are extremely difficult to construct, and solving these models can exceed the computational capabilities of existing reliability prediction tools.

The problem of constructing large, accurate Markov models is currently solved by using a more succinct form of system description and generating the Markov model from this description. Several reliability prediction tools, such as HARP [5, 6], SHARPE [7], and SAVE [8] do this. Models for large systems may also be constructed by first describing system behavior using extended stochastic Petri nets, and then transforming the nets into Markov or semi-Markov processes [3, 5]. However, this may also result in very large models.

Several approaches have been devised to aid in solving Markov models that have large state spaces: structural decomposition, graph pruning, behavioral decomposition and randomization. Behavioral decomposition separates the Markov model according to the relative magnitude of the state-transition rates. All fast transitions are solved separately and the results are used to modify the transition rates of the remaining model. Behavioral decomposition is the approach used in two reliability prediction tools commonly used by system developers: CARE III [9, 10] and HARP. In contrast, SHARPE [7] uses a mixture of modeling techniques to avoid the state space problem. SHARPE uses Markov chains to model subsystems, and the results of these subsystem models are used as inputs to a combinatorial solution of the overall system reliability. SAVE [8] uses randomization to solve the Markov chain, bounds can be found on the error due to using only a limited number of terms in an iterative solution. This technique, coupled with clever storage management techniques allows the solution of state spaces of size 10^5 or greater. Despite the development of such approaches, large state spaces pose a problem as system developers try to predict the reliability of even larger and more complex systems. The Markov models of these systems can be so large that the usefulness of the above approaches is reduced.

Another problem in using Markov models is the difficulty of constructing models that include the coverages associated with each of the error handling regions (i.e., error containment boundaries) present in the design hierarchy. Each error handling region in the design attempts to detect errors present in that region and to recover from errors following detection. If errors propagate out of a "lower" region, or the "lower" region is unable to recover from the error, then the use of another set of mechanisms from a "higher" region will be required to detect or recover from the error. If important system parameters (especially coverages) associated with the individual levels in the error handling hierarchy are not adequately included in the model (e.g., the coverages either incorrectly modeled or not modeled at all), then the model does not provide a good approximation of the system. As a result, inaccurate reliability predictions will occur. Furthermore, in a fault-tolerant design there is a hierarchical treatment of errors, and not all errors are treated the same way at each level.

Recovery is also hierarchical. Also, the handling of errors following detection occurs simultaneously at different levels.

Proposed Approach

This paper presents an approach to avoiding the large state space problem and allows the coverages associated with the hierarchy of error-handling regions to be included in the model. Our solution to these problems involves the numerical solution of a hierarchy of semi-Markov[1] models for analyzing the reliability of complex, fault-tolerant designs. This approach does not depend on any combinatorial solutions and allows the flexibility and advantages of the Markov technique to be used at all levels in the modeling hierarchy. This approach is also simple enough to allow complex models to be solved on a personal computer. Section 2 presents the basic approach, and Section 3 describes SNARC (Semi-Markov Numerically Approximated Reliability Calculations), a new reliability prediction tool which incorporates the new technique. Section 4 presents the steps taken to validate the tool using comparisons with analytical models, other tools and Monte-Carlo simulation. Finally in Section 5, conclusions are presented and some future research is discussed.

2   The Hierarchical Use of Semi-Markov Models

Normally, design levels are referred to as high or low, depending on the complexity of the components in the level. The more complex the components, the higher the level. A component in any level L_i is equal to the network of components taken from level L_{i+1} below it. A system is hierarchical when there is a one-to-one mapping between the components in level L_i and the separate subsystems in level L_{i+1}. At each level, the components of a hierarchical system are self-contained and stable modules. The existence of stable intermediate modules greatly facilitates the design of hierarchical systems.

The notion of hierarchical organization has had a great impact on computer system design. Because computer systems are enormously complex, hierarchical organization has allowed designers to proceed from higher to lower design levels, progressing to successively greater levels of detail. For the most part,

---

[1]In semi-Markov models, the state-holding times do not have to be exponentially distributed; in Markov models, the state- holding times must be exponentially distributed.

the levels in the design hierarchy are defined by the physical boundaries of each subsystem (e.g., chip, board, and so on).

Hierarchical organization can also be applied to handle the complexity of Markov models used for reliability prediction. One can proceed downward from the highest level to the lowest level of the design converting each level into a set of semi-Markov models, one for each separate subsystem in the level. Each semi-Markov model contains only the information important to the subsystem (i.e., the cumulative failure rates of components and coverages) at that level in the design, suppressing unnecessary information about lower levels. Modeling the design as a hierarchy of semi-Markov models avoids the state space problem, since the number of states does not grow exponentially with the number of components, but only linearly. The hierarchical approach allows the coverage values associated with each subsystem's error handling mechanisms at level L_i to be included as parameters in the appropriate subsystem model at level L_i. No other current technique allows this hierarchy of coverages to be modeled. In addition, the models at each level can be easier to construct and verify, since these submodels correspond to the subsystem design at the various levels.

In order to solve a model at any given level, parameters from the models at the lower level are required. Thus, a technique is needed to correctly map these parameters from the lower level models to the models at the next higher level.

The proposed technique to combine the models at different levels is to first take the same time slice from time t to $t + \Delta t$ through all levels of the hierarchy. Then, the reliability for all of the models at the lowest level is calculated. Next, the approximate increase in the cumulative failure rate (CFR) for that time slice of these same models is calculated and "inserted" as parameters in the next higher level models. This iterative procedure is followed until the reliability of the highest level model is calculated. The entire process is repeated for n time slices over the time interval that the reliability of the system is to be determined. The reliability of the highest level model at the last time slice will be the reliability of the entire system for the given total time interval.

The approximate increase in the cumulative failure rate for each model in the interval from time t to $t + \Delta t$ is determined by calculating $[R_i(t + \Delta t) - R_i(t)]/R_i(t)$, where $R_i(t)$ is the reliability of design level i at time t (see Appendix). The amount of error in the approximation is a function of the size

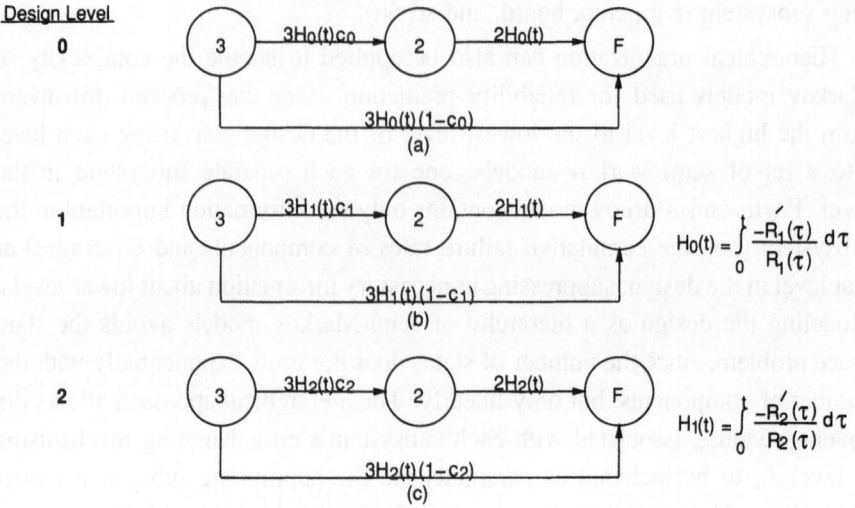

Figure 1: Markov chain of system with three design levels

of time step dt used in solving the model due to the integration technique used.

In order to illustrate the use of this modeling approach for a system that contains three design levels, consider the Markov chain in Figure 1.

Level zero (Figure 1(a)) represents the top of the design hierarchy, and level two (Figure 1(c)) represents the bottom of the design hierarchy. Each design level L_i is composed of three components, of which two must work. The model at any lower level represents a more detailed model of a component at the next higher level.

The models are combined from the lowest design level up to the highest design level for each time slice, starting from time zero and ending with the total mission time. The reliability of design level two (Figure 1(c)) at time slice (t to $t + \Delta t$) is determined by solving its semi-Markov model. The increase in the level one CFR in the time interval from t to $t + \Delta t$ is calculated by

$[R_2(t + \Delta t) - R_2(t)]/R_2(t)$, where $R_2(t)$ is the reliability of design level two at time t. Next, the increase in the level one CFR is used to determine the reliability of design level one (Figure 1(b)) at time $t + \Delta t$ by solving its semi-Markov model. Then the increase in the level zero CFR is approximated for the same time interval by

$[R_1(t + \Delta t) - R_1(t)]/R_1(t)$.

The increase in the level zero CFR is used in the semi-Markov model for design level zero (Figure 1(a)) to calculate the reliability of design level zero, $R_0(t + \Delta t)$, and thus the entire system at time $t + \Delta t$. At a given point in time, therefore, one can proceed from the lowest level, up the hierarchy, to obtain the reliability of the highest level of the design. The computation and storage requirements are minimal, and complex designs can be evaluated readily on a personal computer.

The accuracy of this technique's final result is a function of the number of the time slices used to solve the models. The more time slices used to solve the models, the more accurate the approximation.

3 SNARC description

SNARC models a hierarchical set of subsystems, where the highest level subsystem is treated as the system model. Each subsystem model can be composed of a set of elements that can be i) constant-failure-rate elements, or ii) elements that are lower level subsystems, or iii) a mixture of both. SNARC subsystems fall into one of four types:

A. Self-Contained Models

Type 1. Subsystems with redundant, constant failure rate elements.

These subsystems are composed of a redundant set of elements, but the individual elements are not internally redundant. The reliability $R(t)$ and instantaneous failure probability $R(t) \times dt$ of these subsystems are computed using Constant Transition Rate (Markov) Models.

Type 2. Subsystems that have empirically specified time-varying failure rates.

The failure rate is determined as a function of time for these subsystems from a list of variable pairs specifying a time and failure rate: (time (t_i), FR(i)). The instantaneous failure rate is determined by the straight line interpolation:

$$FR(t) = FR(t_i) + \frac{(t-t_i)}{(t_{i+1}-t_i)}[FR(t_{i+1}) - FR(t_i)],$$
where $t_i \leq t \leq t_{i+1}$

By listing two time-$FR(t)$ pairs with the same time, it is possible to create instantaneous changes in failure rate - the interpolation uses the first value at times below t, and the second value at times after t. The instantaneous probability of failure $dR(t) = R(t) \times FR(t) \times dt$ and the reliability $R(t+1) = R(t) - dR(t)$ are computed for Type 2 models directly from the instantaneous failure rate.

B. Models with Elements Based on Previous Models

Type 3: Subsystems with embedded elements from previous models.

These subsystems contain a redundant configuration of elements. Some of the elements are previously modeled subsystems and that therefore have time-varying failure rates. The time-varying failure rates are due to the internal redundancy of embedded subsystem elements. (It is important to note that in addition to embedded subsystem elements, these models may also include constant-failure-rate elements.) This type of subsystem is modeled with a semi-Markov process in which the transition rates vary as a function of time. The time-varying transition rates are re-computed at each step of a numerical integration and contain a factor of $(dR[M_i(t)]/dt)/R[M_i(t)]$, where $R[M_i(t)]$ is the reliability of a previous model at the current time step.

Type 4. Subsystems with non-redundant elements.

In this case the failure of one element will disable the subsystem. Therefore the reliability of this type of subsystem is simply the product of the element reliabilities. Elements can be any of the four types of subsystems.

A SNARC Model

A SNARC model contains a list of subsystem specifications, which are evaluated numerically in a round-robin fashion. In order to best explain this we use a SNARC modeling example. A hypothetical but representative architecture is shown in Figure 2, and its SNARC description is given below:

This model has six subsystems and four levels. It is assumed that every chip has a failure rate of 100 FITs (i.e. 10^{-7} failures/hour).

1. Subsystem 1 (Type 1, Level 1) is a memory module that consists of a redundant 41-chip memory array and a 2-chip non-redundant interface. Thirty nine memory chips must work, and two are spares. If a memory chip fails, it is replaced with probability (coverage) of 0.998. The memory module also has two non-redundant interface chips whose failure will disable the module. It has the following transition matrix:

```
s1-->s2 = 100*41*.998              S1  (OK)
s1-->s4 = 100*2*1 + 100*41*.002 S2  (ONE RAM FAILED)
s2-->s3 = 100*40*.998              S3  (TWO RAMS FAILED)
s2-->s4 = 100*2*1 + 100*40*.002 S4  (MODULE FAILED)
s3-->s4 = 100*2*1 + 100*39*1
```

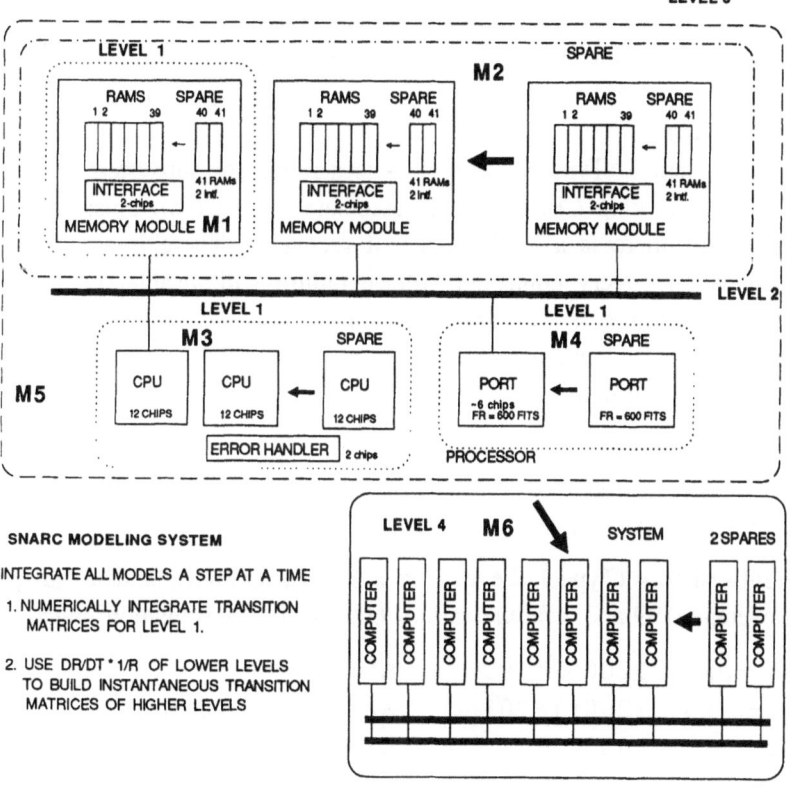

Figure 2: A SNARC Modeling Example

2. Subsystem 2 (Type 3, Level 2) is a memory subsystem that consists of three memory modules. Two must work, and coverage on first failure = C. The value C will be varied as a parameter in following modeling examples.

This subsystem model is entered into SNARC as a 3-state model with the following transitions:

```
s1-->s2 = ITR(M1)*3*C            S1(OK)
s1-->s3 = ITR(M1)*3*(1-C)        S2(ONE MEMORY GONE)
s2-->s3 = ITR(M1)*2*1            S3(FAIL)
```

At every time step as the system is evaluated by numerical integration, the instantaneous transition rate (ITR) is computed for model 1 (the memory module) and is substituted for in the matrix above. This model (2) is then integrated for the same step. ITR(M1) has value:

ITR(M1) = (dR[M1(t)]/dt) / R[M1(t)] where dt is approximated as the integration timestep.

3. Subsystem 3 (Type 1, Level 1) - is a processor subsystem consisting of three self-checking CPUs with an associated cache (two must work). If one CPU fails there is a .995 chance of recovery. There is also a 2-chip non-redundant recovery controller, whose failure will disable the processor subsystem. Each processor consists of 6 chips.

This subsystem is entered into SNARC as a 3-state Markov model with the following transition matrix:

```
s1-->s2   100*18*.995                   S1(OK)
s1-->s3   100*2*1 + 100*18*.005         S2(ONE CPU GONE)
s2-->s3   100*12*1 + 100*2*1            S3(FAIL)
```

4. Subsystem 4 (Type 1, Level 1) is a set of standby redundant bus ports, with coverage 0.99 and a failure rate of 600 FITS. The transition matrix is:

```
s1-->s2   1200*1*.99    S1(OK)
s1-->s3   1200*1*.01    S2(ONE I/O GONE)
s2-->s3   600*1*1       S3(FAIL)
```

5. Subsystem 5 (Type 4, Level 3) is a computer module. Its reliability is simply the product of reliabilities of Models 2,3, and 4.

6. Subsystem 6 (Type 3, Level 4) is a multicomputer made up of 10 computers of which 8 must work. The coverage value, C, will also be varied as a parameter in the following examples. This is represented by a four-state transition model:

```
s1-->s2 = ITR(M5)*10*C      S1(OK)
s1-->s4 = ITR(M5)*10*(1-C)  S2(ONE COMPUTER FAILED)
s2-->s3 = ITR(M5)*9 *C      S3(TWO COMPUTERS FAILED)
s2-->s4 = ITR(M5)*9 *(1-C)  S4(SUBSYSTEM FAILED)
s3-->s4 = ITR(M5)*8 *1
```

At every step as the system is evaluated by numerical integration, the instantaneous probability of failure is computed for Model 5 (the computer module). This value:

ITR(M5) = (dR[M5(t)]/dt))/R[M5(t)] is substituted in the matrix above and evaluated for one step.

SNARC Input

Models are input to SNARC as a text file that lists each subsystem model, its type, and the associated parameters. State transitions are specified as follows:

```
Si --> Sj  BASERATE  \#ELEMS  COV  TRANS
```

The first two entries are the states for this transition, and the transition rate is computed as the product of the other entries. They are:

BASERATE - Fail. rate of single component (FITs)
#ELEMS   - Number of Elements
COV   - Coverage
TRANS   - A transient rate multiplier

For transitions (in Type 3 models) that use the time-varying transition rates from previous models a negative number -i is entered into the BASERATE column to indicate which previous model to use. In this case, ITR(Mi) is substituted for BASERATE. The entries #ELEMS, COV, and TRANS default to 1 if not explicitly entered. This breakdown into BASERATE, #ELEMS, COV, and TRANS is not strictly necessary but it makes the subsystem model specifications easier to read. As another aid to readability, multiple lines may be included for the same transition, and their rates will be added.

For the various subsystem types, the following parameters are included:

Type 1: Number of States and State Transitions
Type 2: List of time-failure-rate pairs to interpolate
        tabulated rates
Type 3: Number of States and State Transitions
        (some with -i BASERATE)
Type 4: No. of Subsystem Reliabilities in the reliability
        product and the Subsystems to be included

Table 1 shows the SNARC input for the model described above with C = 0.95 in Subsystems 2 and 6. In the table, lines are defined by their starting character as follows:

s - System Description - number of subsystem models

m - new model descriptor - model number, type, and number of states or product terms

t - transition matrix entry

l - time-failure rate list for type 2 subsystems

c - comment - ignored by SNARC but useful for reading.

The SNARC Program

The SNARC modeling system is written in Turbo Pascal, and executes on IBM-compatible PCs. It consists of a modeling program and a simple editor for creating model files. The editor is designed to automatically justify data and include comment lines to improve readability. The modeling program requires less than 30 Kbytes of program storage. Within the 64Kbyte data size restrictions of a DOS-PC, it is possible to model 11 subsystem types with up to 20-states each. As can be seen from the model presented above (that contains 64 subsystems and requires 6 subsystem types in the model) complex systems can be modeled within these limitations. By changing a few parameters and recompiling on a larger workstation, much larger models can be handled. (Due to limitations of the current program, only three subsystems with time-varying transition rates can be embedded in a higher level subsystem. This number will be increased in later revisions of SNARC.) The models described above have also been executed with the individual chips modeled as subsystems with empirically specified time-varying failure rates. This has allowed us to include

Table 1. A SNARC Entry File

```
c   ** SNARC SESSION NAME = Multicomputer Model **
c Number of Subsystems=
s        6
c **SUBSYSTEM NAME = memory_module
c  New SubSys Model\#      Type      \#States/Terms
m        1                1 (MM)      4
c  S1 ---> S2  BASERATE  \#ELEMS    COV        TRANS
t   1       2     100       41      .998
t   2       3     100       40      .998
t   3       4     100       39
t   1       4     100       41      .002
t   2       4     100       40      .002
t   1       4     100        2
t   2       4     100        2
t   3       4     100        2
c **SUBSYSTEM NAME = mem-sys
c  New SubSys Model\#      Type      \#States/Terms
m        2                3 (SM)      3
c  S1 ---> S2  BASERATE  \#ELEMS    COV        TRANS
t   1       2      -1        3      .95
t   1       3      -1        3      .05
t   2       3      -1        2
c **SUBSYSTEM NAME = Processor
c  New SubSys Model\#      Type      \#States/Terms
m        3                1 (MM)      3
c  S1 ---> S2  BASERATE  \#ELEMS    COV        TRANS
t   1       2     100       18      .995
t   1       3     100       18      .005
t   2       3     100       12
t   1       3     100        2
t   2       3     100        2
c **SUBSYSTEM NAME = I/O Port
c  New SubSys Model\#      Type      \#States/Terms
m        4                1 (MM)      3
c  S1 ---> S2  BASERATE  \#ELEMS    COV        TRANS
t   1       2    1200        1      .99
t   1       3    1200        1      .01
t   2       3     600        1       1
c **SUBSYSTEM NAME = Computer Module
c  New SubSys Model\#      Type      \#States/Terms
m        5                4 (PR)      3
p   Rm=R2 *R3 *R4
c **SUBSYSTEM NAME = multicomputer
c  New SubSys Model\#      Type      \#States/Terms
m        6                3 (SM)      4
c  S1 ---> S2  BASERATE  \#ELEMS    COV        TRANS
t   1       2      -5       10      .95
t   1       4      -5       10      .05
t   2       3      -5        9      .95
t   2       4      -5        9      .05
t   3       4      -5        8
```

Table 2. Time, Step Size and Convergence of Multicomputer Model

No. of	** Time = 0 to 10 Years **		
Integ.	Time	Reliability	Deviation
Steps	Required	(10 Years)	
2,000	53 Seconds	.979853	
1,000	26 Seconds	.979853	.000000
500	13 Seconds	.979854	.000001
250	6 Seconds	.979855	.000002
100	2 Seconds	.979858	.000005
50	1 Second	.979863	.000010

bathtub curves to deal with infant mortality and wearout, and to model external environmental effects.

The core of the modeling program is a Runge-Kutta numerical integrator that evaluates all of the differential equations represented by the transition matrices one step at a time, in round-robin fashion. The low- level models are integrated one step, and their results are used to build the time-varying portions of the matrices for higher level (type 3) subsystems. Then the higher level subsystems are integrated for that step.

In executing the modeling program, the user is asked to specify the model file, the time-interval over which reliability is to be computed, and an output interval. Two operating modes are provided: auto converge, and manual step size specification. In the autoconverge mode, the user specifies a number of digits for convergence. The SNARC program computes the model with decreasing step sizes until the results agree within the specified convergence limits, and then it produces the output. The manual mode allows the user to compare results with differing step sizes.

The models tend to converge rapidly with decreasing step size due to the low-failure rates of subsystem elements compared with the time intervals of the models. Table 2 shows the computation time, step size and convergence of the multicomputer model above with C= .95) when executed on a 20 MHZ 80386 processor without a floating point coprocessor.

Table 3. Comparison of Reliability Calculations with Analytical Solution

Year	Coverage = 1		Coverage = 0.95	
	Analytical	SNARC	Analytical	SNARC
1	1.000000	1.000000	0.999965	0.999965
2	1.000000	1.000000	0.999918	0.999918
3	1.000000	1.000000	0.999857	0.999857
4	1.000000	1.000000	0.999780	0.999780
5	1.000000	1.000000	0.999685	0.999685
6	1.000000	1.000000	0.999568	0.999569
7	1.000000	1.000000	0.999427	0.999428
8	0.999999	0.999999	0.999257	0.999258
9	0.999998	0.999998	0.999053	0.999055
10	0.999996	0.999996	0.998811	0.998813

4 Validation of SNARC

In addition to verifying the correctness of the mathematical foundations of a complex reliability evaluation tool, the basic assumptions made in the tool and its implementation should be validated. This can be done in three different ways. First, problems with known analytical solutions can be solved using the tool and the results compared. Secondly, the tool can be compared with existing tools using models that can be handled by both tools. Finally, for models which no other tool can solve, the solution can be validated using simulation techniques. This section describes the steps taken to validate SNARC using each of these ways.

Model with known analytical solution

The system in Figure 1 was used as the example. The solution produced by SNARC is compared with the analytical solution for both perfect coverage and a coverage of 0.95 at each design level, in Table 3. The failure rate of the components at level 2 is 1×10^{-6} failures/hour.

The slight difference between some of the analytical calculations and the SNARC results are due to errors introduced by the simple integration technique used in SNARC.

When using tools that cannot model coverages throughout the entire design hierarchy, it was found that they would produce extremely inaccurate predictions of design reliability. An example was CRAFTS, a reliability prediction tool based on ARIES [11, 12]. The CRAFTS models used for design levels zero and one were reliability block diagrams, since constructing a Markov model of the system for use in CRAFTS is precluded by the large state space of the model. While CRAFTS produced the same result as the analytical model (and SNARC) for perfect coverage, CRAFTS predicted an unreliability 54 times greater when the coverage was 0.95. This further demonstrates the usefulness of SNARC.

Comparison with other tools

In this section, SNARC is used to model the multicomputer described in the last section (Figure 2), and the results compared with that of CRAFTS. These models used the fixed coverages for the low-level subsystems (1,3,4) and varied the coverage "C" for the high-level subsystems (2, 6) from one down to 0.95.

Table 4 presents the reliability of this system after 10 years, and shows the results of CRAFTS and SNARC. SNARC was able to solve this model on a personal computer in thirteen seconds. For $C = 1$, both CRAFTS and SNARC gave identical results. Additionally, the SNARC models were evaluated where the coverages, C, at the computer memory subsystem and multicomputer levels, were 0.99, and 0.95. However, CRAFTS was unable to reflect the coverage values at those design levels. Again, the inclusion of coverage values at higher design levels has significantly affected the reliability prediction of the design. We were unable to use CARE III and HARP for comparison, since we were unable to get CARE III and HARP to model this system. The inability of the tools to model the multicomputer resulted from either exceeding the tools' capabilities or the computing resources on which the tools ran.

Use of Monte-Carlo simulation

Since existing reliability evaluation tools are unable to incorporate coverages at different levels in the hierarchy, Monte Carlo simulations of the same multicomputer were run to validate that the numerical technique of SNARC was generating the correct results for all the models. Table 5 shows the results of SNARC and the Monte Carlo simulations for coverages of 1 and 0.95. The

Table 4. Multicomputer Reliability Evaluation

	CRAFTS	SNARC		
Coverage:	1.00	1.00	0.99	0.95
Year				
1	0.999999	0.999999	0.999788	0.998836
2	0.999990	0.999990	0.999546	0.997550
3	0.999963	0.999963	0.999261	0.996127
4	0.999900	0.999900	0.998917	0.994549
5	0.999780	0.999780	0.998492	0.992790
6	0.999574	0.999574	0.997957	0.990817
7	0.999247	0.999247	0.997274	0.988590
8	0.998756	0.998756	0.996401	0.986061
9	0.998051	0.998051	0.995283	0.983171
10	0.997971	0.997970	0.993861	0.979854

Monte Carlo and SNARC reliability results differ by less than 0.1%. The Monte Carlo results provide support that the numerical technique, as implemented in SNARC, generates the correct solutions.

5   Conclusions and Future Work

We have presented a numerical technique for solving a hierarchy of semi-Markov models to analyze the reliability of large, closed, fault-tolerant, on-board computer systems. SNARC, a tool implementing the technique was described, and was shown to avoid the large state space problem. SNARC was used to solve the semi-Markov models of several example systems (that have large corresponding Markov models) on a personal computer in a short period of time (e.g., thirty seconds or less).

Currently, there are two limitations to the hierarchical approach. First, if the fault and error handling mechanisms of subsystem j at design level L_{i+1} are unable to detect and recover from a transient fault and its resulting errors, the failure of subsystem j is treated as permanent at level L_i. Second, the models for every level, except the lowest, suffer the well-known limitations associated with non-homogeneous Markov models. These limitations are the inability to

Table 5. Comparison of SNARC results with Monte-Carlo Simulation

Year	Coverage = 1		Coverage = 0.95	
	Mon. Carlo	SNARC	Mon. Carlo	SNARC
1	1.000000	0.999999	0.998717	0.998836
2	0.999983	0.999990	0.997517	0.997550
3	0.999983	0.999963	0.996150	0.996127
4	0.999950	0.999900	0.994733	0.994549
5	0.999867	0.999780	0.993017	0.992790
6	0.999617	0.999574	0.990867	0.990817
7	0.999333	0.999247	0.986667	0.988590
8	0.998733	0.998756	0.986700	0.986061
9	0.997967	0.998051	0.983867	0.983171
10	0.997300	0.997970	0.980717	0.979854

model cold or warm spares, or repairable systems [13, 14].

Work to improve the technique and tools is continuing in two areas. First, modifications are under way to make the SNARC program easier for designers to use. Also, a graphical interface would allow users to easily input models. Techniques need to be incorporated to correctly model the use of "warm" spares throughout the design hierarchy. Techniques must be developed that allow the hierarchical modeling of transients and the capability to model near-coincident faults throughout the hierarchy. Different Fault/Error Handling Models (FEHMs) must be included in SNARC to enable users to have the ability to select the FEHM that best approximates the fault and error handling mechanisms at each level of the design hierarchy.

Acknowledgements

We wish to thank Venu Shamapant for performing the Monte Carlo simulations, Mark Joseph and Joe Bannister for their useful comments, Patrick Kennedy for his assistance in the analytical calculation of the example triad's reliability, and the developers of CRAFTS for providing the use of their tool for this work.

Appendix

Approx. increase in $H_i(t)$ from time t to t + Δt

We know that the cumulative failure rate $H_i(t) =$
$\int_0^t h_i(\tau)d\tau = \int_0^t -R_i'(\tau)/R_i(\tau)d\tau$.

The integral
$\int_0^t -R_i'(\tau)/R_i(\tau)d\tau = \int_0^{\Delta t} -R_i'(\tau)/R_i(\tau)d\tau +$
$\int_{\Delta t}^{2\Delta t} -R_i'(\tau)/R_i(\tau)d\tau + \int_{2\Delta t}^{3\Delta t} -R_i'(\tau)/R_i(\tau)d\tau + \ldots$
$+ \int_{(N-1)\Delta t}^{N\Delta t} -R_i'(\tau)/R_i(\tau)d\tau,$
where NΔt = t.

We can approximate $\int_t^{t+\Delta t}[R_i(\tau + \Delta\tau) - R_i(\tau)]/[\Delta\tau R_i(\tau)]d\tau$ by
$[R_i(t + \Delta t) - R_i(t)]/R_i(t)$, when Δt is very close to zero.

Substituting into the above equations, we obtain
$H_i(t) = \int_0^t h_i(\tau)d\tau \approx$
$\{[R_i(0+\Delta t)-R_i(\Delta t)]/R_i(0)\}+\{[R_i(0+2\Delta t)-R_i(0+\Delta t)]/R_i(0+\Delta t)\}$
$+\ldots + \{[R_i(0 + (N\Delta t)) - R_i(0 + (N-1)\Delta t))]/R_i(0 + (N-1)\Delta t)\}.$

Therefore, we can approximate the increase in the cumulative failure rate $H_i(t)$, from time t to $t + \Delta t$ by determining $[R_i(t + \Delta t) - R_i(t)]/R_i(t)$, and we can approximate $H_i(t)$ by summing all the incremental increases in $H_i(t)$ from time 0 to the end of mission life time.

References

[1] W. Carter and J. Abraham, "Design and evaluation tools for fault-tolerant systems," in *Proceedings of the AIAA Computers in Aerospace VI Conference*, pp. 70–77, October 1987.

[2] A. M. Johnson and M. Malek, "Survey of software tools for evaluating reliability, availability, and serviceability," *ACM Computing Surveys*, vol. 20, pp. 227–269, December 1988.

[3] R. Geist and K. Trivedi, "Reliability estimation of fault-tolerant systems: Tools and techniques," *Computer*, pp. 52–61, July 1990.

[4] W. G. Bouricius, W. C. Carter, and P. R. Schneider, "Reliability modeling techniques for self-repairing computer systems," in *Proceedings of the 12th ACM National Conference*, pp. 295–309, August 1969.

[5] J. Dugan, K. Trivedi, M. Smotherman, and R. Geist, "The hybrid auto-mated reliability predictor," *AIAA Journal of Guidance, Control, and Dynamics*, pp. 319–331, May–June 1986.

[6] K. Trivedi, R. Geist, M. Smotherman, and J. Dugan, "Hybrid modeling of fault-tolerant systems," *Computers and Electrical Engineering, An International Journal*, vol. 11, no. 2–3, pp. 87–108, 1985.

[7] R. Sahner and K. Trivedi, "A hierarchial combinatorial-markov method for solving complex reliability models," in *Proceedings ACM/IEEE Fall Joint Computing Conference*, November 1986.

[8] A. Goyal, W. Carter, E. de Souza e Silva, S. Lavenberg, and K. Trivedi, "The system availability estimator," in *Proceedings of the 16th IEEE Fault-Tolerant Computing Symposium*, pp. 84–89, July 1986.

[9] S. Bavuso, J. Brunelle, and P. Peterson, "Care iii hands-on demonstration and tutorial," Technical Memorandum 85811, NASA, May 1984.

[10] S. Bavuso, P. Peterson, and D. Rose, "Care iii model overview and user's guide," Technical Memorandum 85810, NASA, June 1984.

[11] Y. Ng and A. Avizienis, "A unified reliability model for fault-tolerant computers," *IEEE Transactions on Computers*, vol. C-29, pp. 1002–1011, November 1980.

[12] S. Makam and A. Avizienis, "Aries 81: A reliability and life-cycle evaluation tool for fault-tolerant systems," in *Proceedings of the IEEE 12th Fault-Tolerant Computing Symposium*, pp. 267–274, June 1982.

[13] K. Trivedi and R. Geist, "A tutorial on the care iii approach to reliability modeling," Contractor Report 3488, NASA, December 1981.

[14] *HARP: The Hybrid Automated Reliability Predictor Introduction and User's Guide.*

Fault Injection Simulation: A Variance Reduction Technique for Systems with Rare Events

AAD P. A. VAN MOORSEL, BOUDEWIJN R. HAVERKORT,
IGNAS G. NIEMEGEERS

University of Twente

Department of Computer Science

Tele-Informatics and Open Systems group

P.O. Box 217, 7500 AE Enschede, The Netherlands

Abstract

In this paper a new technique, called Fault Injection Simulation (FIS), is presented that is suited for deriving results for steady-state measures in discrete event dynamic systems which are strongly influenced by rarely occurring events. FIS is based on decomposition of the observations in those that are affected and those that are not affected by these events. If methods are available FIS can be used as a (partly) analytical technique, else as a pure fast simulation technique. It is shown that under intuitively appealing assumptions FIS gives an unbiased estimator and variance reduction. Furthermore it is shown how to adjust FIS during the simulation to obtain maximum variance reduction. FIS is discussed in the context of rarely occurring failures in networks for real-time applications. Comparisons with other fast simulation techniques are made and results of FIS for an M/M/1-queue with rarely occurring service breakdowns are included.

Key Words: Discrete event simulation, variance reduction techniques, rare events, performability evaluation, real-time systems.

1 Introduction

Fault tolerant distributed computer and communication systems are often used in environments in which they have to fulfill certain real-time constraints. Especially in safety critical applications the dependability characteristics of a system have to support the real-time performance. To establish the real-time characteristics it is necessary to integrate performance and dependability modelling and evaluation. This is called *performability* analysis [1]. Performability modelling has been most successfully applied for gracefully degradable computer systems by using Markov reward models [2]. In this approach the relation of the performability model with the underlying performance and dependability model is exploited. For many systems and for many measures of interest this type of modelling is not suited, so another performability modelling approach is necessary. In this paper we introduce *Fault Injection Simulation* (FIS), a modelling approach which can be used to evaluate highly dependable systems with very small failure rates. In this respect one can for instance think of the very small token loss probabilities in the high speed FDDI token ring.

FIS decomposes a realization of a Discrete Event Dynamic System [3] in a fault affected and a non-affected part. For these two parts steady-state measures are derived and weighted to calculate the desired overall steady-state measure. The individual measures can be obtained by analytical or numerical means or by simulation. In many cases analytical or numerical methods will not be available in which case FIS is a pure fast simulation technique. In this paper we will concentrate on the use of FIS as a pure simulation method.

When one wants accurate simulation results in a reasonable time span for systems that are strongly influenced by rarely occurring events the use of variance reduction techniques (e.g. [3], [4]) is inevitable. Otherwise far too much time is needed before enough of these rare events are simulated. It seems FIS can compete with other known techniques in ease of applicability and amount of variance reduction.

This paper is organized as follows. In Section 2 we informally introduce FIS. In Section 3 we formalize FIS and make assumptions under which FIS gives an unbiased estimator. To show the usefulness of FIS we will derive results for its variance reduction possibilities in Section 4. In Section 5 we discuss on the actual usage of FIS and on how one can optimize the variance reduction during the simulation. In Section 6 we compare FIS to related techniques. An example of the use of FIS for a simple M/M/1 model with server breakdowns is given in Section 7. Section 8 concludes the paper.

2 Fault Injection Simulation

In this section we introduce Fault Injection Simulation (FIS). FIS is a form of decomposition of observations [5] of a discrete event simulation [3] which is visualized in Figure 1.

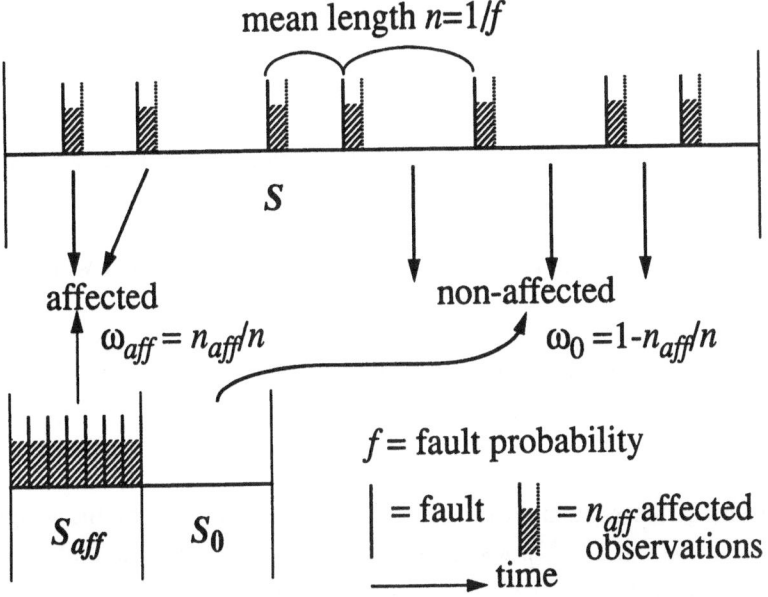

Figure 1: Fault Injection Simulation

FIS is based on the idea that the effect of a fault on the observations fades out. After a fixed number of observations, denoted n_{aff}, the observations are considered to be no longer affected by the occurred fault and to be identical to those in a fault free system. This forms the basis of the following three steps of FIS:

A. Division of observations:

When S is the realization of a system with probability f that a fault occurs between two successive observations, then we divide the observations in S in two sets:

1. *Affected observations* which are considered to be affected by an occurred fault. These are the n_{aff} observations directly following a fault occurrence;

2. *Non-affected observations* which are considered not to be affected by a fault.

B. Separate simulation:

The two sets are simulated separately from each other by the simulation run S_{aff} and S_0 such that

1. the *injected simulation* S_{aff} gives a realization of a system with a fault occurring (*injected*) deterministically every n_{aff} observations;

2. the *fault free simulation* S_0 gives a realization of a system without fault occurrences.

Both S_{aff} and S_0 are steady-state simulations.

C. Weighting:

The results of the simulation runs S_{aff} and S_0 are weighted by the weighting factors ω_{aff} and ω_0. Let $\widehat{Y}_{aff}$ and $\widehat{Y}_0$ respectively be the estimators obtained by S_{aff} and S_0, then the estimator $\widehat{Y}_{FIS}$ obtained by FIS is

$$\widehat{Y}_{FIS} = \omega_{aff}\widehat{Y}_{aff} + \omega_0\widehat{Y}_0. \tag{1}$$

The weighting factors represent the fraction of the overall number of observations that are affected respectively non-affected. This means:

$$\omega_{aff} = \frac{n_{aff}}{n} \quad \text{and} \quad \omega_0 = 1 - \frac{n_{aff}}{n},$$

with $n = 1/f$, the mean number of observations between two successive faults.

3   Unbiasedness of FIS estimators

In Section 3.2 we show that under a so-called *decomposition assumption* FIS gives an unbiased estimator. Therefore we start in Section 3.1 with defining the different stochastic processes in FIS.

3.1   Stochastic processes in FIS

We will look at three simulations: the direct simulation S, the injected simulation S_{aff} of affected observations, and the fault free simulation S_0 of non-affected observations. The direct simulation S delivers the realization $Y_1, \cdots, Y_l$,

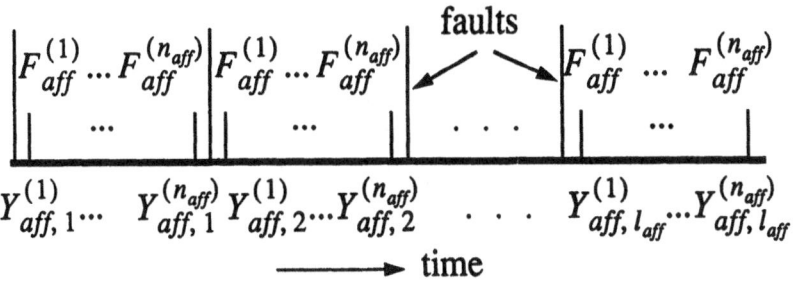

Figure 2: Affected observations

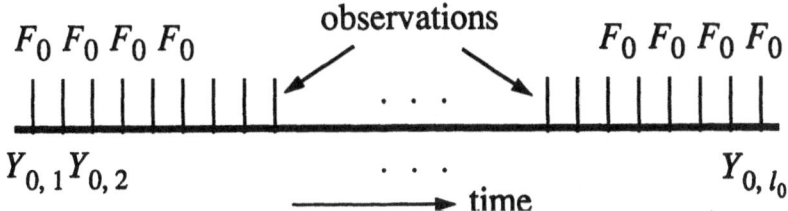

Figure 3: Non-affected observations

with all observations distributed identically to Y, a random variable with cumulative distribution function (cdf) F (see upper part of Figure 4). In other words, we assume that the original system is a stationary stochastic process. The estimator obtained by S is:

$$\widehat{Y} = \frac{1}{l} \sum_{i=1}^{l} Y_i. \tag{2}$$

The injected simulation S_{aff} delivers the realization $Y_{aff,1}^{(1)}, \ldots, Y_{aff,1}^{(n_{aff})}, \ldots, Y_{aff,l_{aff}}^{(1)}$, $\ldots, Y_{aff,l_{aff}}^{(n_{aff})}$. The j-th observation ($j = 1, \cdots, n_{aff}$) after a fault is distributed identically to $Y_{aff}^{(j)}$, a random variable with cdf $F_{aff}^{(j)}$ (see Figure 2). The estimator $\widehat{Y}_{aff}$ obtained by S_{aff} is the mean of all observations:

$$\widehat{Y}_{aff} = \frac{1}{n_{aff} l_{aff}} \sum_{i=1}^{l_{aff}} \sum_{j=1}^{n_{aff}} Y_{aff,i}^{(j)}. \tag{3}$$

The *fault free simulation* S_0 delivers the realization $Y_{0,1}, \cdots, Y_{0,l_0}$ with all observations distributed identically to Y_0, a random variable with cdf F_0 (see

Figure 3). The estimator $\widehat{Y}_0$ obtained by S_0 is:

$$\widehat{Y}_0 = \frac{1}{l_0} \sum_{i=1}^{l_0} Y_{0,i}. \tag{4}$$

As in Section 2 we take as *weighting factors*:

$$\omega_{aff} = \frac{n_{aff}}{n} \quad \text{and} \quad \omega_0 = 1 - \frac{n_{aff}}{n}.$$

Consequently the *FIS estimator* $\widehat{Y}_{FIS}$ (1) equals:

$$\widehat{Y}_{FIS} = \omega_{aff}\widehat{Y}_{aff} + \omega_0\widehat{Y}_0 = \frac{n_{aff}}{n}\widehat{Y}_{aff} + (1 - \frac{n_{aff}}{n})\widehat{Y}_0. \tag{5}$$

We now make an important assumption which makes it possible to relate FIS to direct simulation. It is the fundamental property we assume a system to have:

Assumption 1. Decomposition assumption

The observations in the direct simulation S are all distributed identically to the random variable Y such that

$$Y = \begin{cases} Y_0, & \text{with probability } 1 - n_{aff}/n, \\ Y_{aff}^{(i)}, & \text{with probability } 1/n, \text{ for } i = 1, \cdots, n_{aff}, \end{cases} \tag{6}$$

with Y_0 and $Y_{aff}^{(i)}$, $i = 1, \cdots, n_{aff}$, as specified above.

The idea of Assumption 1, which will be used to show the unbiasedness of the FIS estimator, is depicted in Figure 4. It implies that the observations in S are such that S is built out of a part that can be simulated by S_{aff} and a part that can be simulated by S_0. In Section 5 we will discuss the impact and the heuristic justification of this partition.

3.2   Unbiasedness of FIS estimator

We show that under Assumption 1 the estimator obtained with FIS is an unbiased estimator of $\mathrm{E}Y$, the steady-state mean of the measure of interest in the original system (see (2)). First note that from the fact that S is the realization of a stationary stochastic process it follows directly that $\widehat{Y}$ in (2) is unbiased, i.e. $\mathrm{E}\widehat{Y} = \mathrm{E}Y$.

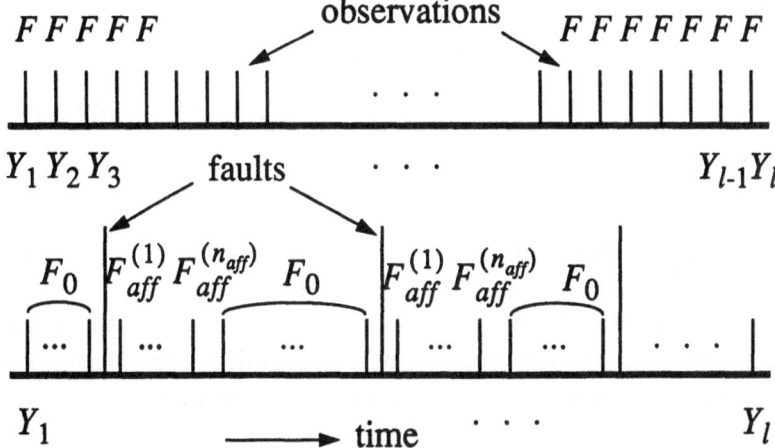

Figure 4: Observations direct simulation, respectively not conditioned and conditioned on fault occurrences

Theorem 1. Under Assumption 1, FIS gives an unbiased estimator of $\mathrm{E}Y$.
Proof. We proof this theorem by showing that the FIS estimator $\widehat{Y}_{FIS}$ in (5) and the direct estimator $\widehat{Y}$ in (2) have the same expectation, i.e. $\mathrm{E}\widehat{Y}_{FIS} = \mathrm{E}\widehat{Y}$.

First, for the direct estimator, by its unbiasedness and by Assumption 1, it follows that:

$$\mathrm{E}\widehat{Y} = \mathrm{E}Y = (1 - \frac{n_{aff}}{n})\mathrm{E}Y_0 + \frac{1}{n}\sum_{j=1}^{n_{aff}} \mathrm{E}Y_{aff}^{(j)}. \tag{7}$$

Secondly, for the FIS estimator in (5) we have that:

$$\mathrm{E}\widehat{Y}_{FIS} = (1 - \frac{n_{aff}}{n})\mathrm{E}\widehat{Y}_0 + \frac{n_{aff}}{n}\mathrm{E}\widehat{Y}_{aff}. \tag{8}$$

The two terms in the right part of (8) can be shown to be equal to those in the right part of (7) by using the facts of Section 3.1. Via (4) and the fact that $\mathrm{E}Y_{0,i} = \mathrm{E}Y_0$ for $i = 1, \cdots, l_0$ it follows for the first term that:

$$\mathrm{E}\widehat{Y}_0 = \frac{1}{l_0}\sum_{i=1}^{l_0} \mathrm{E}Y_{0,i} = \mathrm{E}Y_0. \tag{9}$$

Via (3) and the fact that $\mathrm{E}Y_{aff,i}^{(j)} = \mathrm{E}Y_{aff}^{(j)}$ for $j = 1, \cdots, n_{aff}$, $i = 1, \cdots, l_{aff}$, it follows for the second term that:

$$n_{aff}\mathrm{E}\widehat{Y}_{aff} = \frac{n_{aff}}{n_{aff}l_{aff}}\sum_{i=1}^{l_{aff}}\sum_{j=1}^{n_{aff}} \mathrm{E}Y_{aff,i}^{(j)} = \sum_{j=1}^{n_{aff}} \mathrm{E}Y_{aff}^{(j)}. \tag{10}$$

Substituting (9) and (10) in the right side of (8) gives the equality of (7) and (8), in other words:

$$\mathrm{E}\widehat{Y}_{FIS} = \mathrm{E}\widehat{Y}.$$

Now, because of the fact that $\widehat{Y}$ is unbiased, it follows that the FIS estimator is unbiased.

4   Variance Reduction

In this section we derive results for the variance of the FIS estimator in relation with the variance of the direct estimator. To do this we use results known from Stratified Sampling which is discussed first in Section 4.1. We give an expression for the variance of the FIS estimator in Section 4.2, for the direct estimator in Section 4.3 and we compare them in Section 4.4.

4.1   Stratified Sampling

Stratified Sampling [6] is a technique well known in sampling theory. In this technique a population of N units is partitioned in L subpopulations of respectively $N_1, N_2, \cdots, N_L$ units. A sample is drawn from each subpopulation and the results, denoted $y_1, y_2, \cdots, y_L$, are multiplied by the weighting factors $\omega_1, \omega_2, \cdots, \omega_L$ with $\omega_i = N_i/N, i = 1, \cdots, L$. The obtained estimator $\widehat{Y}$ then equals

$$\widehat{Y} = \sum_{i=1}^{L} \omega_i y_i.$$

From this description of Stratified Sampling the similarity with the three steps in FIS in Section 2 becomes clear. The main difference is that in Stratified Sampling all the N units in the population are considered to be independent, whereas FIS is typically concerned with observations that are correlated. In the following subsections we will obtain results about the variance of the FIS estimator by creating batches which are considered to be independent, thus making it possible to derive results for these batches similar to those for independent units in Stratified Simulation [6].

4.2   Variance of the FIS estimator

We will derive an expression for the variance by applying the method of *batch means* (e.g. [3], [5]). A realization is split in groups of successive observations,

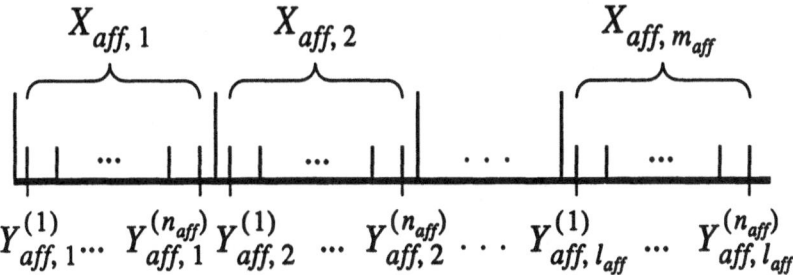

Figure 5: Affected batches

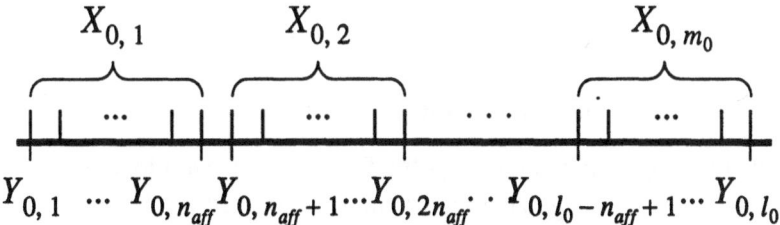

Figure 6: Non-affected batches

the so-called *batches*, that are considered to be independent of each other. The number of observations in a batch is called the batch size, and we will take as batch size n_{aff} for all simulations we consider. The motivation for this choice is that n_{aff} is chosen such that the influence of the injected fault is negligible small after n_{aff} observations (see Section 2).

To determine the variance of the FIS estimator we first look at the injected simulation, then at the fault free simulation and finally we combine them to obtain the variance of the FIS estimator.

We divide the observations in S_{aff} in m_{aff} *affected batches* (see Figure 5):

$$X_{aff,i} = \frac{1}{n_{aff}} \sum_{j=1}^{n_{aff}} Y_{aff,i}^{(j)}, \quad i = 1, \cdots, m_{aff}, \quad (11)$$

all distributed identically to X_{aff}, a r.v. with cdf F_{aff}, mean $\mathbf{E}_{aff}$ and variance $\mathbf{v}_{aff}^2$.

We divide the observations in S_0 in m_0 *non-affected batches* (see Figure 6):

$$X_{0,i} = \frac{1}{n_{aff}} \sum_{j=1}^{n_{aff}} Y_{0,(i-1)n_{aff}+j}, \quad i = 1, \cdots, m_0, \quad (12)$$

all distributed identically to X_0, a r.v. with cdf F_0, mean E_0 and variance v_0^2. Notice that $m_0 \times n_{aff} = l_0$, and $m_{aff} = l_{aff}$.

It can easily be verified that taking the average of all the observations as an estimator is equivalent to taking the average of the batches. So (5) also can be defined in terms of batches as:

$$\widehat{Y}_{FIS} = \frac{n_{aff}}{n} \frac{1}{m_{aff}} \sum_{i=1}^{m_{aff}} X_{aff,i} + (1 - \frac{n_{aff}}{n}) \frac{1}{m_0} \sum_{i=1}^{m_0} X_{0,i}. \tag{13}$$

From (13) and the assumed independence of all the batches we get by (11) and (12) that the *variance* V_{FIS}^2 *of the FIS estimator* equals:

$$v_{FIS}^2 = \left(\frac{n_{aff}}{n}\right)^2 \frac{v_{aff}^2}{m_{aff}} + \left(1 - \frac{n_{aff}}{n}\right)^2 \frac{v_0^2}{m_0}. \tag{14}$$

4.3   Variance of the direct estimator

To be able to compare the variance of the FIS estimator with that of the direct estimator we have to recognize affected and non-affected batches in the direct simulation run S. We divide the observations in S in m batches (see the lower part of Figure 7):

$$X_i = \frac{1}{n_{aff}} \sum_{j=1}^{n_{aff}} Y_{(i-1)n_{aff}+j}, \quad i = 1, \cdots, m. \tag{15}$$

To be able to compare the variance of direct simulation with the variance in FIS we assume:

Assumption 2. Batch decomposition assumption

The r.v.'s X_i, $i = 1, \cdots, m$, defined in (15), have the following property:

$$X_i = \begin{cases} X_{aff}, & \text{with probability } n_{aff}/n, \\ X_0, & \text{with probability } 1 - n_{aff}/n, \end{cases} \tag{16}$$

all distributed identically to X, a r.v. with cdf F, mean E and variance v^2.

Figure 7 illustrates that this assumption implies that the observations of two successive batches in S with both affected and non-affected observations are reordered such that they form one affected and one non-affected batch. This can give bias in the variance estimator when a failure occurs in the initial transient period or in the last batch. We neglect this bias as it will be small

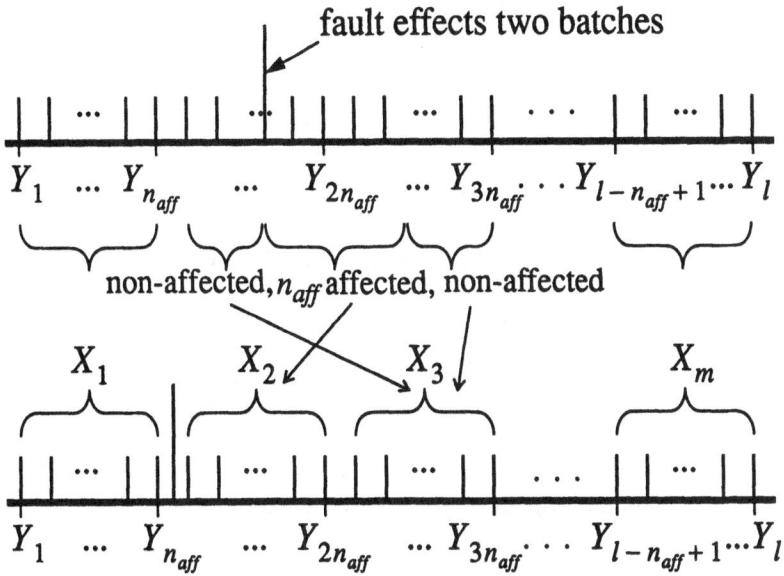

Figure 7: Batches direct simulation

because of the very small fault probabilities. The direct estimator in (2) can be defined in terms of batches as (by (15)):

$$\widehat{Y} = \frac{1}{m} \sum_{i=1}^{m} X_i. \tag{17}$$

From Assumption 2 we obtain by conditioning (compare [6]) the following expression for the *variance of the direct estimator* $\mathbf{v}^2(\widehat{Y})$:

$$\mathbf{v}^2(\widehat{Y}) = \frac{1}{m} \left(\frac{n_{aff}}{n} \mathbf{v}_{aff} + (1 - \frac{n_{aff}}{n}) \mathbf{v}_0 \right)^2 + \frac{1}{m} \frac{n_{aff}}{n} (1 - \frac{n_{aff}}{n}) (\mathbf{v}_{aff} - \mathbf{v}_0)^2$$

$$+ \frac{1}{m} \frac{n_{aff}}{n} (1 - \frac{n_{aff}}{n}) (\mathbf{E}_{aff} - \mathbf{E}_0)^2. \tag{18}$$

4.4   Comparison variance FIS estimator and direct estimator

Now we have expressions for the variance of the FIS estimator in (14) and the direct estimator in (18) we can compare them. In this section we will see that the variance reduction obtained by FIS depends on the fraction of batches dedicated to the injected simulation S_{aff} and to the fault free simulation S_0 respectively. Therefore we define an allocation fraction β, and derive an optimal value for this fraction. Suppose we simulate in FIS a total of m batches of which m_{aff} in S_{aff} and m_0 in S_0 ($m = m_{aff} + m_0$ and m_{aff}, $m_0 > 0$). We now define the allocation fraction β ($0 < \beta < 1$) as the fraction of affected batches out of the total number of m batches. In other words:

$$m_{aff} = \beta m \quad \text{and} \quad m_0 = (1 - \beta)m. \tag{19}$$

Substituting (19) in (14), we obtain for the FIS estimator's variance $v_{FIS}^2(\beta)$ the following expression[1]:

$$v_{FIS}^2(\beta) = \left(\frac{n_{aff}}{n}\right)^2 \frac{v_{aff}^2}{\beta m} + \left(1 - \frac{n_{aff}}{n}\right)^2 \frac{v_0^2}{(1 - \beta)m}. \tag{20}$$

It can be shown that the variance of the estimator obtained by FIS is minimal for the so-called *optimal allocation fraction* β_{opt}, with

$$\beta_{opt} = \frac{\frac{n_{aff}}{n} v_{aff}}{\frac{n_{aff}}{n} v_{aff} + (1 - \frac{n_{aff}}{n}) v_0}. \tag{21}$$

The variance then equals:

$$v_{FIS}^2(\beta_{opt}) = \frac{1}{m} \left(\frac{n_{aff}}{n} v_{aff} + (1 - \frac{n_{aff}}{n}) v_0\right)^2. \tag{22}$$

So, the maximum possible reduction FIS can give is, by subtracting (22) from (18):

$$\frac{1}{m} \frac{n_{aff}}{n} (1 - \frac{n_{aff}}{n})((v_{aff} - v_0)^2 + (E_{aff} - E_0)^2). \tag{23}$$

See for a proof of a similar statement [6].

5   When and how to use FIS

In this section we comment on the use of FIS. First we discuss for what systems, fault types and performance measures FIS can be a useful technique. Then we discuss how to benefit maximally of the variance reduction possibilities of FIS by looking at adjusting FIS during simulation and at possible hybrid use of FIS.

---

[1]β is added in $V_{FIS}^2(\beta)$ to point out the dependence of the FIS estimator on the allocation fraction.

When to use FIS

FIS is useful in discrete event dynamic systems with rarely occurring events that have great influence on the steady-state measure of interest. This amounts to the validity of Assumption 1 and the high value of the mean and variance of the affected batches compared to those of fault free batches. We discuss these two points.

The FIS estimator and the direct estimator are both unbiased when Assumption 1 holds (Theorem 1). This assumption provides the link between the two estimators and is based on the heuristic arguments in Section 2. The basic thought is that the influence of an occurred fault on the observations will be faded out after n_{aff} observations and that from that point on the original system behaves as a fault free system until the next fault occurrence. Theoretically though the effect of an occurred fault can never disappear completely. The problem we face here is identical to the problem of deciding on the length of the initial transient phase of a simulation run (see e.g. [7]); how large should n_{aff} be chosen such that the effect of the fault is negligible? Further investigations should be done to find methods that circumvent the above mentioned purely heuristic arguments.

The usefulness of FIS is clear through its variance reduction capabilities, i.e. the difference between (18) and (22) for the chosen allocation fraction β. When we take for β the optimal allocation fraction the reduction is given in (23) which immediately shows that reduction increases when the differences in mean and variance do. Thus, when a measure is strongly influenced by the occurred fault, i.e. when mean and/or variance of the affected batches are much larger than those of fault free batches it pays out to use FIS. This is for instance the case for measures concerning the tail of some distribution (e.g. waiting time distribution) but far less for e.g. its mean.

FIS can only be applied for very rarely occurring faults. Otherwise it seems impossible to agree on any decomposition of the original system for which Assumption 1 holds. For these faults the value of f, the fault probability, has to be known and at least some of the faults have to occur in steady-state of the fault free system. So, we can also use FIS for fast simulation of systems with bursty fault occurrences provided that the first fault of a burst occurs in steady-state of the fault free system. We can also simulate multiple types of rarely occurring faults by doing more than two different simulations.

How to use FIS

To begin with we note that because of the fact that the FIS estimator (5) is a function in $n = 1/f$, FIS gives results for a range of small fault probabilities f by only one simulation. Further, FIS has the nice property that an optimal application of FIS can be obtained very easily by estimating the optimal allocation fraction (Section 4.4). This can be done during the simulation run by estimating the variance of affected and fault free batches, calculating the optimal allocation fraction and adjusting the run lengths of S_{aff} and S_0.

There is no restriction to the solution method that can be used to obtain the desired results for S_{aff} and S_0. So, when analytical or numerical results for the fault free system (or even for the affected observations) are available they can be used. This would make FIS an hybrid simulation technique [4]. It is also possible to use fast simulation techniques, for instance Importance Sampling within the two simulations S_{aff} and S_0.

6   FIS in relation to other techniques

We compare FIS with known related techniques. First we will discuss the relation with Fault Injection and then with other fast simulation techniques from which Importance Sampling is the most prominent one.

Fault Injection [8], [9] has had extensive coverage in the literature. A motivation for applying Fault Injection is to accelerate dependability analysis for highly dependable systems. As FIS, Fault Injection is based on the artificial creation (injection) of a fault. This can be done in an abstract mathematical model or, as is most often done, in a physical model [8]. According to [8] Fault Injection's major goals are to validate the methods and mechanisms used to obtain the desired dependability. Typical interest exists in elements as coverage factors, error latency, machine crash probabilities, etc. In these measures of interest FIS differs from Fault Injection. Our intention is to obtain steady-state measures of a fault tolerant system which have no direct relation with the occurred fault, whereas Fault Injection aims on analyzing transient results. FIS achieves its goal by including a simulation of a fault free system and weighting the results appropriately.

Importance Sampling [10] alters the probability density of the realizations of a stochastic process. A so-called likelihood ratio is calculated and used to make the resulting estimator have the same expectation as the direct estimator. In for instance [11] and [12] the usefulness of Importance Sampling for simulating communication networks is shown. A difference with FIS is the remaining probabilistic nature. FIS injects faults deterministically which is part of the cause of the variance reduction of FIS (compare [7]).

In our case an obvious way to use Importance Sampling is to increase the fault probability f. A problem lies in the numerically stable calculation of the likelihood ratios when the regenerative periods are long. For the same reason Importance Sampling is hard to use within a batch mean approach.

Various other techniques have been developed that can reduce variance, see for instance [3], [4], [10], [13]. We have already introduced *Stratified Sampling* in Section 4.1. FIS might be seen as a form of Stratified Sampling extended for Discrete Event Simulations.

A technique that has some similarities with FIS is described in [4]. The rare event is injected at the beginning of each busy period of the system under consideration. This might be a satisfying approximate approach for some applications. In [14] *Virtual Measures*, a form of conditional sampling based on a similar idea as injecting faults, are constructed which can be useful when the measure of interest is directly related to the occurrence of a fault.

7   A simple FIS implementation

In this section we work out a FIS implementation for a simple M/M/1 queueing model with service breakdowns. We first discuss how we implemented FIS, then we show some results.

Actual implementation of FIS

We look at an M/M/1-queue with arrival rate λ, service rate μ and infinite buffer. After each service completion a service breakdown can occur with probability f, after which an exponentially distributed "repair time" with mean $1/\nu$ follows. We are interested in the steady-state probability of exceeding some response time (*deadline*) α. The response time is the waiting time of a client in the queue plus its service time.

Our simulation starts with FIS for 100 batches of both S_{aff} and S_0. After that we calculate the optimal allocation fraction β_{opt} as well as the number of batches m necessary to get results that are 95% certain within 10% of the estimate. Then the FIS simulation continues with simulating $\beta_{opt}m$ affected batches and

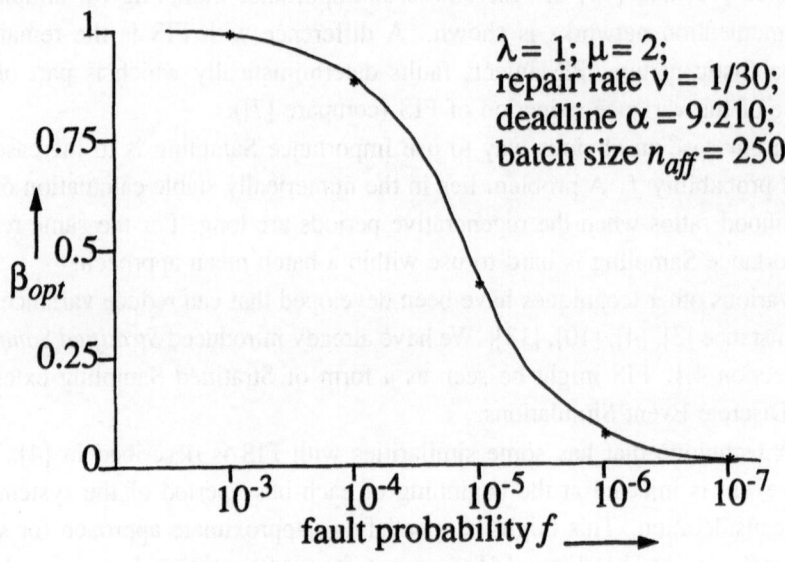

Figure 8: Optimal allocation fraction β_{opt}

$(1 - \beta_{opt})m$ non-affected batches (see Section 4.4). This is repeated until the desired accuracy is obtained.

As stated when we introduced FIS in Section 2 the basic thought behind the method is the fading out of a fault effect after some number (n_{aff}) of observations. When we apply FIS we therefore have to take care to choose an appropriate value for n_{aff} (see Section 5). In our simulation we keep track of the correlation between successive batches to get an indication of their dependency. To check our results we apply FIS with two different values of n_{aff} and compare them.

Results

We show results for the following parameter values: $\lambda = 1$, $m = 2$, repair rate $\nu = 1/30$, fault probability $f = 10^{-4}$, deadline $\alpha = 9.210$ and batch size $n_{aff} = 250$. At the end of this section we comment on the consequences of a change in some of these values.

The results for the two simulations in FIS were: for S_{aff}: $E_{aff} = 44.36$, $v_{aff}^2 = 2862$, and for S_0: $E_0 = 0.0219$, $v_0^2 = 0.0363$ (see Section 4.2). From this the optimal allocation fraction β_{opt}, i.e. the fraction of batches that should

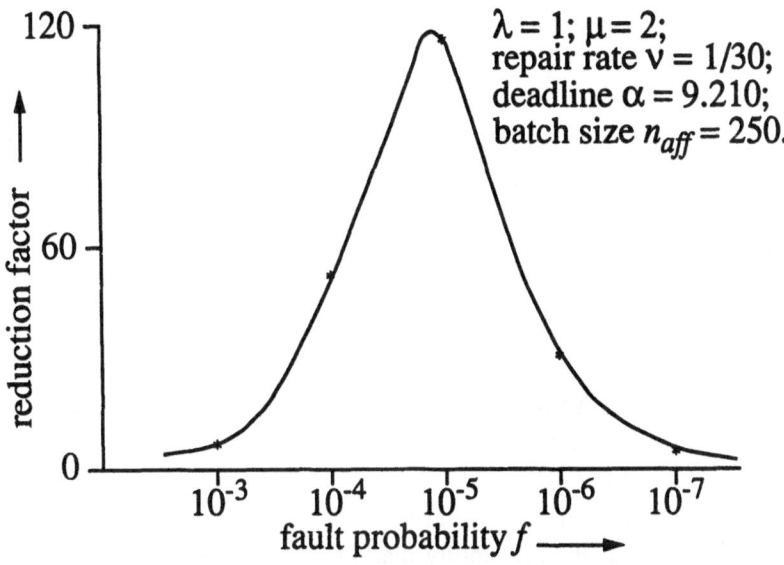

Figure 9: Factor reduction in time

be dedicated to S_{aff}, can be calculated (Section 4.4). This leads to Figure 8. Notice that we can calculate β_{opt} for a range of fault probabilities but that we will continue the simulation with the allocation fraction belonging with $f = 10^{-4}$. The factor of variance reduction, and thus the time saving factor, obtained with these optimal allocation fractions is shown in Figure 9. This factor is equal to the reduction factor in the number of batches that has to be simulated.

We want results for the system with breakdown probability $f = 10^{-4}$, so we take the estimated optimal allocation fraction belonging with that fault probability and simulate until the 95% confidence interval is within 10% of the estimate. This leads to Table 1 in which the estimated probabilities of missing the deadline α are shown with 95% confidence intervals.

The results for batch size $n_{aff} = 500$, for which the effect of the fault may last longer, hardly differ from those of $n_{aff} = 250$. The same holds for a direct simulation we did. This direct simulation took 36 times as much simulation time as FIS with $n_{aff} = 250$. In all three cases we had that for $f = 10^{-4}$ the probability of missing the deadline was $4.5 \times 10^{-3} \pm 0.4 \times 10^{-3}$. The estimate for the correlation between successive batches is 0.06 for $n_{aff} = 250$ and 0.01 for $n_{aff} = 500$.

Table 1: $\overline{F}(\alpha)$ with confidence intervals

f	point estimate with 95% confidence interval
10^{-7}	$9.20 \times 10^{-5} \pm 12.8 \times 10^{-5}$
10^{-6}	$1.32 \times 10^{-4} \pm 1.28 \times 10^{-4}$
10^{-5}	$5.31 \times 10^{-4} \pm 1.32 \times 10^{-4}$
10^{-4}	$4.52 \times 10^{-3} \pm 0.38 \times 10^{-3}$
10^{-3}	$4.44 \times 10^{-2} \pm 0.36 \times 10^{-2}$

Simulation time reduction factors of the order of 100 are no rarity. For instance, for the same model with $\mu = 4/3$, $\alpha = 20.72$ and batch size $n_{aff} = 500$ we obtained a reduction factor of almost 200 for $f = 3.7 \times 10^{-6}$.

For the discussed model we could have taken known analytical results for the fault free model as the probability that a response time exceeds $\alpha = 9.210$ is 10^{-4}. For batch size $n_{aff} = 250$ this would give $\text{E}_0 = 2.5 \times 10^{-2}$ and $v_0^2 = 0$. With FIS the same number of faults then would be simulated in a fraction n_{aff}/n of the time necessary with direct simulation.

8   Conclusions

We have presented a rather straightforward method, Fault Injection Simulation (FIS), that can be used for the computation of performability characteristics of communication systems which are strongly influenced by very rarely occurring faults. Among the advantages of FIS we have:

- Large variance reduction possibilities, the single reason to use FIS in the first place;

- An optimal application of FIS can be calculated during the simulation;

- The possibility to use FIS as an hybrid, i.e. partly analytical, technique when analytical results for the fault free system are available;

- Extrapolation of results; one simulation gives results for a range of fault probabilities;

- Straightforward applicability; the method is intuitively appealing and easy to implement.

Of course FIS also has its limitations. One should keep in mind that the method is based on heuristic arguments, the notion of the fading out of the influence of a fault should be considered with care. We conclude with discussing some extensions of FIS. It is possible to evaluate systems with different rarely occurring faults by running more than two separate simulations. Furthermore one can mix different variance reduction techniques, e.g. within the fault free simulation part of FIS one can apply Importance Sampling to further speed up the simulation, etc. Based on the first results presented in this paper FIS seems to be a flexible, promising variance reduction method for studying the performability influence of rare failures.

Acknowledgment

We thank Erik van Doorn for his helpful comments during the development of the Fault Injection Simulation technique.

References

[1] J.F. Meyer, "On evaluating the performability of degradable computing systems", *IEEE Trans. Comp.*, vol. C-29, pp. 720-731, Aug. 1980.

[2] R.M. Smith, K.S. Trivedi and A.V. Ramesh, "Performability analysis: Measures, an algorithm and a case study", *IEEE Trans. Comp.*, vol. C-37, pp. 406-417, April 1988.

[3] G.S. Fishman, *Principles of Discrete Event Simulation*, John Wiley & Sons, New York, 1978.

[4] V.S. Frost, W.W. Larue Jr. and K.S. Shanmugan, "Efficient techniques for the simulation of computer communication networks", *IEEE J. Select. Areas Commun.*, vol. SAC-6, pp. 146-157, Jan. 1988.

[5] P.D. Welch, "The statistical analysis of simulation results", in *Computer Performance Modeling Handbook*, S.S. Lavenberg Ed., New York, Academic Press, 1983, pp. 267-329.

[6] W.G. Cochran, *Sampling Techniques*, New York, John Wiley & Sons, 1977.

[7] K. Pawlikowski, "Steady-state simulation of queueing processes: A survey of problems and solutions", *ACM Comp. Surveys*, vol. 22, pp. 123-170, June 1990.

[8] J. Arlat, Y. Crouzet and J.C. Laprie, "Fault injection for dependability validation of fault-tolerant computing systems", in *Proc. FTCS-19*, pp. 348-355, 1989.

[9] J.H. Barton, E.W. Czeck, Z.Z. Segall and D.P. Siewiorek, "Fault injection experiments using FIAT", *IEEE Trans. Comp.*, vol. 39, pp. 575-582, April 1990.

[10] J.P.C. Kleynen, *Statistical Techniques in Simulation I and II*, New York, Marcel Dekker, 1974.

[11] A. Goyal, P. Shahabuddin, Ph. Heidelberger, V.F. Nicola and P.W. Glynn, "A unified framework for simulating Markovian models of highly dependable systems", *IBM Res. Rep. RC 14772*, Nov. 1989 (to appear in *IEEE Trans. Comp.*).

[12] J. Walrand, "Quick simulation of queueing networks: An introduction", in *Computer Performance and Reliability*, G. Iazeolla, P.J. Courtois and O.J. Boxma, Eds., Amsterdam, North-Holland, 1988, pp. 275-286.

[13] S.M. Ross, *Introduction to Probability Models*, Fourth Edition, San Diego, CA, Academic Press, 1989.

[14] G. Carter and E.J. Ignall, "Virtual measures: A variance reduction technique for simulation", *Mgt. Sc.*, vol. 21, pp. 607-616, Feb. 1975.

Limits of Parallelism in Fault-Tolerant Multiprocessors

VICTOR F. NICOLA, AMBUJ GOYAL

IBM Thomas J. Watson Research Center

P.O. Box 704

Yorktown Heights, NY 10598, USA

Abstract

Using simple and realistic models we show that, in the presence of failures, there is a limit to the amount of effective parallelism in a multiprocessor system, even if the application is infinitely parallelizable. We consider the execution of a given application on a multiprocessor system with fault-tolerance capabilities, and use its expected completion time as a measure of computational efficiency. We show that more parallelism may not necessarily decrease the expected task completion time, and if it does, there exists an optimal number of processors beyond which the expected task completion time starts to increase monotonically. This optimal number may also be interpreted as a limit on the effective parallelism in a multiprocessor system. This phenomenon occurs regardless of the fault-tolerance mechanism used in the system; however, the optimal number of processors depends on that mechanism. We consider four different methods of fault-tolerance:

1. Restarts
2. Triple modular redundancy (TMR) with restarts
3. Checkpointing, centralized and distributed
4. Resilient applications, e.g., iterative or hill-climbing methods.

For each of the above fault-tolerance methods, we give an expression for the expected completion time, as a function of the number of processors in the system, and provide numerical results to illustrate the limit of parallelism.

Key Words: Fault-tolerance, parallel processing, restarts, checkpointing, resilient applications.

1 Introduction

One of the main reasons to build a parallel processing system is to reduce the run-time of a large computation. For example, the GF11 computer [1] has been designed to solve the Quantum Chromo Dynamics (QCD) application which has approximately 10^{17} floating-point operations for a reasonable size QCD experiment. The GF11 computer, which has 576 processors and a total computing power of 11 gigaflops, takes one year to perform a single QCD experiment. Another example of a large running computation is Complex Queries which may involve multi-way joins on relational tables of hundreds of gigabytes in size. The Teradata machine [2] has been designed to solve such complex queries in parallel. Specifically, examples from real applications show that a 144 processor Teradata machine may take more than a day to solve many of these large queries. Also, in real-time avionic and space applications, parallelism is used to speed up long computations. In these systems, a delayed response may lead to a catastrophic failure.

In the design phase, when building these large parallel processors, we must consider the possibility of failures during system operation. Many parallel processing systems have been designed without any consideration for fault-tolerance. Adding fault-tolerance as an after-thought is either not feasible or ends up adding too much cost to the machine. In this paper, we show using simple and realistic models that, in the presence of failures, there is a limit to the amount of effective parallelism in a multiprocessor system. In fact, an application which has a finite computation time on a uniprocessor system could potentially take a much longer time to complete on a multiprocessor system. Therefore, explicit consideration for fault-tolerance while designing a multiprocessor system is of utmost importance.

The basic idea behind the "limits of parallelism" can be explained as follows. As the number of processors (or parallelism) increases, the computation time decreases, at best linearly. On the other hand, the overall failure rate of the system (i.e., the rate at which the first processor fails) increases linearly with the number of processors. Therefore, if the computation time decreases at a rate which is less than linear (even if it is fractionally less), then for a sufficiently large computation and/or number of processors, the failure rate function would become the dominating function. At this point, the loss in computational efficiency due to increased failures offsets the gain obtained by more parallelism. Thus, the completion time would eventually start increasing with the number of processors. This is true regardless of what hardware and software redundancy mechanisms we use to mask failures. These mechanisms, however, would delay the occurrence of this phenomenon, i.e., a larger number of processors can now be used in a multiprocessor system without encountering the "failure domination" effect.

Depending upon the fault-tolerance techniques used and the size of the computation, the limits of parallelism occur at different points for different computational models. The goal of this paper is to use simple models of completion time to determine these limits. Notice that the completion time is a task oriented measure [3, 4], as opposed to system oriented performability measures considered by others [4, 5, 6]. We do not focus on the derivation of the expected completion time equations for the various models considered; these have been derived elsewhere in the literature [3, 7]. However, we do parametric studies using these existing results to illustrate our ideas.

In Section 2 we consider parallelism in a failure free environment. We define the service time requirement as a function of the number of processors and introduce two computational models. We also define the completion time of a given task. In Section 3 we consider parallelism in the presence of failures without fault-tolerance capabilities. In this case, the task is lost if not completed before any failure. With fault-tolerance capabilities, the task is guaranteed to complete. In this case, we are interested in the expected task completion time as a measure of computational efficiency. For different methods of fault-tolerance; namely, restarts, triple modular redundancy (TMR) with restarts, checkpointing and resilient applications, in Section 4 we give expressions for the expected task completion time in terms of the number of processors in the system and numerically determine the optimal number of processors. We conclude in Section 5.

2 Parallelism with no Failures

In this section we consider the parallel execution of a task on a multiprocessor system under the assumption of no failures; this is to examine the effect of parallelism in isolation. The computational efficiency gained by parallelism depends on how much parallelism can be exploited when executing a task and how much overhead is involved in coordinating the parallel execution of a task. Note that even with maximum parallelism (i.e., the task is equally divided among all available processors), the gain in computational efficiency is reduced, and may be offset, due to the increased overhead of parallelism.

2.1 The Computational Model

Consider the execution of a given task with a fixed computational requirement (e.g., a fixed number of instructions to be executed). We define its service time requirement on any system (uniprocessor system or multiprocessor system) as the time required to execute the task alone (i.e., with no other tasks in the system) with no failures. Since the task is executed alone on the system its service time requirement is deterministic. Throughout this paper we consider a task with a deterministic service time requirement on a uniprocessor system given by x. On a multiprocessor system with no failures, assuming that the task can be divided among all processors, the service time requirement of the task is a function fcxm , where M is the number of processors in the system. Ideally, $f_c(x, M)$ decreases linearly with M; this is not the case for most realistic applications due to the several factors, such as the non-uniform division of the task among all processors and the overhead involved in coordinating the parallel computation. In this paper we consider two models of computation; namely, the power model and the logarithmic model.

  i) The power model:
In this model $f_c(x, M)$ is given by

$$f_c(x, M) = x/M^p, \tag{1}$$

where p is a constant ($0 \leq p \leq 1$), i.e., $f_c(x, M)$ decreases (with M) as fast as the p power of M increases (with M). Usually, $p < 1$ (i.e., $f_c(x, M)$ decreases less than linearly with M) and increases with the efficiency of the parallel computation. Therefore, the power model can be used as a parameterization of a broad class of parallel applications. In our experiments we consider different values of p.

  ii) The logarithmic model:
In this model $f_c(x, M)$ is given by

$$f_c(x, M) = x/(1 + \log_p M). \tag{2}$$

Here p is the base of the logarithm function. For most applications, p is naturally equal to 2, which is also our choice throughout this paper. Parallel "divide and conquer" algorithms, to find the maximum or the minimum or sort the elements of an array, are examples of such applications.

  In Figure 1 we plot the normalized service time requirement $f_c(x, M)/x$ as a function of M, the number of processors in the system, for the logarithmic

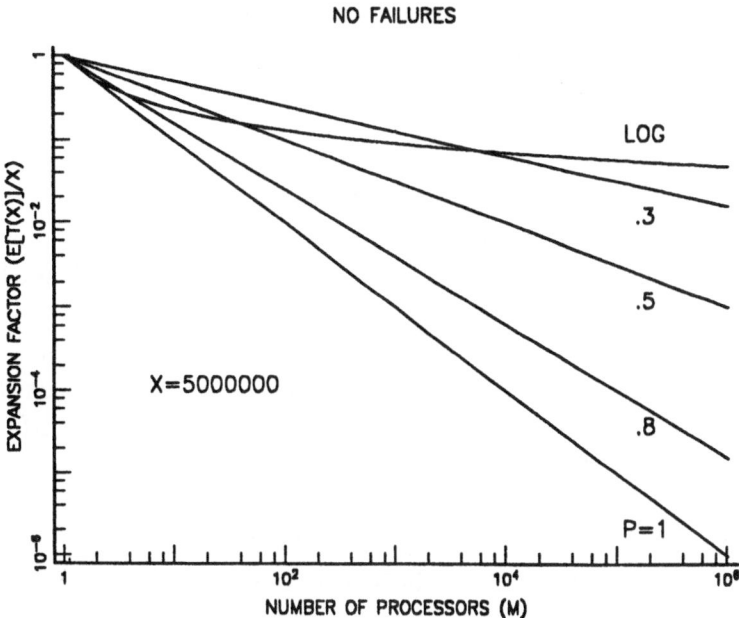

Figure 1: Normalized service time requirement $f_c(x, M)/x$.

and the power models of computation. We have selected a value for x commensurate with the QCD application mentioned in Section 1. For example, if we assign $x = 5000000$ hours, then ideally, i.e., assuming a linear computational model ($p = 1$), 500 processors will require 10000 hours (approximately one year) to solve the problem. Notice that, asymptotically (for large M), $f_c(x, M)$ from equation 2 (the logarithmic model) decreases with M slower than $f_c(x, M)$ from equation 1 (the power model) for any value of $p > 0$.

In the following section we define the completion time of a task and the expansion factor which is used as a measure of computational efficiency in the presence of failures.

2.2 The Completion Time

The completion time of a given task is defined to be the time elapsed between the beginning and the completion of its execution on the system. It is identical to the service time requirement when the task is executed alone on the system under a failure free environment. In the presence of failures, the completion time of a task on a multiprocessor system is a random variable denoted by

$T(x, M)$, where x is the service time requirement on a uniprocessor system and M is the number of processors in the system. In this paper we consider the completion time of a given task when executed alone on the system (i.e., with no multiprogramming), in the presence of random failures. In particular, we choose the expected task completion time, $E(T(x, M))$, as the measure of interest. We also define the expansion factor as the ratio between the expected task completion time and its service time requirement on a uniprocessor system, i.e., $e(x, M) = E(T(x, M))/x$. The expansion factor $e(x, M)$ is used as a measure of the computational efficiency when executing a given task on a given system. Obviously, there is efficiency loss in the presence of failures and no parallelism, i.e., $e(x, M) \geq 1$. On the other hand, with parallelism and in the absence of failures, there is usually efficiency gain, i.e., $e(x, M) < 1$, unless the overhead of parallelism is so large that it is more efficient to execute the task on uniprocessor system. While more parallelism usually improves the computational efficiency, this is not necessarily true in the presence of processor failures. Our goal in this paper is to study the gains and limitations of parallelism in the presence of processor failures. It is assumed that the failure of each processor occurs according to a Poisson process at a rate γ. Typical processor failure rates are of the order of one in 100000 hours, i.e., $\gamma = .00001$.

In the presence of failures, fault-tolerance is necessary in most applications in order to avoid or reduce the possibility of losing a task or getting incorrect results in the presence of failures. We briefly discuss the case with no fault-tolerance in Section 3, and then move to the more interesting case with fault-tolerance in Section 4.

3   Parallelism with no Fault-tolerance

In this section we briefly discuss the parallel execution of a task on a multi-processor system without fault-tolerance capabilities, i.e., any processor failure results in a non- recoverable task failure. Here, a task restart is also not possible. In this environment we are interested in measures, such as the probability of completing a task before any failure, and its expected completion time given that it com- pletes before any failure. Since the failure of each processor occurs according to a Poisson process at a rate γ, the overall rate of processor failures is γM , i.e., it increases linearly with the number of processors in the system.

Again, let x be the service time requirement of a given task on a uniprocessor system, then its service time requirement on a multiprocessor system is $f_c(x, M)$. For the task to complete successfully, there must be no failures during the first $f_c(x, M)$. time units. Since the overall failure process is Poisson with rate γM, the probability of successfully completing the task is given by

$$P(T(x, M) < \infty) = e^{-\gamma M f_c(x,M)}. \tag{3}$$

For the power model of computation equation 3 reduces to $e^{-\gamma x M^{1-p}}$. For the logarithmic model of computation equation 3 reduces to $e^{-\gamma x M/(1+\log M)}$.

Given that a task completes before any failure, its completion time is deterministic given by $f_c(x, M)$, which yields the conditional expansion factor

$$e(x, M \mid T(x, M) < \infty) = f_c(x, M)/x, \tag{4}$$

which is the normalized service time requirement.

In Figure 2, for different computational models, we plot the probability of task completion in the presence of failures as a function of the number of processors M. We set $x = 100000$ hours and $\gamma = .00001$ per hour. The curve marked "LOG" is for the logarithmic model of computation with $p = 2$. The other curves are for the power model of computation with different values of p. As expected, unless $p = 1$, the probability of a task completion before a failure decreases considerably with M, particularly for small values of p.

4   Parallelism with Fault-tolerance

In many critical and commercial applications the loss or failure of a task is not acceptable in a normally operating system. Therefore, it is necessary that fault-tolerant techniques be used in these systems to restore the correct operation after a failure and guarantee a successful completion of the running task. For example, the system may be equipped with a repair/reconfiguration service after which the interrupted task is resumed or restarted from its beginning. Often, failures result in a loss of work, and resuming a task from the point of failure is not possible. However, checkpointing is a commonly used technique, particularly in long running applications, to avoid total loss of work after failures. With checkpointing, relevant information is saved periodically in a safe storage, so that, after a failure, the interrupted task may be restarted from a previous checkpoint rather than from its beginning. In resilient applications, such as iterative methods and hill-climbing techniques, some kind of

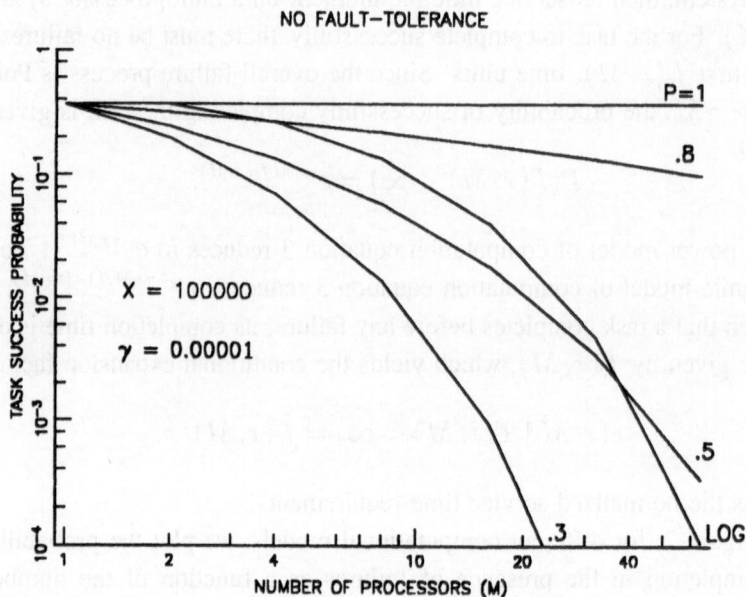

Figure 2: Probability of task completion without fault-tolerance.

"checkpointing" is inherently built-in; a failure in these applications does not necessarily require restarting from the beginning after repair.

With fault-tolerance, a task executing on the system will eventually complete. However, it should be noted that in some real-time applications, a successful completion of a task may involve timing constraints that must be satisfied in the course of its execution. In our treatment we do not consider such time constraints, we are only interested in the expected task completion time (or the expansion factor) as a measure of computational efficiency.

In this section, four methods of fault-tolerance are discussed; namely, restarts, triple modular redundancy (TMR) with restarts, checkpointing and resilient applications. Analytical results are used to study the first three methods, while simulation is used to evaluate the fourth method.

4.1 Restarts

In this fault-tolerance method, the interrupted task is restarted from its beginning after repair following a failure. Recall that, in the absence of failures, the service time requirement of a task on a multiprocessor system is given by $f_c(x, M)$. In the presence of failures, the time it takes to complete the task is the first time an uninterrupted interval of length $f_c(x, M)$. is encountered. Since the overall failure process is Poisson at a rate γM, the probability of no failure during the time interval $f_c(x, M)$. is given by $e^{\gamma M f_c(x,M)}$. It follows that the expected number of restarts (or failures) before completing the task is given by $(e^{\gamma M f_c(x,M)} - 1)$. Again, because failures occur according to a Poisson process, we can determine the expected time to complete the task by dividing the expected number of failures by the overall failure rate, which yields $(e^{\gamma M f_c(x,M)} - 1)/\gamma M$. Let R be the expected repair time following a failure, then the expected task completion time including repair time is given by

$$E(T(x, M)) = (\frac{1}{\gamma M} + R)(e^{\gamma M f_c(x,M)} - 1). \qquad (5)$$

For the power model of computation, equation 5 becomes

$$E(T(x, M)) = (\frac{1}{\gamma M} + R)(e^{\gamma x M^{1-p}} - 1). \qquad (6)$$

For the logarithmic model of computation, equation 5 becomes

$$E(T(x, M)) = (\frac{1}{\gamma M} + R)(e^{\gamma x \frac{M}{1+\log M}} - 1). \qquad (7)$$

In Figure 3 we plot the expansion factor $e(x, M) = E(T(x, M))/x$ as a function of the number of processors M, for the logarithmic and power models of computation. We set $x = 100000$ hours, $\gamma = .00001$ per hour and $R = 1.0$ hours. Notice that at $M = 1$, the expansion factor is greater than one, this is because of the presence of failures. Depending on the power p of the computational model, we see that increased parallelism may not improve the computational efficiency; this is particularly true for small values of p. However, for sufficiently large p, the expansion factor initially decreases until it reaches a minimum then starts to increase. The same observation also holds for the logarithmic model of computation. This means that depending on the service time requirement x, the computational model and the failure rate, parallelism may not necessarily improve the computational efficiency, and if it does, then there exists an optimal value for M which maximizes the computational efficiency (or minimizes the expansion factor). It is interesting, however, to note that in the ideal case, i.e., $p = 1$, the expansion factor decreases linearly with M.

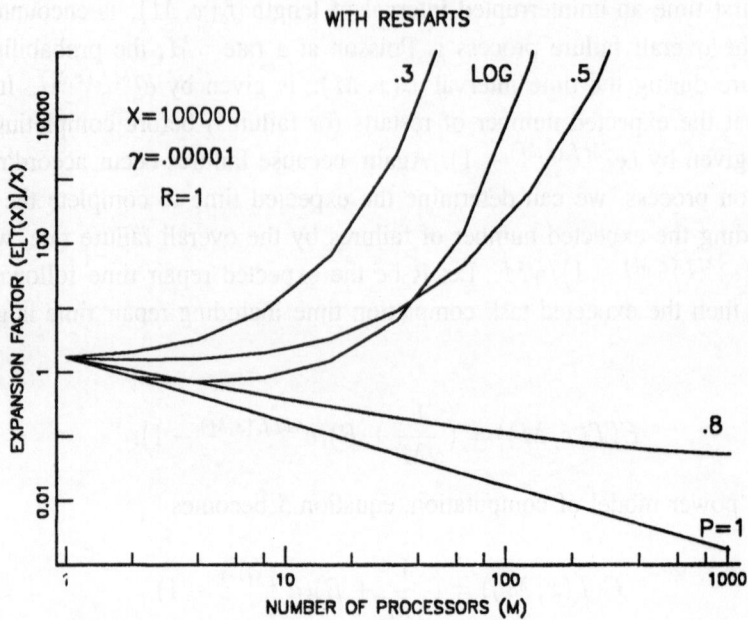

Figure 3: Computational efficiency with restarts in a standard configuration.

4.2 TMR with Restarts

In this fault-tolerance method, triple modular redundancy (TMR) [8] is used to increase the reliability of each node (processor) in the system. Each processor in the standard (original) configuration is replaced by one node, which is composed of three redundant processors and a voting mechanism. The voting mechanism selects the correct output as long as two or more (out of three) processors produce the correct output. In other words, the failure of one processor (Poisson process with a rate γ) does not bring a node down. In the mean time, this failed processor is being repaired at a rate μ (assuming the repair time is exponentially distributed). If a second processor fails in the same node while the first is in repair, then the node is considered failed. This node failure corresponds to a processor failure in the standard configuration which, in a fault-tolerant system, is followed by a repair time of a mean R. Figure 4 is a continuous-time Markov chain (CTMC) representing the failure process of a node. The broken line corresponds to a repair transition, following a node failure, after which the execution of the interrupted task is restarted from the beginning.

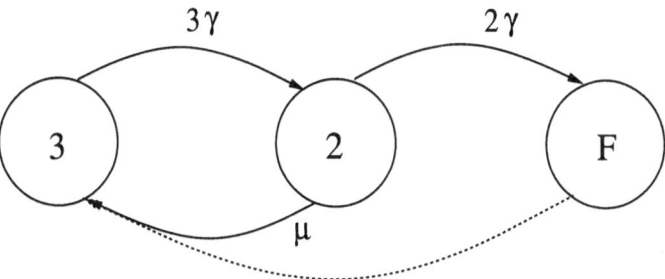

Figure 4: CTMC representing the failure of a TMR node.

Starting with three processors operational, the mean time to a node failure (MTTF) can be obtained from the CTMC in Figure 4 [8]

$$MTTF_{TMR} = (5\gamma + \mu)/6\gamma^2. \tag{8}$$

Since $\gamma \ll \mu$, using a known result from the theory of "process thinning" [9], the distribution of the time to a node failure can be very well approximated by an exponential distribution with a rate $\gamma_{TMR} = 1/MTTF_{TMR}$. This result is very useful, since now we can use the same model of Section 4.1 to obtain the expected task completion time on a TMR multiprocessor system configuration. Therefore, equation 5 holds for the considered TMR configuration, with γ replaced by γ_{TMR} and M replaced by m_{TMR} (=$M/3$), the number of TMR nodes in the system.

It should be noted that the TMR configuration increases the reliability of each individual node by orders of magnitude, but it reduces potential parallelism only by a factor of 3. However, it is not obvious when the TMR configuration (with higher node reliability and less parallelism) is more efficient than the standard configuration (with lower node reliability and more parallelism). Clearly, for a sufficiently small failure rate, the standard configuration is more efficient; this suggests that the TMR configuration could be more efficient for a sufficiently large failure rate. For the purpose of comparison with the standard configuration, in Figure 5 we plot the expansion factor as a function of the total number of processors in the system with the TMR configuration. We set $\mu = 1.0$ per hour, and use the same parameter values as those used to generate Figure 3. For this realistic choice of parameters, we see a significant

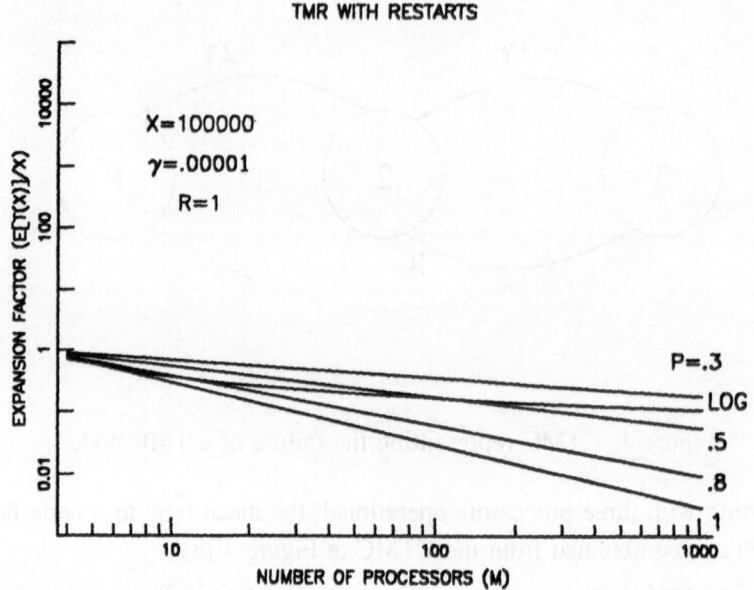

Figure 5: Computational efficiency with restarts in a TMR configuration ($x =$ 100000 hours).

improvement in computational efficiency by using the TMR configuration over the standard configuration.

In Figure 6, we show that also with the TMR configuration there is a limit on parallelism beyond which the computational efficiency decreases. We plot the expansion factor as a function of M, for a larger service time requirement $x = 5000000$ hours (as required in QCD application), all other parameters are the same as used in Figure 5. Similar to the standard multiprocessor system configuration with restarts, for any $p < 1$, the expansion factor may not decrease with M, and if it does, then it will eventually reach a minimum and then starts to increase monotonically.

4.3 Checkpointing

Checkpointing is a technique commonly used along with fault-tolerance, particularly in long-running applications, to avoid a total loss of work after a failure. A vast literature exists on practical as well as modelling aspects of checkpointing, some references may be found in [7]. With checkpointing, sufficient information is saved periodically in a safe storage, so that the interrupted task

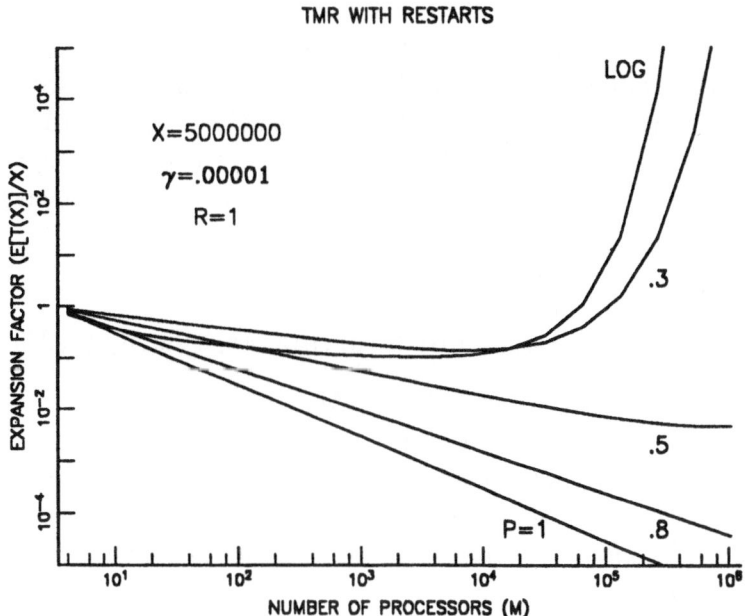

Figure 6: Computational efficiency with restarts in a TMR configuration ($x =$ 5000000 hours).

is resumed from the last checkpoint after repair following a failure. Therefore, only work done since the last checkpoint is lost.

Models of checkpointing in a program are considered in [7, 10]. In [7] checkpoints are assumed to occur during normal operation and also during reprocessing after failures. The duration of a checkpoint saving operation as well as the repair time after a failure are generally distributed with means C and R, respectively. The time between two successive checkpoints (excluding checkpointing and repair times) is exponentially distributed with a mean $1/\alpha$. In [7] they derive the distribution of the completion time of a task, executing on a uniprocessor system, with a given service time requirement x. Their expression for the expected task completion time holds also for our multiprocessor system model after replacing γ by γM, x by $f_c(x, M)$ and C by $C(M)$. $C(M)$ is the mean duration of a checkpoint as a function of M. The resulting expression for the expected task completion time in a multiprocessor system is given by (see [7])

$$E(T(x, M)) =$$
$$((1 + \alpha C(M) + \gamma MR)/\alpha)((\alpha + \gamma M)f_c(x, M) + \ln(A(f_c(x, M)))),$$

with $A(x) = (\alpha + \gamma Me^{-(\alpha + \gamma M)x})/(\alpha + \gamma M)$.

We consider two functions for $C(M)$ corresponding to centralized and distributed checkpointing. In centralized checkpointing, the status of each processor is saved in a central safe storage. The saving operations are performed sequentially. Therefore, the duration of a checkpoint is proportional to M, and we use the following linear model

$$C(M) = C_0 + C_1 M. \tag{9}$$

In distributed checkpointing, the status of each processor is saved in a local safe storage. The saving operations are performed in parallel. Therefore, the duration of a checkpoint is the maximum of M saving operations. Assuming that the saving time is exponentially distributed, then the duration of a checkpoint is proportional to $\sum_{i=1}^{M} 1/i$, which is well approximated by a logarithmic function. Therefore, we use the following model

$$C(M) = C_0 + C_1 \log M. \tag{10}$$

In the above two models for $C(M)$ we set the constants C_0 and C_1 to one.

For a uniprocessor system and sufficiently large service time requirement x, in [7] they give a simple approximation for the optimal checkpointing rate

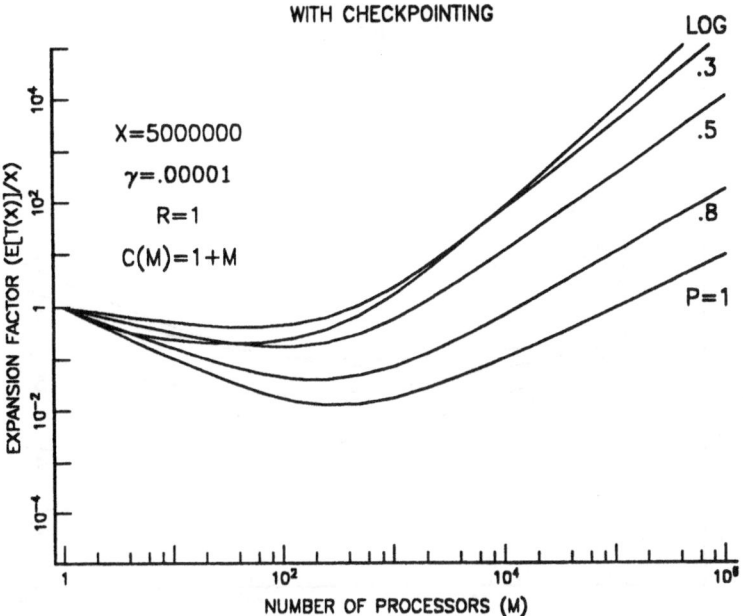

Figure 7: Computational efficiency with centralized checkpointing.

$\alpha^* \simeq \sqrt{\gamma(1 + \gamma R)/C}$ which minimizes the expected task completion time. Since in a multiprocessor system the model parameters depend on M, so does the optimal checkpointing rate $\alpha^*(M)$. In our experiments, as we vary M we also vary the checkpointing rate to its approximate optimal value which is given by the same expression for α^* with γ replaced by γM and C replaced by $C(M)$

$$\alpha^*(M) \simeq \sqrt{\gamma M(1 + \gamma M R)/C(M)}. \tag{11}$$

In Figures 7 and 8 we plot the expansion factor, for a task running on a multiprocessor system, as a function of M with centralized and distributed checkpointing, respectively. We set the parameters as follows: $x = 5000000$ hours, $\gamma = 0.00001$ per hour, $R = 1.0$ hours. For the power as well as the logarithmic model of computation, we see that there is an optimal M for which the expansion factor is minimized. The optimal number of processors increases with the power of the computational model p. Notice also that, for the same computational model, the expansion factor with distributed checkpointing is lower and less sensitive to M than that with centralized checkpointing.

It is also interesting to compare checkpointing with restarts when $p = 1$.

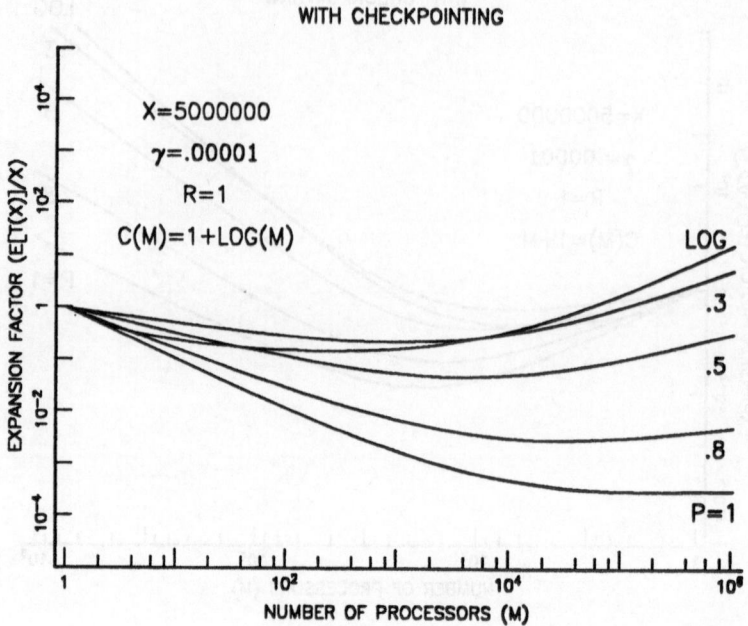

Figure 8: Computational efficiency with distributed checkpointing.

Since, for restarts, the expansion factor decreases linearly with M (Figures 5 and 6), it is not intuitively clear why, for checkpointing, it (eventually) increases with M (Figures 7 and 8). This behavior is caused by the checkpointing overhead (duration) which increases with M according to our model. In fact, for a constant checkpoint duration (independent of M), we would observe a behavior similar to the case with restarts, i.e., the expansion factor would decrease monotonically with M.

4.4 Resilient Applications

Resilient applications, such as iterative methods and hillclimbing techniques, inherently possess some kind of fault-tolerance (or "checkpointing"). When the execution of a resilient algorithm is interrupted by the occurrence of a failure, and if the system is repairable, the algorithm need not restart from the beginning (initial starting guess); the last available iteration point could be used as a restarting point, thus avoiding a total loss of work. This is a very useful property of resilient applications when executed in a failure prone environment.

The execution of a resilient task in the presence of failures can be modelled

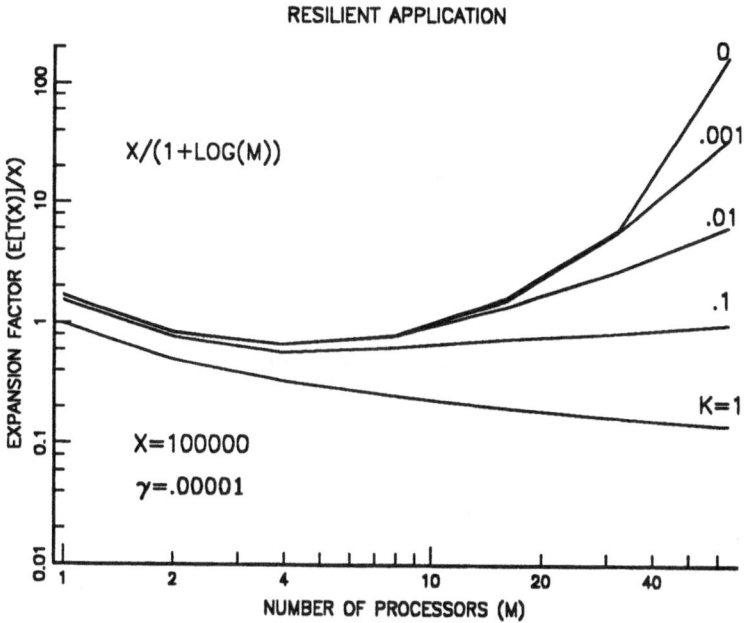

Figure 9: Computational efficiency in a resilient application (the logarithmic model).

by assuming that upon the occurrence of a failure only a fraction K ($0 \leq K \leq 1$) of the work done since either the last failure or the beginning of the task (whichever is the later) is preserved (i.e., not lost). The analysis of the completion time of a resilient task is a difficult problem which is being investigated, but not considered in this paper. Instead, we use simulation to estimate the expected completion time of a resilient task in a multiprocessor system.

In Figures 9 and 10 we plot the expansion factor, in a resilient application executing on a multiprocessor system, as a function of M for the logarithmic and power models, respectively. The parameters are set as follows: $x = 100000$ hours, $\gamma = 0.00001$ per hour. Without loss of generality, repair is assumed instantaneous, i.e., $R = 0.0$. Again, the figures indicate that there is an optimal M, which increases with K, for which the expansion factor is minimized and beyond which the effect of failures starts to lower the computational efficiency. Notice that the expansion factor first increases slowly with decreasing K (from 1 downwards), then it increases sharply as K approaches 0. This suggests that the expected task completion time increases exponentially with decreasing K.

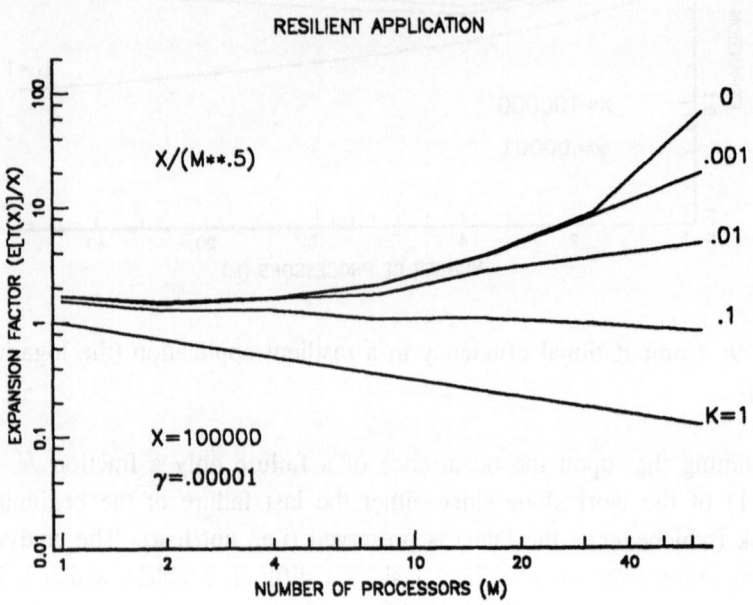

Figure 10: Computational efficiency in a resilient application (the power model).

5 Conclusions

In this paper, we have demonstrated that, in the presence of failures, increasing parallelism does not necessarily increase the computational efficiency. This is due to the fact that the overall failure rate increases linearly with the number of processors in the system. Therefore, unless the gain in computational efficiency due to parallelism is linear (ideal case), the loss due to failures will eventually dominate and cause the computational efficiency to drop monotonically with the number of processors. In other words, in the presence of failures, there is a limit on the effective parallelism in a multiprocessor system. This is true regardless of the fault-tolerance mechanism used in the system. However, the optimal number of processors, which minimizes the expected completion time of a given application, depends on the fault-tolerance mechanism used. We have considered four such mechanisms; namely, restarts, triple modular redundancy (TMR) with restarts, checkpointing and resilient applications. For each of these mechanisms, we have given an expression for the expected task completion time in terms of the number of processors in the system and numerically determined the limit beyond which the effectiveness of parallelism reduces monotonically.

Acknowledgement

We wish to thank Stephen S. Lavenberg for his interest and careful reading of the manuscript.

References

[1] J. Beetem, M. Denneau, and D. Weingarten, "The gf11 supercomputer," in *Proceedings of the 12th Annual International Symposium on Computer Architecture*, IEEE Computer Society, 1985.

[2] P. M. Neches, "Hardware support for advanced data management systems," *Computer*, vol. 17, no. 11, pp. 29–40, 1984.

[3] V. G. Kulkarni, V. F. Nicola, and T. K. S., "The completion time of a job on multi-mode systems," *Advances in Applied Probability*, vol. 19, pp. 932–954, 1987.

[4] V. F. Nicola, A. Bobbio, and K. S. Trivedi, "A unified performance reliability analysis of a system with a cumulative down time constraint,"
Microelectronics and Reliability, 1991. to appear.

[5] A. Goyal and A. N. Tantawi, "Evaluation of performability for degradable computer systems," *IEEE Transactions on Computers*, vol. 36, no. 6,
pp. 738–744, 1987.

[6] J. F. Meyer, "On evaluating the performability of degradable computer
systems," *IEEE Trans. on Computers*, vol. 29, no. 8, pp. 720–731, 1980.

[7] V. G. Kulkarni, V. F. Nicola, and T. K. S., "Effects of checkpointing
and queueing on program performance," *Communications in Statistics:
Stochastic Models*, vol. 6, no. 4, pp. 615–648, 1990.

[8] K. S. Trivedi, *Probability and Statistics with Reliability, Queueing and
Computer Science Applications*.   Englewood Cliffs, NJ: Prentice-Hall,
1982.

[9] J. Keilson, *Markov Chain Models: Rarity and Exponentiality*.   Springer-
Verlag, 1979.

[10] A. Duda, "The effects of checkpointing on program execution time," *Information Processing Letters*, vol. 16, pp. 221–229, 1983.

Correlated Failures

Correlated Hardware Failures in Redundant Systems

JOANNE BECHTA DUGAN

Department of Computer Science

Duke University

Durham, NC 27706, USA

Abstract

Fault tolerant systems make extensive use of hardware redundancy in order to improve the system dependability. Models used to assess the dependability of these systems usually assume that failures in the redundant units are independent, even though it is generally believed that this assumption is optimistic. Correlation between failures occurs for a variety of reasons, ranging from design flaws to environmental interference. In this paper we investigate the effect of a postulated correlation coefficient on the predicted availability of a redundant system, when compared with a system in which the failures are assumed to occur independently. The purpose of this study is to show the sensitivity of an availability estimate to correlation, and to urge designers and measurers of redundant systems to consider this phenomenon.

To facilitate this study, we develop a technique for incorporating the concept of correlation between failures into a Markov model of a redundant system. This methodology uses a correlation coefficient to determine the (constant) rate of occurrence of the second and subsequent faults in a Markov model of a system. We solve models of two, three and four component systems subject to permanent, intermittent and transient correlated faults, and show that a correlation coefficient as small as 0.01 can increase the unavailability of a system by several orders of magnitude. To demonstrate the calculation of the necessary correlation coefficients from observed data, we analyze the results of a VAXcluster measurement-based analysis using the proposed methodology.

1 Introduction

Designers of fault tolerant systems often use redundant hardware to improve dependability. One justification for the use of redundant components is the belief that failures are random events that affect the individual components independently (or at least nearly so). It is generally conceded that the independence assumption is optimistic, but the effect of small amounts of correlation are assumed to be negligible. However, recent measurements of a VAXcluster system have shown that correlated failures are significant [1].

Most dependability analysis techniques for modeling hardware failures in redundant systems assume that failures occur independently [2, 3, 4, 5]. From the point of view of a system designer or measurer, correlation is not well understood, and techniques for incorporating correlation in analytical models are not sufficiently developed [6].

In this paper we develop a technique for considering correlation between the faults that affect redundant components of a system. We show that even a very small correlation coefficient can dramatically affect the predicted availability of a redundant system. Correlation between component failures in fault tolerant systems can arise from several sources. If the system is exposed to a harsh environment, several components can be affected simultaneously. Identical redundant components may be subject to the same design or manufacturing flaw. We define a correlation coefficient ρ_i between the ith and $(i + 1)$st fault, and use this coefficient to determine λ_i, the rate at which the ith fault occurs. The λ_i values can then be used in a standard Markov model of the system.

There are two previous works which are related. In [7], Krishna and Singh present an integrated model that accounts for the effects of independent failures and environmentally induced failures. They do not define a correlation coefficient, but rather formulate a model of the environment such that when the environment is in a malicious state, the hardware failure rate is increased. They then propose the use of time redundancy (slack time in the schedule) so as to tolerate the effects of environmental disturbances.

In [8], Malaiya introduces two measures to describe dependence of intermittent failures, *duration linear correlation* and *transition linear correlation*. Duration linear correlation relates the time period during which faults are active, while transition linear correlation relates the closeness in time instants at which faults become active. The work described here generalizes Malaiya's concept of duration linear correlation. First, we extend the technique to consider more than two components, by defining ρ_i values for the occurrence of the ith fault. (Malaiya considers correlation between two components, and does not discuss correlation between multiple components.) Second we consider the correlation of transient, intermittent and permanent faults using a single methodology, and define a cohesive methodology to include all three types. Finally, we analyze the results of a measurement-based analysis of a VAXcluster system [1] to show how the correlation coefficient can be calculated from observed data, and to help determine what "typical" correlation coefficients might be.

2 Methodology

In this section we present the general methodology used to define the correlation coefficient, and use the correlation coefficient to determine the arrival rates for the second and subsequent faults in a system. Suppose that $A(t)$ and $B(t)$ are indicator variables for some conditions $\mathcal{A}$ and $\mathcal{B}$, that is,

$$A(t) = \begin{cases} 0 & \text{if condition } \mathcal{A} \text{ is } \textit{TRUE} \text{ at time } t \\ 1 & \text{if condition } \mathcal{A} \text{ is } \textit{FALSE} \text{ at time } t \end{cases}$$

and

$$B(t) = \begin{cases} 0 & \text{if condition } \mathcal{B} \text{ is } \textit{TRUE} \text{ at time } t \\ 1 & \text{if condition } \mathcal{B} \text{ is } \textit{FALSE} \text{ at time } t \end{cases}$$

For example, A and B might indicate whether particular components or subsets of components have failed. We discuss a general correlation coefficient for A and B and will apply this correlation to several particular cases we wish to study.

2.1 Defining the correlation coefficient

Define m_A as the time average of $A(t)$ and m_B as the time average of $B(t)$.

$$m_A = \lim_{t \to \infty} \frac{1}{t} \int_0^t A(t)$$

$$m_B = \lim_{t \to \infty} \frac{1}{t} \int_0^t B(t)$$

The *steady-state* linear correlation between the two variables is given by

$$\rho_{AB} = \frac{E[(A - m_A)(B - m_B)]}{\sqrt{(E[(A - m_A)^2])(E[(B - m_B)^2])}} \tag{1}$$

Since A and B are either zero or one, $A^2 = A$, $B^2 = B$, and $E[AB]$ reduces to $Prob\{A = 1 \wedge B = 1\}$. Further, the assumption that the processes indicated by A and B are ergodic implies that the time averages are equal to

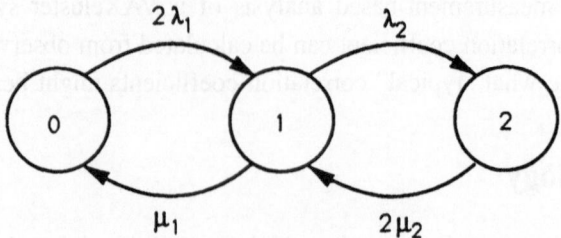

Figure 1: Behavior of intermittent faults in a two-component system

the ensemble averages, that is, $m_A = Prob[A = 1]$ and $m_B = Prob[B = 1]$. Equation 1 then reduces to

$$\rho_{AB} = \frac{Prob\{A \wedge B\} - Prob\{A\}Prob\{B\}}{\sqrt{(Prob\{A\} - Prob^2\{A\})(Prob\{B\} - Prob^2\{B\})}} \qquad (2)$$

where we have used the shorthand notation $Prob\{A\}$ to mean $Prob\{A = 1\}$. If the two variables are independent, then $Prob\{A \wedge B\} = Prob\{A\}Prob\{B\}$ and $\rho_{AB} = 0$.

2.2   Relating the correlation coefficient to the component failure rates

The methodology being used to relate the correlation coefficient to the failure rates is most easily described by considering a simple two-component system. This methodology is based on that in [8]. Consider a system with two identical components that are subject to intermittent faults. An intermittent fault is one that cycles between being active (capable of producing errors) and benign. An example would be a loose wire that sometimes makes the connection and sometimes does not, or a component that is subject to an environment that stresses its tolerance. A Markov chain diagram representing the behavior of the components with respect to intermittent faults is shown in figure 1.

The state indices for the Markov chain represent the number of components that are behaving incorrectly. The λ_i transition rate represents the rate at which the ith fault becomes active in the system, and μ_i represents the rate at which the ith fault goes benign. If the failures are independent, then $\lambda_1 = \lambda_2$ and $\mu_1 = \mu_2$, and the "usual" Markov process for a two component system results.

The goal of this section is to determine $\lambda_1, \lambda_2, \mu_1$ and μ_2 given the correlation coefficient, and the rates at which a failure becomes active (λ) and goes benign (μ) for a single component in isolation. To reduce the number of factors being considered, we assume that the active-to-benign rates (μ_1 and μ_2) are independent of the correlation coefficient. That is, we assume that $\mu_1 = \mu_2 = \mu$. Now the problem reduces to determining λ_1 and λ_2, given λ and ρ_{AB}, where A indicates the status of one component and B indicates the status of the other. Since the components are identical, it doesn't matter which way we assign the two components X_1 and X_2 to A and B; what ρ_{AB} represents is the correlation between faults in the two components.

Let P_i ($0 \leq i \leq 2$) be the steady state probability of being in state i for the Markov chain shown in figure 1. The values for the P_i are given by (assuming that $\mu_1 = \mu_2 = \mu$)

$$\begin{aligned} P_0 &= \mu^2/(\lambda_1\lambda_2 + 2\lambda_1\mu + \mu^2) \\ P_1 &= 2\lambda_1\mu/(\lambda_1\lambda_2 + 2\lambda_1\mu + \mu^2) \\ P_2 &= \lambda_1\lambda_2/(\lambda_1\lambda_2 + 2\lambda_1\mu + \mu^2) \end{aligned}$$

For this system, $Prob\{A\} = Prob\{B\} = (\frac{1}{2}P_1 + P_2)$ and $Prob\{A \wedge B\} = P_2$. Replacing the probabilities for A and B in equation 2 with the appropriate expression of state probabilities from the Markov chain yields [8]

$$\rho_{AB} = \frac{\mu}{\lambda_1 + \mu} - \frac{\mu}{\lambda_2 + \mu} \tag{3}$$

In the uncorrelated case $\lambda_1 = \lambda_2$ causing $\rho_{AB} = 0$.

In order to determine both λ_1 and λ_2 from the correlation coefficient ρ_{AB} an additional equation is needed. (The first is given in equation 3.) We can obtain a second equation by relating the failure rates for the correlated system to the failure rate of the single component (λ). We can relate the correlated and uncorrelated models by setting the rate for the first failure to equal the uncorrelated failure rate. Substituting λ for λ_1 in equation (3) and then solving for λ_2 results in the following.

$$\begin{aligned} \lambda_1 &= \lambda \\ \lambda_2 &= \frac{\mu(\lambda + \lambda\rho_{AB} + \mu\rho_{AB})}{\mu - \lambda\rho_{AB} - \mu\rho_{AB}} \end{aligned} \tag{4}$$

2.3 A two component modeling example

Suppose that we have a two component subject to permanent, intermittent and transient faults. The permanent faults are instantly and perfectly detected, after which the system is repaired. Transient and intermittent faults behave similarly,

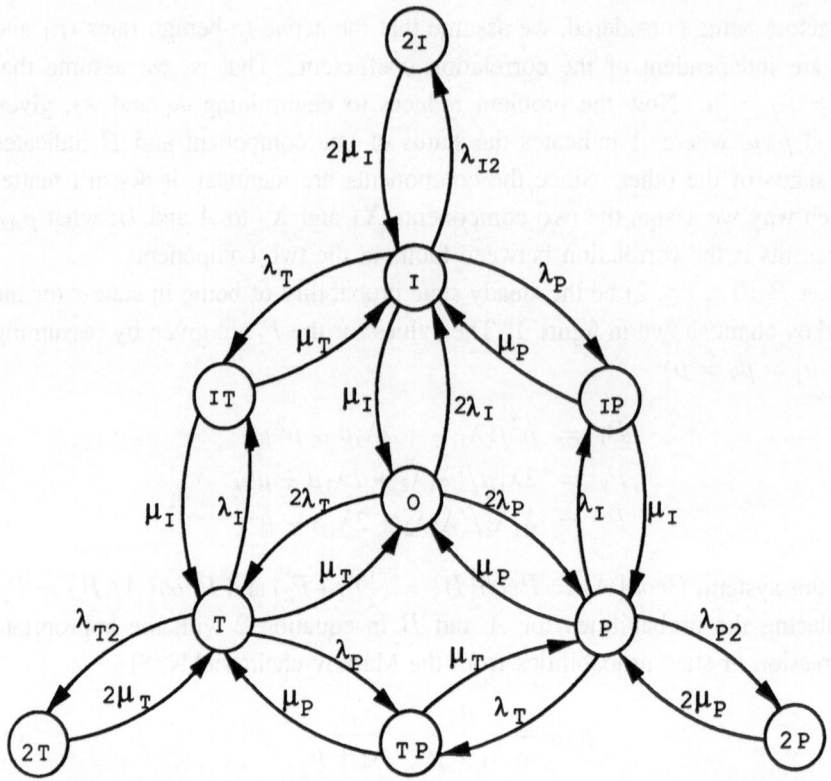

Figure 2: Markov model of two component system subject to transient, permanent and intermittent faults

in that the fault is present for some period of time after which it goes benign (in the case of intermittents) or disappears (in the case of transients, which are generally caused by some temporary external environmental interference). Transients, intermittents and repairable permanent faults are all modeled in the same manner; the fault types differ in the relative values of the parameters. Transients generally arrive much more frequently than intermittents and permanents. Transients and intermittents disappear at much faster rates than permanents are repaired.

The system is operational as long as one of the two components is operational; the availability of the system is given by the sum of the steady state probabilities that one or both components are operational.

Figure 2 shows the Markov model of the duplex system subject to transient, intermittent and permanent faults. The state labeled 0 is the state where both

Symbol	Meaning	Value (per hour)
λ_P	Arrival rate for uncorrelated permanent faults	10^{-4}
λ_T	Arrival rate for uncorrelated transient faults	10^{-3}
λ_I	Arrival rate for uncorrelated transient faults	10^{-4}
μ_P	Repair rate for permanent faults	0.5
μ_T	Disappearance rate for transient faults	100
μ_I	Rate at which intermittent faults go benign	10
λ_{P2}	Arrival rate for second permanent fault	depends on ρ
λ_{T2}	Arrival rate for second transient fault	depends on ρ
λ_{I2}	Arrival rate for second intermittent fault	depends on ρ

Table 1: Parameters for two component system

ρ_{AB}	0	0.0001	0.001	0.01
λ_{P2}	10^{-4}	1.5×10^{-4}	10^{-3}	5.15×10^{-3}
λ_{T2}	10^{-3}	1.0111213	0.1011021	1.10012×10^{-2}
λ_{I2}	10^{-4}	0.101112	1.01102×10^{-2}	1.10012×10^{-3}
Unavailability	4.84×10^{-8}	7×10^{-8}	4.28×10^{-7}	2.27×10^{-6}

Table 2: Effect of correlation on unavailability of two component system

components are operational. States P and 2P represent the situation where one and two permanent failures exist, respectively. Similarly, I, 2I and T, 2T represent the existence of one and two intermittent and transient faults, respectively. States IT, IP and TP represent the coexistence of two faults of different types. To keep the model simple, we conservatively assume that only faults of the same type are correlated. For example, the second permanent fault is correlated only with the first permanent fault. All other events are independent. A description of the parameters and their baseline numerical values is given in table 1.

To determine the arrival rates for the second failure of the same type as the first (λ_{P2}, λ_{T2} and λ_{I2}), we use equation 4. The values for these parameters for several values of the correlation coefficient, as well as the steady-state unavailability of the system are given in table 2. For simplicity we assume the same correlation coefficient for all three fault types, although the methodology can certainly handle the more general case. In this table we see that, even for very small values of the correlation coefficient, the unavailability increases dramatically.

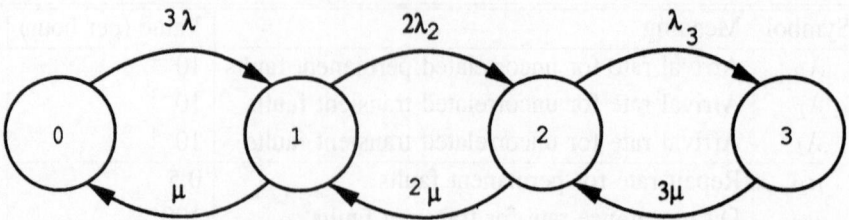

Figure 3: Markov model of three component system with one fault type

3 Correlation in larger systems

In the previous section we defined the methodology for modeling correlation in a two component system. In this section we extend the methodology to systems with three or four components, and describe how the methodology extends to larger systems. To extend the methodology we must answer the question, what does it mean for three (or more) random variables to be correlated, considering that the random variables represent a set of identical components? Most texts that deal with correlation define a correlation matrix which lists correlation components for pairwise combinations of variables. Since we are interested in defining the correlation between the first and second faults that occur, and then between the second and third faults, etc., this approach is unsuitable. Instead, to analyze a system with n total components, we wish to define $n - 1$ correlation coefficients, where ρ_i will represent the correlation between the ith and $(i + 1)$st fault in the system. We then want to use this set of correlation coefficients to determine the λ_i values, the rate at which the ith fault occurs in the system. For each ρ_i value we define A and B variables, and, repeatedly using equation (2), we obtain a set of i equations for determining the λ_i values. We demonstrate this methodology by analyzing a three-component and a four-component system, and compare the results of the correlated system with that of an uncorrelated system.

3.1 Correlation among three components

A three component system with one fault type can be in any of four states, representing the existence of zero, one, two or three faults (see figure 3). The state probabilities for this system are given by

$$P_0 = \frac{\mu^3}{\mu^3 + 3\mu^2\lambda + 3\mu\lambda\lambda_2 + \lambda\lambda_2\lambda_3}$$

$$P_1 = \frac{3\mu^2\lambda}{\mu^3 + 3\mu^2\lambda + 3\mu\lambda\lambda_2 + \lambda\lambda_2\lambda_3}$$

$$P_2 = \frac{3\mu\lambda\lambda_2}{\mu^3 + 3\mu^2\lambda + 3\mu\lambda\lambda_2 + \lambda\lambda_2\lambda_3}$$

$$P_3 = \frac{\lambda\lambda_2\lambda_3}{\mu^3 + 3\mu^2\lambda + 3\mu\lambda\lambda_2 + \lambda\lambda_2\lambda_3} \tag{5}$$

In order to determine λ_2 (the arrival rate of the second fault) and λ_3 (the arrival rate of the third fault), two equations are needed. We define two correlation coefficients, ρ_1 is the correlation between the first and second faults, and ρ_2 is the correlation between the second and third faults. The relationship of these coefficients and the desired failure rates proceeds from two applications of equation 2, where the indicator A represents the condition (failed or not) of a single component, and where a separate indicator B_i is defined for each ρ_i. For ρ_1, indicator B_1 is true if exactly one other component (other than that indicated by A) is faulty. For ρ_2, indicator B_2 is true if exactly two other components (other than that indicated by A) are failed.

Considering ρ_1, the correlation coefficient between the first and second failures, label the indicator variables for the three identical components with X_1, X_2 and X_3. To use equation 2, let A be X_1, and let B_1 be one when $X_2 + X_3 = 1$, that is, when one other component has failed. Then B_1 and A can be used together to represent, respectively, the first and second failures to occur in the system. To use equation 2 we define $Prob\{A\}$, $Prob\{B_1\}$ and $Prob\{A \wedge B_1\}$ in terms of the state probabilities given by equation 5.

$$Prob\{A\} = \frac{1}{3}P_1 + \frac{2}{3}P_2 + P_3$$

$$Prob\{B_1\} = \frac{2}{3}P_1 + \frac{2}{3}P_2$$

$$Prob\{A \wedge B_1\} = \frac{2}{3}P_2 \tag{6}$$

The correlation coefficient between the second and third failures, ρ_2, is similar to the ρ_1 presented above. For this coefficient, A is again X_1, while B_2 is true when both X_2 and X_3 have failed. Then B_2 and A can be used together to represent, respectively, the second and third failures to occur in the system.

We again use the state probabilities from equation 5 in the definition of the general correlation coefficient (equation 2), using the following probabilities.

$$Prob\{A\} = \frac{1}{3}P_1 + \frac{2}{3}P_2 + P_3$$

$$Prob\{B_2\} = \frac{1}{3}P_2 + P_3$$

$$Prob\{A \wedge B_2\} = P_3 \tag{7}$$

Once we have the values for the correlation coefficients, we use these to determine the λ_i values. Unfortunately, it is not a simple matter to invert equations (6) and (7) to obtain expressions for λ_2 and λ_3 in terms of ρ_1 and ρ_2 as in the two component case. However, one can repeatedly guess at the appropriate values for λ_i that result in the desired ρ_i values. We do this by assigning values to λ_2 and λ_3, determining the state probabilities via equation (5), using the state probabilities to determine the event probabilities in equations (6) and (7), and substituting these values in equation (2) to find the resulting correlation coefficient. The λ_2 and λ_3 values are changed as needed until the desired correlation coefficient values are obtained. In practice, it takes only two or three minutes to interactively determine appropriate λ_i values.

Figure 4 shows the Markov model of a three component system subject to permanent, intermittent and transient faults. The baseline parameters used for this system are the same as those used for the two-component system presented earlier. In this model, we assumed that the correlation coefficient for the first and second faults was equal to the correlation coefficient for the second and third faults ($\rho_1 = \rho_2$). The failure rates determined by the correlation coefficients and the resulting system unavailability are listed in table 3. In this case, we can see an increase of four orders of magnitude in the unavailability of the system (the probability that three components are failed simultaneously) when the correlation coefficient is 0.01. The increases for smaller values are also significant.

3.2   Correlation among four components

Figure 5 shows the Markov model of a four component system subject to one fault type. The state probabilities for this system are given by

$$P_0 = \frac{\mu^4}{\mu^4 + 4\lambda\mu^3 + 6\lambda\lambda_2\mu^2 + 4\lambda\lambda_2\lambda_3\mu + \lambda\lambda_2\lambda_3\lambda_4}$$

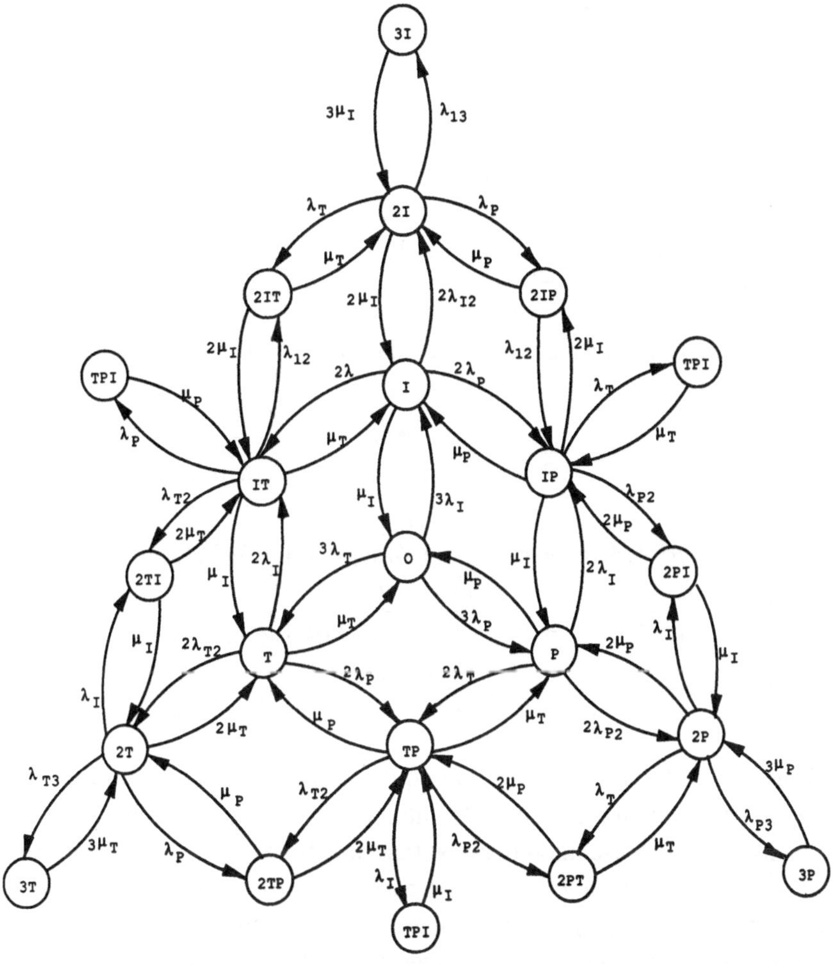

Figure 4: Markov model of three component system subject to transient, permanent and intermittent faults

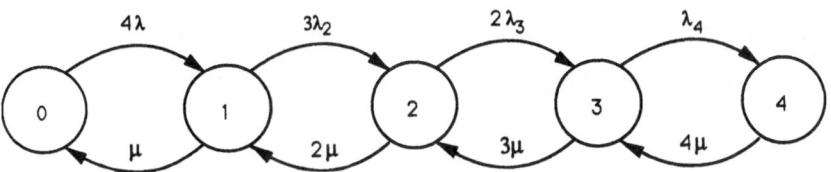

Figure 5: Markov model of four component system with one fault type

$\rho_1 = \rho_2$	0	0.0001	0.001	0.01
λ_{P2}	10^{-4}	1.35×10^{-4}	4.47×10^{-4}	3.47×10^{-3}
λ_{P3}	10^{-4}	3.155×10^{-3}	1.72×10^{-2}	6.45×10^{-2}
λ_{T2}	10^{-3}	8.05×10^{-3}	7.05×10^{-2}	6.75×10^{-2}
λ_{T3}	10^{-3}	1.129	3.845	13.1
λ_{I2}	10^{-4}	8.05×10^{-4}	7.1×10^{-3}	6.785×10^{-2}
λ_{I3}	10^{-4}	1.13×10^{-1}	3.85×10^{-1}	1.3
Unavailability	1.06×10^{-11}	3.63×10^{-10}	6.71×10^{-9}	1.97×10^{-7}

Table 3: Effect of correlation on unavailability of three component system

$$P_1 = \frac{4\lambda\mu^3}{\mu^4 + 4\lambda\mu^3 + 6\lambda\mu^3 + 6\lambda\lambda_2\mu^2 + 4\lambda\lambda_2\lambda_3\mu + \lambda\lambda_2\lambda_3\lambda_4}$$

$$P_2 = \frac{6\lambda\lambda_2\mu^2}{\mu^4 + 4\lambda\mu^3 + 6\lambda\lambda_2\mu^2 + 4\lambda\lambda_2\lambda_3\mu + \lambda\lambda_2\lambda_3\lambda_4}$$

$$P_3 = \frac{4\lambda\lambda_2\lambda_3\mu}{\mu^4 + 4\lambda\mu^3 + 6\lambda\lambda_2\mu^2 + 4\lambda\lambda_2\lambda_3\mu + \lambda\lambda_2\lambda_3\lambda_4}$$

$$P_4 = \frac{\lambda\lambda_2\lambda_3\lambda_4}{\mu^4 + 4\lambda\mu^3 + 6\lambda\lambda_2\mu^2 + 4\lambda\lambda_2\lambda_3\mu + \lambda\lambda_2\lambda_3\lambda_4} \tag{8}$$

We use the state probabilities from equation 8, along with definitions of ρ_1 (correlation between first and second faults), ρ_2 (correlation between second and third faults) and ρ_3 (correlation between third and fourth faults) to determine λ_2, λ_3 and λ_4 (where λ_i is the rate of occurrence of the ith fault). The individual ρ terms are defined using equation 2 where $\mathcal{A}$ indicates the condition of a single component X_1, and where a $\mathcal{B}_i$ is defined for each ρ_i.

For the coefficient of correlation between the first and second faults, ρ_1, we define $\mathcal{B}_1$ to represent the failure of exactly one of the components X_2, X_3 and X_4. The ρ_i will then be used to determine the arrival rate of the second failure in the system, that is, the failure rate of a component given that exactly one component (of the other three) has failed. To define ρ_1 we use the following, defined in terms of the state probabilities given by equation 8.

$$Prob\{A\} = \frac{1}{4}P_1 + \frac{1}{2}P_2 + \frac{3}{4}P_3 + P_4$$

$$Prob\{B_1\} = \frac{3}{4}P_1 + \frac{1}{2}P_2$$

$$Prob\{A \wedge B_1\} = \frac{1}{2}P_2 \tag{9}$$

To determine the correlation between the second and third failures, ρ_2, for the four component system, B_2 represents the failure of exactly two of the components other than X_1 (which is represented by A). This results in the following definitions, in terms of the state probabilities given by equation 8.

$$Prob\{B_2\} = \frac{1}{2}P_2 + \frac{3}{4}P_3$$
$$Prob\{A \wedge B_2\} = \frac{3}{4}P_3 \tag{10}$$

Finally, the correlation between the third and fourth failures in a four component system, ρ_3, uses B_3, which is true when all three of X_2, X_3 and X_4 have failed. Thus, to use equation 2, we define

$$Prob\{B_3\} = \frac{1}{4}P_3 + P_4$$
$$Prob\{A \wedge B_3\} = P_4 \tag{11}$$

As with the three component system, we can determine the appropriate values for the λ_i that yield the desired ρ_i by repeated guessing. For a four component system subject to permanent, transient and intermittent faults, we use the same baseline parameters as for the two and three component systems presented earlier. The arrival rate for the second and subsequent failures, as well as the resulting unavailability, are given in table 4. The Markov model for this system had 35 states and 120 transitions, and so is too big to present pictorially. The increase in the unavailability for the system with four components is more dramatic than the others. A correlation coefficient of 0.0001 increases the unavailability by 4 orders of magnitude; a coefficient of only 0.01 corresponds to an increase in unavailability of 7 orders of magnitude.

3.3 Correlation among more than four components

Extending this methodology to systems consisting of more than four identical components is relatively straightforward. For the n identical components, define $n - 1$ correlation coefficients ρ_i, where ρ_i represents the correlation between the ith and $(i + 1)$st failure in the system. To determine the ρ_i, pick an

$\rho_1 = \rho_2 = \rho_3$	0	0.0001	0.001	0.01
λ_{P2}	10^{-4}	1.29×10^{-4}	3.9×10^{-4}	3.045×10^{-3}
λ_{P3}	10^{-4}	1.91×10^{-3}	1.064×10^{-2}	3.89×10^{-2}
λ_{P4}	10^{-4}	5.35×10^{-2}	1.39×10^{-1}	2.94×10^{-1}
λ_{T2}	10^{-3}	6.8×10^{-3}	5.9×10^{-2}	5.9×10^{-1}
λ_{T3}	10^{-3}	7.1×10^{-1}	2.5	8
λ_{T4}	10^{-3}	15.6	29.7	63
λ_{I2}	10^{-4}	6.8×10^{-4}	5.9×10^{-3}	5.9×10^{-2}
λ_{I3}	10^{-4}	7.1×10^{-2}	2.43×10^{-1}	7.89×10^{-1}
λ_{I4}	10^{-4}	1.55	3.02	5.95
Unavailability	2.34×10^{-15}	2.93×10^{-11}	1.40×10^{-9}	8.64×10^{-8}

Table 4: Effect of correlation on unavailability of four component system

State	Expected State Prob.	Observed State Prob.
0	0.955378	0.9638
1	0.043754	0.0324
2	0.000859	0.0011
3	0.000006	0.0002
4	6.13×10^{-8}	0.0007
5	2.41×10^{-10}	0.0016
6	5.25×10^{-13}	0.0
7	4.90×10^{-16}	0.0

Table 5: Comparison of expected and observed state probabilities for VAX-cluster system

arbitrary component, and let A monitor its condition. Define a $\mathcal{B}_i$ which is true when exactly i of the n components (other than that one that $\mathcal{A}$ is monitoring) have failed. Determine the probabilities that $\mathcal{A}$ and each of the $\mathcal{B}_i$ are true, and use these in equation 2 to determine the ρ_i. The ρ_i are then used to determine the λ_i, the arrival rate for the ith failure.

4   Measured Correlation

In [1], the authors present a measurement based analysis of real error data collected from a DEC VAXcluster multicomputer system. Their results indicate that "correlated errors and correlated failures are significant." In this section we analyze some of their results to determine the correlation coefficients in their observed data. In section 5.2 of [1], the authors construct a semi-Markov failure model for the VAXcluster based on the failure event data. The model has eight states, where F_i is the state with i failed machines (of seven total). In table 5, we compare the observed occupation probabilities for the states (taken from table 7 in [1]) with the expected occupation probabilities if the failures are independent. The authors observed no occurrences of six or seven joint failures.

In order to assess whether the postulated correlation coefficients that were used in earlier sections of this paper were realistic, we calculate the correlation coefficients associated with the observed data in the VAXcluster study. We calculate ρ_1, ρ_2, ρ_3 and ρ_4 from the observed data using the state probabilities listed in table 5 in four instances of equation 2. The calculation of each observed correlation coefficient requires the definition of $Prob\{A\}$, $Prob\{B\}$, and $Prob\{A \wedge B\}$. The $Prob\{A\}$ terms for all coefficients is the same, as it is the probability that one particular component (X_1) has failed.

$$Prob\{A\} = \frac{1}{7}F_1 + \frac{2}{7}F_2 + \frac{3}{7}F_3 + \frac{4}{7}F_4 + \frac{5}{7}F_5 + \frac{6}{7}F_6 + F_7 \qquad (12)$$

For ρ_1, the correlation coefficient between the first and second failures, $\mathcal{B}_1$ is the event that exactly one other unit, other than X_1 has failed.

$$Prob\{B_1\} = \frac{6}{7}F_1 + \frac{6}{21}F_2$$

$$Prob\{A \wedge B_1\} = \frac{6}{21}F_2$$

For ρ_2, the correlation coefficient between the second and third failures, $\mathcal{B}_2$ is the event that exactly two other units, other than X_1 have failed.

$$Prob\{B_2\} = \frac{15}{21}F_2 + \frac{15}{35}F_3$$

$$Prob\{A \wedge B_2\} = \frac{15}{35}F_3$$

For ρ_3, the correlation coefficient between the third and fourth failures, $\mathcal{B}_3$ is the event that exactly three other units, other than X_1 have failed.

$$Prob\{B_3\} = \frac{20}{35}F_3 + \frac{20}{35}F_4$$

$$Prob\{A \wedge B_3\} = \frac{20}{35}F_4$$

Coefficient	Value
ρ_1	0.0097
ρ_2	0.0336
ρ_3	0.2165
ρ_4	0.3696

Table 6: Observed correlation coefficients for VAXcluster system

Finally, for ρ_4, the correlation coefficient between the fourth and fifth failures, $\mathcal{B}_4$ is the event that exactly four other units, other than X_1 have failed.

$$Prob\{B_4\} = \frac{15}{35}F_4 + \frac{15}{21}F_5$$

$$Prob\{A \wedge B_4\} = \frac{15}{21}F_5$$

Repeatedly inserting these values into equation (2) yields the correlation coefficients listed in table 6. The observed values from the VAXcluster system are in some cases significantly higher than those used in the example calculations. The authors noted in [1] that the observed occupation probability for state F_5 was six orders of magnitude larger than what would be expected under the independence assumption.

5   Conclusions

The assumption that failures occurring in redundant units in fault tolerant systems are independent can be dramatically optimistic, even if the failures are only slightly correlated. In this paper we presented a methodology for assessing the effect of even small levels of correlation on the unavailability of a fault-tolerant system. In example models of systems consisting of two, three or four components subject to permanent, intermittent and transient faults, we have shown that even extremely small values of the correlation coefficient reduce the predicted availability of the system by orders of magnitude. The increase in unavailability is more dramatic for larger systems. For a (small) fixed correlation coefficient, a system with more components is more available than one with fewer components, even though the increase in unavailability is larger as the system size increases. As the correlation coefficient increases, the marginal benefit of an additional active redundant component decreases more rapidly. That is, the additional active component has a lesser effect on the availability of a system as the correlation coefficient increases. Once the correlation coefficient gets large enough, the system with more components can actually become less reliable than one with a lower level of redundancy. For the systems modeled in this paper, for a correlation coefficient of 0.1, the four unit system was less reliable than the one with only three units. If the units are perfectly correlated ($\rho_i = 1$) then the most reliable system will be the one with the fewest components.

6 Acknowledgements

Most of this work was performed while the author was a visiting scientist at the Center for Digital Systems Research of the Research Triangle Institute. The author is grateful to Bob Baker of the Research Triangle Institute, Don Lee of the Aerospace Corporation and Malathi Veeraraghavan of AT&T Bell Labs for their helpful comments and discussions.

References

[1] D. Tang, R. Iyer, and S. Subramani, "Failure analysis and modeling of a VAXcluster system," in *Proc. IEEE International Symposium on Fault-Tolerant Computing, FTCS-20*, pp. 244–251, 1990.

[2] A. Costes, J. E. Doucet, C. Landrault, and J. C. Laprie, "SURF: A program for dependability evaluation of complex fault-tolerant computing systems," in *Proceedings IEEE Symposium on Fault-Tolerant Computing, FTCS-11*, pp. 72–78, June 1981.

[3] J. B. Dugan, K. S. Trivedi, M. K. Smotherman, and R. M. Geist, "The hybrid automated reliability predictor," *AIAA Journal of Guidance, Control and Dynamics*, vol. 9, pp. 319–331, May-June 1986.

[4] A. Goyal, W. C. Carter, E. de Souza e Silva, S. S. Lavenberg, and K. S. Trivedi, "The system availability estimator," in *Proceedings IEEE International Symposium on Fault-Tolerant Computing, FTCS-16*, pp. 84–89, 1986.

[5] S. V. Makam and A. Avizienis, "ARIES 81: A reliability and life-cycle evaluation tool for fault-tolerant systems," in *Proceedings IEEE Interna-*

tional Symposium on Fault-Tolerant Computing, FTCS-12, pp. 267–274, June 1982.

[6] J. B. Dugan, "On measurement and modeling of computer systems dependability: A dialog among experts," *IEEE Transactions on Reliability*, October 1990.

[7] C. Krishna and A. Singh, "Modeling correlated transient failures in fault-tolerant systems," in *Proceedings IEEE International Symposium on Fault-Tolerant Computing, FTCS-19*, pp. 374–381, 1989.

[8] Y. K. Malaiya, "Linearly correlated intermittent failures," *IEEE Transactions on Reliability*, vol. R-31, pp. 211–215, June 1982.

Impact of Correlated Failures on Dependability in a VAXcluster System

DONG TANG, RAVISHANKAR K. IYER

Center for Reliable and High-Performance Computing

Coordinated Science Laboratory

University of Illinois at Urbana-Champaign

1101 W. Springfield Ave., Urbana, IL 61801, USA

Abstract

This paper addresses the issue of correlated failures and their impact on system dependability. Measurements are made on a VAXcluster system and validated analytical models are proposed to calculate availability and reliability for simple systems with correlated failures. A correlation analysis of the VAXcluster data shows that errors are highly correlated across machines (average correlation coefficient $\rho = 0.62$) due to sharing of resources. The measured failure correlation coefficient, however, is not high (0.15). Based on the VAXcluster data, it is shown that models that ignore correlated failures can underestimate unavailability by orders of magnitude. Even a small correlation significantly affects system unavailability. A validated analytical model, to calculate unavailability of 1-out-of-2 systems with correlated failures, is derived. Similarly, reliability is also significantly influenced by correlated failures. The joint failure rate of the two components, λ_f, is found to be the key parameter for estimating reliability of 1-out-of-2 systems with correlated failures. A validated relationship between ρ and λ_f is also derived.

1 Introduction

Measurement-based analysis of computer system failures is valuable for understanding actual failure characteristics and developing representative analytical models. The dependence of failures on workload was found by measurement-based analyses [1, 2, 3, 4] and later was modeled analytically [5, 6]. This paper addresses the issue of correlated failures by using measurements from a 7-machine VAXcluster system [7]. Most analytical dependability models of computer systems assume that failures on the different components in the modeled system are independent (with a few exceptions such as [8, 9]. If component failures are correlated, model specifications become complicated. In addition to the individual component parameters (failure rates, recovery rates, etc.), interactions among components must also be considered. As a result, both parameters and transitions in a Markov model may be greatly increased. Although the increased model complexity can be handled by modern dependability evaluation tools, there is little knowledge about the parameters needed to specify interactions among components in the model. For example, the modeling tool SAVE [10] allows users to specify the failure probabilities of other components when one component fails, thereby taking correlated failures into account for the modeled system. However, these probabilities are likely to be unknown. Thus, measurement-based analysis is necessary for providing information for modeling such complicated systems.

In our previous analysis [7], we found that correlated errors and failures existed significantly in a measured VAXcluster system. This paper quantifies error and failure correlations and investigates their impact on dependability. The measured results are compared against analytical models assuming failure independency to analyze the effect of correlations on availability and reliability estimation. New models, based on the correlation coefficient and the multimachine joint failure rate, are proposed to calculate availability and reliability for simple systems with correlated failures. The models are validated using the VAXcluster data.

In the next section we briefly review the VAXcluster measurement. Section 3 presents a correlation analysis for both errors and failures on the different machines in the VAXcluster. We analyze the effect of correlated failures on availability in Section 4, and the influence of correlated failures on reliability in Section 5. Section 6 summarizes the major results of this study.

2 Measurement

The measured system was a VAXcluster [11], a distributed computer system consisting of seven VAX machines and four mass storage controllers connected by the Computer Interconnect bus. The source of the error-data for this study was the ERROR_LOG files produced by the VAX/VMS operating system running on the seven VAXcluster machines (Earth, Europa, Jupiter, Leonardo, Mars, Mercury, and Saturn). The data collection period was 250 days (Dec.

Class	Software	Machine	Device	Unknown	Operator
Freq.	14	5	25	9	38
Percent	15.4	5.7	27.5	9.9	41.8

Table 1: Data Statistics in Joint Failures by Error Classes

9, 1987 - Aug. 15, 1988). Based on the information provided by the error log files, errors were classified into seven types according to the subsystems or devices they occurred: software, CPU, memory, disk, tape, network, and unknown. In addition, CPU and memory errors were grouped into *machine* errors, and disk, tape, and network errors were grouped into *device* errors. The *failure* was identified by a reboot log following error reports. The details of the measurement can be found in [7].

Before investigating the impact of correlated failures on dependability, we made an attempt to identify causes for correlated failures based on the failure data. For convenience of discussion, we define the *joint failure* to be a set of two or more failure events (including recovery or repair times) which occur on the different machines and intersect with each other in time. Table 1 shows statistics of the failure data included in joint failures by error class, where "Unknown" denotes an unknown reason failure (two or more reboots occurred successively), and "Operator" means the operator requested system shutdown due to system problems.

By the table it is seen that causes of about 50% failures are not known (Unknown and Operator). For the remaining part, more than half of the failures were due to errors from shared resources. Software errors also led to a significant percentage of the failures. Because the VAX machines were working in a common interconnect environment, a hardware or software problem may affect multiple machines. Thus, the reliability of network-related hardware and software is particularly important for a distributed system.

3 Correlation Analysis

In this section we first explain how the measured data is fitted into a matrix form for correlation analysis. Then we conduct a correlation analysis to quantify error and failure relationships between each pair of the VAXcluster machines.

3.1 Data Matrices

The correlation coefficient, $Cor(x_1, x_2)$, between the random variables x_1 and x_2 is defined as [12]

$$Cor(x_1, x_2) = \frac{E[(x_1 - \mu_1)(x_2 - \mu_2)]}{\sigma_1 \sigma_2}, \tag{1}$$

where μ_1 and μ_2 are the means of x_1 and x_2, and σ_1 and σ_2 the standard deviations of x_1 and x_2, respectively. When computing correlation coefficients, we use estimates of these parameters from samples.

For the VAXcluster, x_1 and x_2 represent errors or failures on machines 1 and 2, respectively. Since there are seven machines in the VAXcluster, we have seven variables $(x_1, \ldots, x_7)$. In order to compute $Cor(x_i, x_j)$ for each pair of machines, we need to built a sample matrix for these variables based on the raw data. The whole measured period (250 days) was divided into 72000 equal intervals of five minutes. A 72000 x 7 data matrix was then constructed in the following way. The columns of the matrix represent machines. The rows of the matrix represent time intervals. Element (i, j) of the matrix was set to the number of errors occurring within interval i on machine j. Multiple error observations of the same type within the same interval were counted as a single error, which is consistent with the rule used by the coalescing algorithm in our previous analysis.[1] Each interval could have at most seven errors since there are seven error types. Thus, each element of the matrix has an integer value in the range (0,7). It is obvious that the seven columns of the matrix are samples of the random variables $(x_1, \ldots, x_7)$ representing errors on the seven machines.

Similarly, data matrices for each error type (CPU, disk, etc.) can be constructed. The elements of the matrix for each error type can only take (0,1) values since only two states (normal and error) are possible. If we use $D, D_1, D_2, \ldots, D_7$ to represent the all, software, CPU, memory, disk, tape, network, and unknown error matrices, respectively, then they have the following

---

[1] Our previous study [7] used the 5-minute interval as a criterion to coalesces repeated error reports for the same problem because error hazard curves showed that error bursts (repeated error reports) occurred within 5 minutes. For the same reason we choose the 5-minute window as the scope of an error for our correlation analysis. Our experimental results show that the correlation coefficient increases with the increase of the time window, but not very sensitively.

Between	All	Software	CPU	Memory	Disk	Tape	Network
Ear & Eur	0.597	0.000	0.000	–	0.745	0.096	0.668
Ear & Jup	0.590	0.066	0.001	–	0.748	0.087	0.672
Ear & Leo	0.531	0.000	0.001	–	0.695	0.053	0.668
Ear & Mar	0.511	0.000	0.001	–	0.647	0.091	0.674
Ear & Mer	0.457	0.000	0.000	–	0.778	0.090	0.653
Ear & Sat	0.580	0.000	0.000	–	0.728	0.137	0.691
Eur & Jup	0.734	0.182	0.068	0.000	0.826	0.130	0.654
Eur & Leo	0.672	0.000	0.000	0.000	0.779	0.063	0.647
Eur & Mar	0.647	0.000	0.000	–	0.742	0.085	0.689
Eur & Mer	0.516	0.084	0.000	-0.001	0.755	0.095	0.591
Eur & Sat	0.731	0.000	0.000	0.000	0.845	0.117	0.683
Jup & Leo	0.759	0.000	0.000	-0.001	0.866	0.112	0.784
Jup & Mar	0.708	0.000	0.001	–	0.808	0.107	0.800
Jup & Mer	0.529	0.077	0.000	0.012	0.760	0.107	0.666
Jup & Sat	0.805	0.000	0.476	0.000	0.922	0.164	0.790
Leo & Mar	0.675	0.000	0.000	–	0.791	0.054	0.810
Leo & Mer	0.520	0.000	0.000	0.128	0.737	0.029	0.641
Leo & Sat	0.748	0.000	0.000	-0.001	0.883	0.072	0.791
Mar & Mer	0.461	0.000	0.000	–	0.667	0.134	0.655
Mar & Sat	0.704	0.000	0.000	–	0.807	0.145	0.814
Mer & Sat	0.515	0.000	0.000	-0.001	0.757	0.109	0.669
Average	0.619	0.019	0.026	0.014	0.775	0.099	0.700

Table 2: Error Correlation Coefficients for VAXcluster

relationship:

$$D = \sum_{k=1}^{7} D_k. \tag{2}$$

3.2 Error and Failure Correlations

The correlation coefficients between each pair of machines for each type and all errors were computed using the data matrices introduced above. Each time we chose two columns (i and j) from the matrices to compute $Cor(x_i, x_j)$. The results are shown in Table 2.

In general, correlation coefficients are low for software (average=0.02), CPU

(0.03), memory (0.01), and tape (0.10) errors, and high for disk (0.78) and network (0.70) errors. Disk errors have the strongest correlation because errors in the disk subsystem often affect multiple machines due to the sharing and heavy usage of disks. For similar reasons network errors are strongly correlated across machines. Since disk and network errors constitute the majority of all errors, it is not surprising that correlations across machines for the "all" errors are significant (0.62). The only exception is the correlations involving Mercury which are slightly low (0.46-0.52). This is due to a higher proportion of machine errors (CPU and memory errors) on Mercury (27.4% vs. the average 6.3%).

In a similar way to the error correlation computation, failure correlations were also computed with the time window being one minute. In contrast with the error correlations, the failure correlation coefficients (shown in the "ρ" column in Table 3) are low (0.017-0.375). The average value is 0.155. (When we changed the time window to 5 and 10 minutes, the average value, taken to two significant digits, still remained to be 0.16.) This result appears to suggest that the impact of correlated failures on system dependability may be negligible. However, we will show in the next two sections that even these low coefficients significantly influence system availability and reliability for the VAXcluster.

To summarize, it was found that most errors occurring on the different machines are highly correlated (average coefficient = 0.62) due to the strong correlations of the disk (0.78) and network (0.70) errors. In contrast with the error correlation, the failure correlation is low (0.15).

4   Correlation vs. Availability

We have found the correlation of failures across machines to be low. Now we investigate how this orrelation affects system availability, and show the role of the correlation coefficient in estimating unavailability (U_{12}) for 1-out-of-2 systems. We will focus our attention on the 1-out-of-2 system, because the correlation coefficient, ρ, is a measure of the relationship between pair of machines. We will also investigate the impact of correlated failures on availability for the k-out-of-n VAXcluster systems.

4.1 Analytical Relationship between ρ and U_{12}

The published analytical availability models of VAXcluster systems [13, 14] assume that failures of the different subsystems are independent. Thus, if U_1 denotes the unavailability (failure probability) of machine 1 and U_2 the unavailability of machine 2 for a 1-out-of-2 VAXcluster system, the system availability is modeled as $1 - U_1 U_2$ [13]. This "independent model" assumes

$$U_{12} = U_1 U_2, \tag{3}$$

where U_{12} is the unavailability of machines 1 and 2, i.e., the proportion of time that both machine 1 and machine 2 are unavailable.

What is the actual U_{12} if failures of the two computers are correlated? To address this issue, we first derive a relationship between ρ and U_{12}, where ρ is the correlation coefficient of failures on the two machines, and then use this relationship to analyze the VAXcluster.

Let N denote the number of time units in the whole measured period, $N_i(i = 1, 2)$ the number of time units in which machine i fails, and N_{12} the number of time units in which both machines 1 and 2 fail. We have

$$U_i = \frac{N_i}{N}, U_{12} = \frac{N_{12}}{N}. \tag{4}$$

Define the random variable X_i ($i=1,2$) representing the state of machine i in such way:

$$X_i = \begin{cases} 1 & \text{machine } i \text{ in failure state} \\ 0 & \text{machine } i \text{ in normal state} \end{cases} \tag{5}$$

Because X_i is a Bernoulli variable, we know

$$\mu_i = U_i, \sigma_i = \sqrt{U_i(1 - U_i)}, \tag{6}$$

where μ_i is the sample mean of X_i and σ_i the sample standard deviation of X_i. Then the correlation coefficient of X_1 and X_2, ρ, can be estimated as

$$\rho = \frac{E[(X_1 - \mu_1)(X_2 - \mu_2)]}{\sigma_1 \sigma_2} = \frac{E[(X_1 - U_1)(X_2 - U_2)]}{\sqrt{U_1(1 - U_1)U_2(1 - U_2)}}. \tag{7}$$

To calculate $E[(X_1 - U_1)(X_2 - U_2)]$, consider the following four cases of X_1 and X_2:

Case	X_1	X_2	Number of instances
1	1	1	N_{12}
2	1	0	$N_1 - N_{12}$
3	0	1	$N_2 - N_{12}$
4	0	0	$N - N_1 - N_2 + N_{12}$

Thus

$$E[(X_1 - U_1)(X_2 - U_2)]$$
$$= \frac{1}{N}[(1 - U_1)(1 - U_2)N_{12} - (1 - U_1)U_2(N_1 - N_{12}) -$$
$$U_1(1 - U_2)(N_2 - N_{12}) + U_1U_2(N - N_1 - N_2 + N_{12})]$$
$$= U_1U_2 - \frac{N_2}{N}U_1 - \frac{N_1}{N}U_2 + \frac{N_{12}}{N}.$$

By 4,

$$E[(X_1 - U_1)(X_2 - U_2)] = U_{12} - U_1U_2. \tag{8}$$

Substituting 8 into 7, U_{12} is solved as

$$U_{12} = \rho\sqrt{U_1(1 - U_1)U_2(1 - U_2)} + U_1U_2. \tag{9}$$

This formula quantifies the relationship between the failure correlation coefficient ρ and the unavailability U_{12} for a 1-out-of-2 system. It is seen that 3 is a special case of 9 when $\rho = 0$. For convenience of discussion, we will call 3 the *independent* model and 9 the *dependent* model.

4.2   Model Comparison with Real Data

In this subsection we apply both the independent and dependent models to our failure data to analyze the effect of correlated failures on system availability. Since there are 7 machines in the VAXcluster, we have 21 pairs of machines which can be used to analyze a 1-out-of-2 system availability. We measured U_1, U_2, and U_{12} (with time granularity being one second), and calculated U_1U_2, $\rho\sqrt{U_1(1 - U_1)U_2(1 - U_2)}$, U_{12}/U_1U_2 (ratio between the measured and the independent values), and $U_{12}/(\rho\sqrt{U_1(1 - U_1)U_2(1 - U_2)} + U_1U_2)$ (ratio between the measured and the dependent values) for each of these 1-out-of-2 systems. The results are shown in Table 3.

It is seen that in most cases, the measured U_{12} is an order of magnitude greater than the estimate given by the independent model. Their average ratio is approximately 27. On the other hand, the estimate given by the dependent

Machines 1 & 2	ρ	U_1	U_2	U_{12}	U_1U_2	$\rho\sqrt{U_1(1-U_1)U_2(1-U_2)}$	Meas/Ind	Meas/Dep
Ear & Eur	0.238	0.006591	0.007774	0.001769	0.000051	0.001690	34.53	1.02
Ear & Jup	0.137	0.006591	0.004723	0.000791	0.000031	0.000762	25.42	1.00
Ear & Leo	0.024	0.006591	0.001121	0.000065	0.000007	0.000066	8.86	0.90
Ear & Mar	0.375	0.006591	0.005730	0.002337	0.000038	0.002289	61.88	1.00
Ear & Mer	0.185	0.006591	0.010102	0.001577	0.000067	0.001501	23.68	1.01
Ear & Sat	0.235	0.006591	0.010935	0.002060	0.000072	0.001976	28.58	1.01
Eur & Jup	0.066	0.007774	0.004723	0.000430	0.000037	0.000395	11.72	1.00
Eur & Leo	0.102	0.007774	0.001121	0.000307	0.000009	0.000301	35.26	0.99
Eur & Mar	0.289	0.007774	0.005730	0.001974	0.000045	0.001913	44.30	1.01
Eur & Mer	0.184	0.007774	0.010102	0.001703	0.000079	0.001612	21.68	1.01
Eur & Sat	0.157	0.007774	0.010935	0.001541	0.000085	0.001436	18.12	1.01
Jup & Leo	0.136	0.004723	0.001121	0.000319	0.000005	0.000311	60.29	1.01
Jup & Mar	0.219	0.004723	0.005730	0.001150	0.000027	0.001132	42.48	0.99
Jup & Mer	0.017	0.004723	0.010102	0.000155	0.000048	0.000116	3.24	0.94
Jup & Sat	0.079	0.004723	0.010935	0.000612	0.000052	0.000565	11.85	0.99
Leo & Mar	0.127	0.001121	0.005730	0.000324	0.000006	0.000321	50.37	0.99
Leo & Mer	0.030	0.001121	0.010102	0.000099	0.000011	0.000100	8.78	0.89
Leo & Sat	0.024	0.001121	0.010935	0.000084	0.000012	0.000085	6.82	0.86
Mar & Mer	0.212	0.005730	0.010102	0.001661	0.000058	0.001602	28.69	1.00
Mar & Sat	0.270	0.005730	0.010935	0.002180	0.000063	0.002122	34.80	1.00
Mer & Sat	0.144	0.010102	0.010935	0.001602	0.000110	0.001494	14.51	1.00
Average	0.155	0.006711	0.006711	0.001083	0.000043	0.001038	27.42	0.98

Table 3: Comparisons between Measured and Analytical Models

model basically fits the measured U_{12} (the little difference is due to the minor error in estimating ρ caused by using one minute time granularity, instead of one second). The table also makes it clear as to why the independent model is not satisfied. This is because the contribution of U_1U_2 to U_{12} is very small. The dominant term is $\rho\sqrt{U_1(1-U_1)U_2(1-U_2)}$, which quantifies the impact of correlation on availability. It is interesting that the actual unavailability is much greater than would be expected in the independent model, even though the failure correlation coefficient is not so high. In the next subsection we will investigate this further.

4.3 Further Discussion on ρ and U_{12}

For simplicity of argument, we assume that $U_1 = U_2 = U$, i.e., the two subsystems have the same unavailability. It is very common for the modern computer systems to have $U \ll 1$, i.e., $1 - U \approx 1$. Thus, from equation 9,

$$U_{12} \approx \rho U + U^2 = (\rho + U)U. \tag{10}$$

If $\rho \gg U$, then

$$U_{12} = \rho U. \tag{11}$$

That is, U_{12} is approximately a linear function of U, rather than a quadratic function of U as in the independent model 3. Therefore, as long as $\rho \gg U$, the

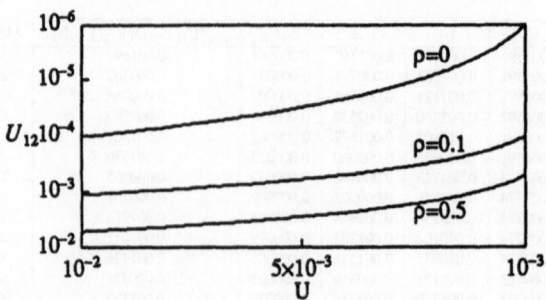

Figure 1: Effect of ρ on U_{12}

difference between the results given by 3 and 9 will be significant. From Table 3 it is seen that ρ (average = 0.155) is usually much greater than U (average = 0.0067), explaining the big difference between U_{12} and $U_1 U_2$.

Figure 1 shows the effect of ρ on U_{12} for different values of ρ (Notice that the logarithmic coordinate is used for U_{12} and both horizontal and vertical coordinates are decreasing). It is seen that when $\rho = 0$, U_{12} is quadratically reduced as U is reduced. However, when $\rho \neq 0$, even though it is very low (0.1), U_{12} is only linearly reduced as U is reduced (As U decreases from 10_{-2} to 10^{-3}, U_{12} decreases approximately from 10^{-3} to 10^{-4}). A 0.5 coefficient could lead to an availability loss of 40 system hours in one year of operation. The significance of this result is that increasing component availability may not be the most efficient way of increasing system availability. Reducing correlated failures of components or adding new redundant components may be a more efficient way to increase system availability.

4.4 Impact on Availability for k-out-of-n Systems

We have seen that correlated failures significantly affect unavailability of the 1-out-of-2 systems. What is the impact of correlations on unavailability for generalized k-out-of-n systems? We answer this question by comparing results from measurements and results from the analytical models which assume independent failures for the k-out-of-7 VAXcluster systems.

Let $N = 1, 2, \ldots, n$ and A_i be the availability of component i in a system with n components. If failures on the different components are independent, the probability that m components are functioning and the remaining $n - m$

Model	6/7	5/7	4/7	3/7	2/7
Meas.	$3.3 \times 1-^{-3}$	2.6×10^{-3}	2.4×10^{-3}	1.7×10^{-3}	3.1×10^{-5}
Indep.	8.9×10^{-4}	9.2×10^{-6}	5.5×10^{-8}	1.8×10^{-10}	3.0×10^{-13}
Ratio	3.8	2.9×10^2	4.4×104	9.3×106	1.0×10^8

Table 4: Unavailability for k-out-of-n VAXcluster Models

components have failed, P_m, is estimated as [12]

$$P_m = \sum_{j_1,\ldots,j_m \in N} A_{j_1} \cdots A_{j_m}(1 - A_{j_{m+1}}) \cdots (1 - A_{j_n}). \qquad (12)$$

The availability of a k-out-of-n model, A_{kn}, and the unavailability of a k-out-of-n model, U_{kn}, can be calculated by

$$U_{kn} = 1 - A_{kn} = 1 - \sum_{m=k}^{n} P_m. \qquad (13)$$

Table 4 compares the measured unavailability with the "independent" unavailability estimated using equation 13, and their ratio for the k-out-of-7 VAXcluster models. Since we are concerned with joint failures, the 7-out-of-7 model is not listed. The 1-out-of-7 model is also not provided because no 7-joint failures are found in our measured data. Comparing Table 4 with Table 3, it can be seen that the impact of correlated failures on availability for the k-out-of-n models is much greater than that for the 1-out-of-2 models. The ratio between the measured and the independent unavailabilities for the 1-out-of-2 systems (Table 3) is only one order of magnitude, but for the k-out-of-7 VAXcluster systems, this ratio is over two orders of magnitude except for the 6-out-of-7 system.

To summarize, we derived a relationship between failure correlation coefficient ρ and system unavailability U_{12} for 1-out-of-2 systems (dependent model) which can accurately estimate system availability when failures of the two components are correlated. It was found that the widely-used independent model underestimates system unavailability by several orders of magnitude for the VAXcluster. This is because unavailability is very sensitive to ρ. It remains an open problem at this stage to determine availability for k-out-of-n systems in which failures on different components are dependent.

5   Correlation vs. Reliability

We have shown that the effect of correlated failures on unavailability is significant. In this section we investigate the impact of correlated failures on reliability. Similar to the discussion in last section, we use 1-out-of-2 systems as examples to analyze the effect of correlations on reliability. Finally, the result for the 4-out-of-7 VAXcluster system is also provided. ·

5.1   Impact on Reliability for 1-out-of-2 systems

Of the seven machines in the VAXcluster two pairs considered as 1-out-of-2 systems are discussed here: 1) Earth and Europa (System 1) which has a correlation coefficient 0.238, and 2) Jupiter and Mercury (System 2) which has a very low correlation coefficient 0.017. For each pair, a measurement-based Markov model was built using real data and reliability function ($R(t)$) was calculated by solving the model [15]. The resulting curves provide *measured* reliability.

We also measured the mean failure rates, λ_1 and λ_2, and the mean recovery rates, μ_1 and μ_2, for the two machines in each pair. These parameters were then used to construct an *independent* Markov model (Figure 2) for each system. The model assumes that failures on the two machines in the system are independent. There are four states in the model: S_0 - normal state; S_i ($i = 1, 2$) - only machine i fails; and S_f — both machines 1 and 2 fail, which is set to absorbing state for computing reliability. We will call the reliability curves computed from this model the *independent* reliability. Figures 3 and 4 show the measured and the independent reliability curves for the two systems, respectively.

It is seen from the figures that for the both systems, there is a large difference between the measured and the independent curves. If we take a 0.5 reliability value to quantify the difference, the ratio of the times for the two curves to drop to 0.5 is 9.5 for System 1, and 9.3 for System 2. The average relative errors of the two independent curves are 420% and 264% (over 200 days), respectively. Although System 2 has an extremely low ρ (0.017), the difference between the measured and the independent curves is still significant.

Thus, it is clear that reliability is also sensitive to correlated failures. However, the reliability calculation directly depends on the joint failure rate of correlated components (transition rate to the joint failure state), not on the correlation coefficient. This is because the correlation coefficient is a measure of the common intervals in which both components fail, rather than a measure

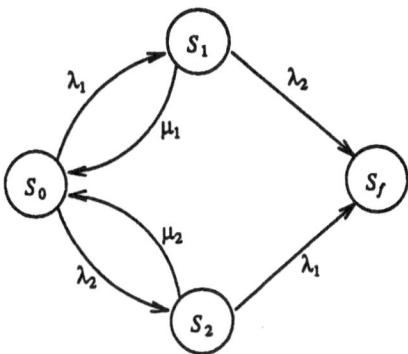

Figure 2: Independent Model for a 1-out-of-2 System

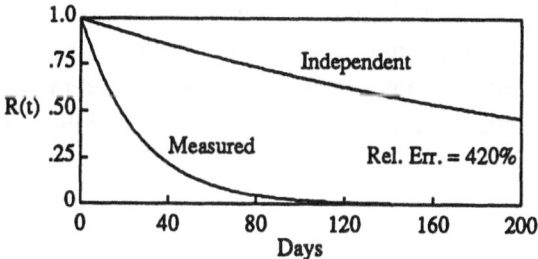

Figure 3: Reliability Comparison for System 1 ($\rho = 0.238$)

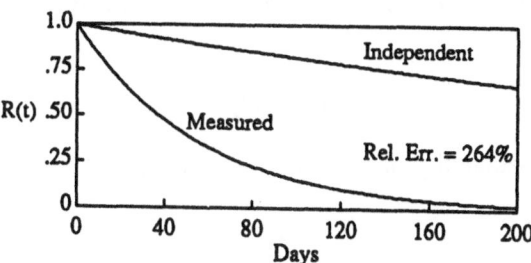

Figure 4: Reliability Comparison for System 2 ($\rho = 0.017$)

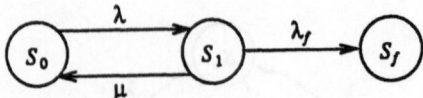

Figure 5: Reduced Model for a 1-out-of-2-system

of failure frequency. The joint failure rate issue is discussed in the next two subsections.

5.2   Role of the Joint Failure Rate

To investigate the role of the joint failure rate in reliability estimation, we reduce the independent model in Figure 2 by merging single failure states, S_1 and S_2, and compare the parameters in the simplified equivalent model with those from the measurement-based model. The reduced 3-state model for a 1-out-of-2 system is shown in Figure 5. In this 3-state model, S_1 denotes that only one machine has failed, while S_0 and S_f are still the normal and the failure states, respectively. Since component failures are assumed to be independent, λ, μ, and λ_f (joint failure rate) can be calculated by the following formulas:

$$\lambda = \lambda_1 + \lambda_2, \tag{14}$$
$$\mu = w_1\mu_1 + w_2\mu_2, \tag{15}$$
$$\lambda_f = w_1\lambda_2 + w_2\lambda_1, \tag{16}$$

where w_i (i=1,2) is the proportion of time in S_1 during which machine i fails. If the occupancy probability of S_1 is much greater than that of S_f (which is assumed by the independent model), w_i can be approximately estimated by

$$w_i = \frac{U_i}{U_1 + U_2}, \tag{17}$$

where U_i is the unavailability of machine i and can be calculated based on λ_i and μ_i [12]:

$$U_i = \frac{\lambda_i}{\lambda_i + \mu_i}. \tag{18}$$

Tables 5 and 6 compare the calculated λ, μ, and λ_f using equations 14 to 18 with the measured λ, μ, and λ_f for System 1 and System 2, respectively. The tables show that the parameter with the greatest error is λ_f, the joint failure rate.

Parameter	λ	μ	λ_f
Reduced Model	0.55/day	37.82/day	0.27/day
Measurement	0.50/day	45.81/day	3.70/day
Relative Error	10%	17%	93%

Table 5: Parameter Comparision for System 1

Parameter	λ	μ	λ_f
Reduced Model	0.27/day	17.68/day	0.13/day
Measurement	0.25/day	16.81/day	1.38/day
Relative Error	7%	5%	90%

Table 6: Parameter Comparision for System 2

That is, in the systems with correlated failures, λ_f is significantly increased. By replacing λ_f in the reduced model with the measured λ_f we achieve a modified model which we will refer to as the *dependent model* (DM). Figures 6 and 7 show the reliability curve computed from the dependent model and the measured reliability curve for System 1 and System 2, respectively.

From the figures we see that the estimated reliability curve given by the dependent model is a good approximation of the measured curve. The average relative errors are 20.7% and 2.2% for the two systems. This fact indicates that λ_f plays an important role in modeling 1-out-of-2 systems with correlated failures and λ and μ are not as significant. That is, when correlated failures are taken into account by analytical reliability models, it is essential to estimate

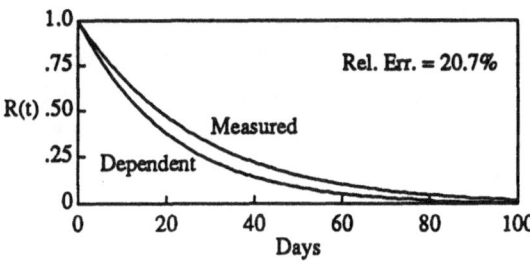

Figure 6: Reliability Estimated by DM for System 1

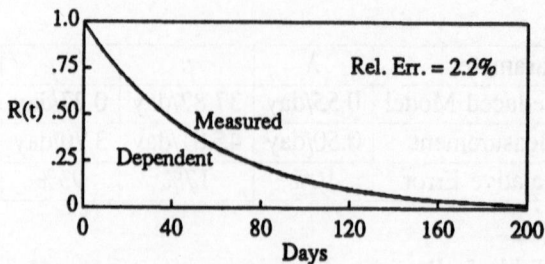

Figure 7: Reliability Estimated by DM for System 2

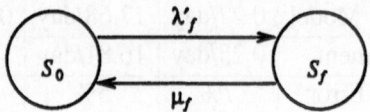

Figure 8: The 2-state Model for a 1-out-of-2-system

joint failure rates correctly.

5.3   Analytical Relationship between ρ and λ_f

So far we have discussed two measures for correlated failures between two components: ρ — correlation coefficient, and λ_f — joint failure rate. We have shown that ρ is useful for availability estimation and λ_f is useful for reliability estimation for 1-out-of-2 systems. What is the relationship between the two parameters? In this subsection, we build a relationship for them.

To obtain the relationship, we further simplify the model in Figure 5 to the 2-state model in Figure 8 for 1-out-of-2 systems. The normal state, S_0, represents three cases: both machines 1 and 2 are working, machine 1(or 2) has failed but machine 2(or 1) is working. The failure state, S_f, represents that both machines 1 and 2 have failed. The λ'_f and μ_f denote the system failure rate (at which both machines fail) and the system recovery rate (at which one or both machines recover), respectively. Note that λ'_f in Figure 8 and λ_f in Figure 6 are different. However, if λ'_f is known, λ_f can be estimated by

$$\lambda_f = \frac{\lambda + \mu}{\lambda}\lambda'_f. \tag{19}$$

Thus, if we can build a relationship between ρ and λ'_f, we actually build a relationship between ρ and λ_f. It is obvious that the two state model in

Parameter	Measured λ_f	Calculated λ_f	Error
System 1	3.70/day	3.64/day	1.6%
System 2	1.38/day	1.46/day	5.8%

Table 7: Validation via Comparison on λ_f

Figure 8 is equivalent to a single machine model in that it has only two states: operational and failed. If we apply equation 18 to this system, the unavailability of the system, U_{12}, can be expressed as

$$U_{12} = \frac{\lambda'_f}{\lambda'_f + \mu_f}. \tag{20}$$

By solving equations 9 and 20, we obtain

$$\lambda'_f = \frac{\mu_f(\rho\sqrt{U_1(1 - U_1)U_2(1 - U_2)} + U_1U_2)}{1 - (\rho\sqrt{U_1(1 - U_1)U_2(1 - U_2)} + U_1U_2)}. \tag{21}$$

This equation gives a relationship between ρ and λ'_f.

From equation 21 we see that when ρ is fixed, λ'_f is proportional to μ_f. That is, it is possible for two systems to have a same ρ and different system failure rates, therefore to have different reliability curves. equations 21 and 19 allow us to estimate λ_f, given the correlation coefficient, the system recovery rate, and the single machine parameters. Table 7 shows the measured λ_f and the calculated λ_f using equations 21 and 19 for System 1 and System 2. The results verify the derived relationship because the measured and the calculated values are very close for the both systems.

5.4 Impact on Reliability for 4-out-of-7 VAXcluster

Our analysis has shown that even for a simple 1-out-of-2 system, the impact of correlation on reliability is significant. Now we extend our investigation to the 4-out-of-7 VAXcluster system. The reason we choose the 4-out-of-7 model is that it is consistent with the VAXcluster Quorum algorithm [16]. The algorithm requires (usually) at least half of the processors in the VAXcluster to function for the VAXcluster to function. (However, the system manager can change Quorum to reduce this requirement).

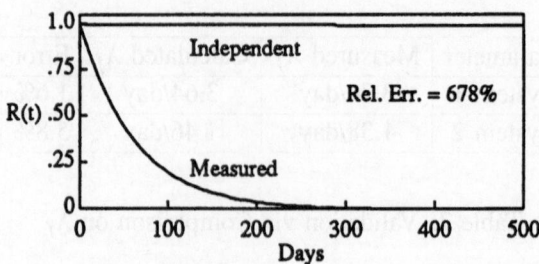

Figure 9: Reliability Comparision for 4/7 VAXcluster

As we did for 1-out-of-2 systems, we built a measurement-based model from the data and an independent model (similar to that in Figure 2) based only on the individual component parameters for the 4-out-of-7 VAXcluster system. Reliability curves were then computed from the two models. The result is shown in Figure 9. It is seen that the difference between the two curves is much greater than in the 1-out-of-2 systems. The average relative error is 678%. This result indicates that the impact of correlated failures on reliability for k-out-of-n systems can be more significant than for 1-out-of-2 systems.

To summarize, we compared the reliability curves from the measurement-based models with those from the analytical models using an independency assumption and found that the difference is significant for k-out-of-n systems. We showed that joint failure rate is a key parameter in estimating reliability for 1-out-of-2 systems with correlated failures. We also gave a validated relationship between correlation coefficient ρ and joint failure rate λ_f.

6   Conclusion

In this paper, we investigated correlated failures in a VAXcluster system. We conducted a correlation analysis for both errors and failures on the different machines in the VAXcluster. We analyzed the impact of correlated failures on availability and reliability by comparing the results from measurements with the results from analytical models using an independency assumption. We also proposed new models to accurately estimate availability and reliability for simple systems with correlated failures.

Results of the correlation analysis showed that a strong correlation existed among network errors (average correlation coefficient = 0.70) and among disk errors (0.78) across all machines. This is explained by the sharing and heavy usage of these resources. The failure correlation coefficient was not high (0.15). However, even this low correlation significantly influenced the system unavailability. Our measurement showed that the widely used models, which ignore correlated failures, can underestimate system unavailability by several orders of magnitude. A new formula was proposed to accurately estimate unavailability of 1-out-of-2 systems with correlated failures and was validated with real data. Reliability was also found to be significantly influenced by correlated failures. There was a large difference between the reliability curves from the measurement-based models and those from the independent analytical models. The joint failure rate of the two components, λ_f, was found to be a key parameter in estimating reliability for 1-out-of-2 systems with correlated failures. A relationship between ρ and λ_f was derived, given the system recovery rate and the individual component parameters. This relationship was also validated with real data.

Acknowledgement

The authors would like to thank Professor Kumar Joag-dev of the Department of Statistics at UIUC for helpful discussion. Thanks are also due to Robert Dimpsey, Inhwan Lee, and Susan Taylor, and to the anonymous referees for their valuable comments on this manuscript. This research was supported in part by the Joint Services Electronics Program under Contract N00014-90-J-1270 and in part by the National Aeronautical and Space Administration under Grant NAG1-613.

References

[1] S. Butner and R. Iyer, "A statistical study of reliability and system load at SLAC," in *FTCS-10*, pp. 207–209, IEEE, October 1980.

[2] X. Castillo and D. Siewiorek, "Workload, performance, and reliability of digital computer systems," in *FTCS-11*, pp. 84–89, IEEE, June 1981.

[3] X. Castillo and D. Siewiorek, "A workload dependent software reliability prediction model," in *FTCS-12*, pp. 279–286, IEEE, June 1982.

[4] R. Iyer and D. Rossetti, "A statistical load dependency model for cpu errors at SLAC," in *FTCS-12*, pp. 363–372, IEEE, June 1982.

[5] J. Dunkel, "On the modeling of workload-dependent memory faults," in *FTCS-20*, pp. 348–355, IEEE, June 1990.

[6] J. Meyer and L. Wei, "Analysis of workload influence on dependability," in *FTCS-18*, pp. 84–89, IEEE, June 1988.

[7] D. Tang, R. Iyer, and S. Subramani, "Failure analysis and modeling of a vaxcluster system," in *FTCS-20*, pp. 244–251, IEEE, June 1990.

[8] J. Dugan, "Correlated hardware failures in redundant systems," in *Preprints of the 2nd IFIP Working Conference on Dependable Computing for Critical Applications*, February 1991.

[9] C. Krishan and A. Singh, "Modeling correlated transient failures in fault-tolerant systems," in *FTCS-19*, pp. 374–381, IEEE, June 1989.

[10] A. Goyal *et al.*, "The system availability estimator," in *FTCS-16*, pp. 84–89, IEEE, June 1986.

[11] N. Kronenberg, H. Levy, and W. Strecker, "Vaxcluster: A closely-coupled distributed system," *ACM Transactions on Computer Systems*, vol. 4, pp. 130–146, May 1986.

[12] K. Trivedi, *Probability and Statistics with Reliability, Queueing, and Computer Science Applications*. Englewood Cliffs, NJ: Prentice-Hall, 1982.

[13] E.E. Balkovich *et al.*, "Vaxcluster availability modeling," *Digital Technical Journal*, pp. 69–79, September 1987.

[14] O. Ibe, R. Howe, and K. Trivedi, "Approximate availability analysis of vaxcluster systems," *IEEE Transactions on Reliability*, vol. 38, pp. 146–152, April 1989.

[15] D. Heimann, N. Mittal, and K. Trivedi, "Availability and reliability modeling for computer systems," *Advances in Computers*, vol. 31, pp. 175–233, 1990.

[16] *VAXcluster Systems Handbook*, April 1986.

Design Diversity

Assuring Design Diversity in *N*-Version Software: A Design Paradigm for *N*-Version Programming

MICHAEL R. LYU

Electrical and Computer Engineering Department

The University of Iowa

Iowa City, IA 52242, USA

ALGIRDAS AVIŽIENIS

Computer Science Department

University of California at Los Angeles

Los Angeles, CA 90024, USA.

Abstract

The *N*-Version Programming (NVP) approach achieves fault-tolerant software units, called *N*-version Software (NVS) units, through the development and use of software diversity. To maximize the effectiveness of the NVP approach, the probability of similar errors that coincide at the NVS decision points should be reduced to the lowest possible value. Design diversity is potentially an effective method to get this result. It has been the major concern of this paper to formulate a set of rigorous guidelines, or a *design paradigm*, for the investigation and implementation of design diversity in building NVS units for practical applications. This effort includes the description of a most recent formulation of the NVP design paradigm, which integrates the knowledge and experience obtained from fault-tolerant system design with software engineering techniques, and the application of this design paradigm to a real-world project for an extensive evaluation. Some limitations of the approach are also presented.

1 Introduction

The *N*-Version Programming (NVP) approach to fault-tolerant software systems involves the generation of functionally equivalent, yet independently developed and maintained software components, called *N*-Version Software (NVS)

[1]. These components are executed concurrently under a supervisory system, called *N*-Version eXecutive (NVX), that uses a decision algorithm based on consensus to determine final output values. Whenever probability of similar errors is minimized, distinct, erroneous results tend to be masked by a consensus decision during NVS execution [2].

NVS systems are gaining acceptance in critical application areas such as aerospace industry, nuclear power industry, and ground transportation industry. The construction of such systems is still, however, done mostly in an *ad hoc* manner. In order to obtain a paradigmatic approach in applying the NVS techniques for fault-tolerant software systems, a joint project [3] (to be called "the Six-Language Project" throughout the text) was initiated at the UCLA Dependable Computing & Fault-Tolerant Systems (DC & FTS) Laboratory and at the Honeywell Commercial Flight Systems Division. This paper will describe an NVP design paradigm which was applied to conduct the Six-Language Project, and discuss evidences and lessons learned from the project to testify and revise the proposed paradigm.

2   A Design Paradigm for *N*-Version Programming

NVP has been defined from the beginning as "the independent generation of $N \geq 2$ functionally equivalent programs from the same initial specification [1]." "Independent generation" meant that the programming efforts were to be carried out by individuals or groups that did not interact with respect to the programming process. The NVP approach was motivated by the "fundamental conjecture that the independence of programming efforts will greatly reduce the probability of identical software faults occurring in two or more versions of the program [1]."

The research effort at UCLA has been addressed to the formation of a set of guidelines for systematic design approach to implement NVS systems, in order to achieve efficient tolerance of *design faults* in computer systems. The gradual evolution of these rigorous guidelines was revealed in several previous research activities [4, 5, 6], which have investigated a total of 81 programs in four different applications. This evolving methodology was most recently formulated in [7] as an NVS *design paradigm* by integrating the knowledge and experience obtained from both software engineering techniques and fault tolerance investigations. The word "paradigm," used in the dictionary sense, means "pattern, example, model," which refers to a set of guidelines and rules with illustrations.

The objectives of the design paradigm are:

1. to reduce the possibility of oversights, mistakes, and inconsistencies in the process of software development and testing;

2. to eliminate most perceivable causes of related design faults in the independently generated versions of a program, and to identify causes of those which slip through the design process;

3. to minimize the probability that two or more versions will produce similar erroneous results that coincide in time for a decision (consensus) action of NVX.

The application of a proven software development method, or of diverse methods for individual versions, remains the core of the NVP process. This process is supplemented by procedures that aim: (1) to attain suitable isolation and independence (with respect to software faults) of the N concurrent version development efforts, (2) to encourage potential diversity among the N versions of an NVS unit, and (3) to elaborate efficient error detection and recovery mechanisms. The first two procedures serve to reduce the chances of related software faults being introduced into two or more versions via potential "fault leak" links, such as casual conversations or mail exchanges, common flaws in training or in manuals, use of the same faulty compiler, etc. The last procedure serves to increase the possibilities of discovering manifested errors before they are converted to coincident failures. Figure 1 describes the current NVP paradigm for the development of NVS.

In Figure 1, the NVP paradigm is composed of two categories of activities. The first category, represented by boxes and single-line arrows at the left-hand side, contains typical software development procedures. The second category, represented by ovals and double-line arrows at the right-hand side, describes the concurrent enforcement of various fault- tolerant techniques under the *N*-version programming environment. Detailed descriptions of the incorporated activities and guidelines are presented in the following sections.

2.1 Determine Method of NVS Supervision and Execution Environment in System Requirement Phase

The NVS execution environment has to be determined early in the system requirement phase to evaluate the overall system impact and to obtain required

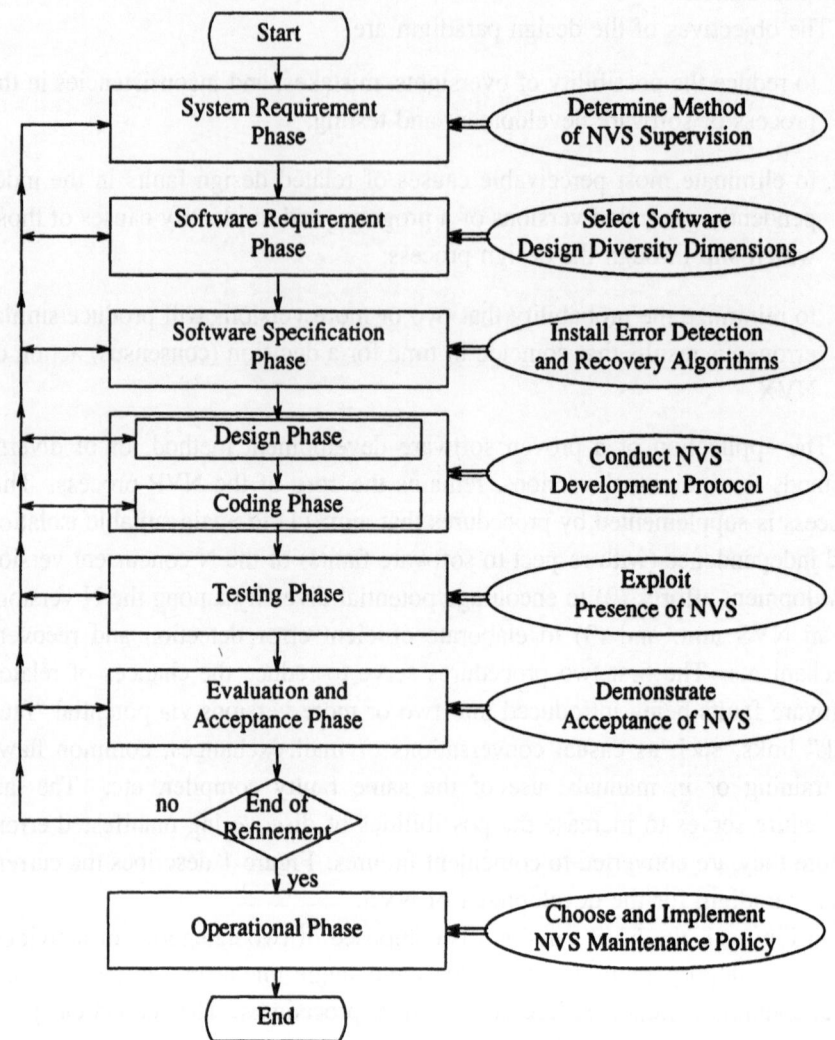

Figure 1: A Design Paradigm for *N*-Version Programming

support facilities.

(1) Decide NVS execution methods and required resources

The overall system architecture might be well defined during system requirement phase, at which time the software configuration items could be properly
identified. This means the number of software versions and their interaction
could be investigated and determined. Due to the cost of software development for multiple versions, the number of versions is not expected to be large at
present. However, from dependability (including reliability and safety) viewpoints, at least two versions are required for single-failure-detection operations,
and at least three versions are required for single-failure- correction/double-
failure-detection operations. The current limitation is the lack of quantitative
methods for an accurate decision.

(2) Develop support mechanisms and tools

Generally speaking, a generic class of NVX forming the NVS execution support environment is favorable. The NVX may be implemented in software, in
hardware, or in a combination of both. The basic functions that the NVX must
provide for NVS execution are: (a) the decision algorithm, or set of algorithms;
(b) assurance of input consistency for all versions; (c) interversion communication; (d) version synchronization and enforcement of timing constraints; (e)
local supervision for each version; (f) the global executive and decision function
for version error recovery; and (g) a user interface for observation, debugging,
injection of stimuli, and data collection during N- version execution of application programs. The nature of these functions was extensively illustrated in
the descriptions of the DEDIX testbed system [2].

(3) Comply with hardware architecture

Special dedicated hardware processors might have to be implemented or pro-
cured in advance for the execution of NVS systems, especially when the NVS
supporting environments need to operate under certain stringent requirements
(e.g., accurate supervision, efficient CPUs, etc.). The options of combining
with hardware fault-tolerance for a hybrid configuration could also be considered [8, 9, 10].

In order to create enough sampling space under the budget constraint, it was decided that six versions would be generated in the Six-Language Project. Representatives from Honeywell extracted the information needed for the Six-Language Project from their original system specification and provided it in a System Description Document (SDD). Moreover, for the purpose of industrial-standard validation and verification, a Model Definition Document (MDD) was also supplied. This document described a mathematical Aircraft Model and a Square Wave Model. The former provided functions within the landing control loop but external to the application program to form closed-loop flight simulations, while the latter applied open-loop testing strategy with various stringent conditions to saturate the executions of the control laws in the application program.

2.2   Investigate Software Design Diversity Dimensions in Software Requirement Phase

The major reason for choosing design diversity is to eliminate the commonalities between the separate programming efforts, as they have the potential to cause related faults among the multiple versions.

(1) Compare random diversity vs. enforced diversity

Different dimensions of diversity can be applied to the building of NVS systems. Design diversity could be achieved either by randomness or by enforcement. The *random diversity*, such as that provided by independent personnel, leaves the dissimilarity to be generated according to individual's training background and thinking process. The diversity is achieved somewhat in an uncontrolled manner by this way. The *enforced diversity*, on the other hand, investigates different aspects in several dimensions, and deliberately requires them to be implemented into different program versions. The purpose of such *required diversity* is to minimize the opportunities for common causes of software faults in two or more versions (e.g., compiler bugs, ambiguous algorithm statements, etc.), and to increase the probabilities of significantly diverse approaches to version implementation.

	implementors	languages	tools	algorithms	methodologies
spec.	higher	higher	lower-	higher+	lower
design	higher+	lower	lower	higher+	higher
coding	higher+	higher+	lower	higher	higher
testing	lower	lower-	higher	lower	higher+

Table 1: A Qualitative Design Diversity Metric

(2) Derive qualitative design diversity metrics

There are four phases in which design diversity could be applied: the specification phase, the design phase, the coding phase, and different implementors, different languages [11], different tools [12], different algorithms [13], and different software development methodologies (including phase-by-phase software engineering, prototyping, computer-aided software engineering, or even the "clean room" approach [14]).

A qualitative design diversity metric is proposed in Table 1. This assessment of diversity is an initial effort based on the experiences gained from previous experiments at UCLA [4, 5, 15], and published work from other sites [12, 13, 16, 17].

This table suggests that in the specification phase, using different implementors, languages or algorithms might achieve higher diversity than applying other dimensions. In the design phase, using different implementors, different algorithms, or different methodologies tends to be more helpful. All dimensions except tools are considered effective in the coding phase. Finally, investigation of different tools or methodologies might be more favorable in the testing phase. Moreover, to compare rows and columns in Table 1, extra granularity is provided by using "+" (indicating "further" for "higher") and "-" (indicating "further" for "lower") signs. For example, diversity by using different implementors in testing (marked "lower") is considered lower than using them in the previous three phases, but that could still be higher than using different languages in testing phase, which is marked "lower-".

(3) Evaluate cost-effectiveness along each dimension

Since adding more diversity implies adding more resources, it is important to evaluate cost-effectiveness of the added diversity along each dimension.

This evaluation will enable trade-off studies between cost and efficiency. Such evaluation might be application-dependent, and thus may need to be elaborated, possibly by several iterations of investigations. It is hypothesized that the main cost of NVP is dominated by the employment of extra implementors. Cost of adding other diversity dimensions should not be significant, especially when the resources in these dimensions are abundant (e.g., languages, hypothesis, though.

(4) Obtain the final choice under particular constraints

After the above investigation, the final combination of diversity could be deter-mined under particular project constraints. Typical project constraints include: cost, schedule, and required dependability performance. At the current stage, however, this decision might have to involve substantial subjective judgements, due to the lack of quantitative measures for design diversity and the resulting cost impacts. As more experiences and evidences are gained by researchers in this field, schemes for achieving an optimal solution prior to a project start might be available in the future.

In the Six-Language Project, it was decided that different algorithms were inappropriate for investigation due to tight accuracy requirements for numerical computations. Based on the availability of computer resources and language-knowledgeable programmers, applying different programming languages was considered as a cost-effective investigation in enforcing design diversity. Six programming languages of various programming style were chosen, consist-ing two widely used conventional procedural languages (C and Pascal), two object-oriented programming languages (Ada and Modula-2), one logic pro-gramming language (Prolog), and one functional programming language (T, a variant of Lisp). It was postulated that different programming languages would force people to think differently about the application problem and the program design, and to use different tools in their programming and testing activities, which could lead to significant diversity in the software development efforts.

2.3   Install Error Detection and Recovery Algorithms in Software Specifi-cation Phase

The specification of the member versions, to be called "V-spec," needs to state the functional requirements completely and unambiguously, while leaving the widest possible choice of implementations to the N programming efforts.

Sufficient error detection and recovery algorithms have to be carefully designed and specified in order to detect related errors that could potentially lead to coincident failures.

(1) Prescribe the matching features needed by NVX

Each V-spec must prescribe the *matching features* that are needed by the NVX to execute the member versions as an NVS unit in a fault-tolerant manner [4]. The V-spec defines: (a) the *functions* to be implemented, the time constraints, the inputs, and the initial state of a member version; (b) requirements for internal *error detection* and *exception handling* (if any) within the version; (c) the *diversity* requirements; (d) the *cross-check points* ("cc-points") at which the NVX decision algorithm will be applied to specified outputs of all versions; (e) the *recovery points* ("r-points") at which the NVX can execute *community error recovery* [18] for a failed version; (f) the choice of the NVX *decision algorithm* and its *parameters* to be used at each cc-point and r-point; (g) the *response* to each possible outcome of an NVX decision, including absence of consensus; and (h) the prevention to the *Consistent Comparison Problem* [19].

(2) Avoid diversity-limiting factors

The specifications for simplex software tend to contain guidance not only "what" needs to be done, but also "how" the solution ought to be approached. Such specific suggestions of "how" reduce the chances for diversity among the versions and should be eliminated from the V-spec. Another potential diversity-limiting factor is the over-specification of cc-points and r-points. The installation of cc-points and r-points enhances error detection and recovery capability, but it imposes extra common constraints to the programs and might tend to limit design diversity. The choice of number of these points and their placements depend on the size of the software, the control flow of the application, the number of variables to be checked and recovered, and the time overhead allowed to perform these operations.

(3) Require the enforced diversity

The V-spec may explicitly require the versions to differ in the "how" of implementation. Diversity may be specified in the following elements of the NVP process: (a) training, experience, and location of programmers; (b) application algorithms and data structures; (c) software development methods; (d) programming languages; (e) programming tools and environments; (f) testing methods and tools.

(4) Protect the specification

The use of two or more distinct V-specs, derived from the same set of user requirements, can put extensive protection against specification errors. Two cases have been practically explored: a set of three V-specs (formal algebraic OBJ, semi-formal PDL, and English) that were derived together [5, 15], and a set of two V-specs that were derived by two independent efforts [20]. These approaches provide additional means for the verification of the V-specs, and offer diverse starting points for version implementors.

In the Six-Language Project, only absolutely necessary information was supplied to the programmers in the software specification. The diagrams describing the major system functions were taken directly from the original SDD, while the explanatory text was shortened and made more concise. A further enhancement to the specification was the introduction of *seven* cross-check points, placed right after each main computation unit, and *one* recovery point at the end of the last computation unit. In total, the instrumented application required 14 external variables (for input functions), 68 intermediate and variables (for recovery function). The resulting specification given to the programmers was a 64-page English document.

Another characteristics of the Six-Language Project is that the "(in)consistent comparison problem" [19] *did not* exist in the application. This was due to two main reasons: (1) We did not vote final results on Boolean or integer values; the final, critical results that required consensus were always real numbers upon which cross-checking could be properly applied for error detection and recovery; (2) Multiple correct values were allowed for intermediate results during computation without limiting the potential for diversity in various implementation; however, as imposed by the specified algorithm, they would converge to the final results with tight numerical precision for an accurate comparison. Although there might be applications appearing unsuitable for NVS investigations, this automatic landing problem, a typical example in aeronautic industry, turned out to be appropriate.

2.4   Conduct NVS Development Protocol in Design and Coding Phase

In this phase, multiple programming teams (P-teams) start to develop the NVS concurrently according to a given software specification. The main concern herein is to maximize the isolation and independence of each version, and to smooth the overall software development efforts.

(1) Derive a set of mandatory rules of isolation

The purpose of imposing rules on the P-teams is to assure the *independent generation* of programs, which means that programming efforts are carried out by individuals or groups that do not interact with respect to the programming process. The rules of isolation are intended to identify and eliminate potential "fault leak" links between the P-teams. The development of the rules is an on-going process, and the rules are enhanced when a previously unknown "fault leak" is discovered and its cause pinpointed. Current isolation rules include: strict prohibition of any discussion of technical work between P-teams, widely separated working areas (offices, computer terminals, etc.) for each P- team, usage of different host machines for the software development, protection of all on-line computer files, and safe deposit of technical documents.

(2) Define a rigorous communication and documentation (C&D) protocol

The C&D protocol imposes rigorous control on the manner in which all necessary information flow and documentation efforts are conducted. The main goal of the C&D protocol is to avoid opportunities for one P-team to influence another P-team in an uncontrollable, and unnoticed manner. In addition, the C&D protocol documents communications in sufficient detail to allow a search for "fault leaks" if potentially related faults are discovered in two or more versions at some later time.

(3) Form a coordinating team (C-team)

The C-team is the keystone of the C&D protocol. The major functions of this team are: (a) to prepare the final texts of the V-specs and of the test data sets; (b) to set up the implementation of the C&D protocol; (c) to acquaint all P-teams with the NVP process, especially rules of isolation and the C&D protocol; (d) to distribute the V-specs, test data sets, and all other information needed by the P-teams; (e) to collect all P- team inquiries regarding the V-specs, the test data, and all matters of procedure; (f) to evaluate the inquiries (with help from expert consultants) and to respond promptly either to the inquiring P-team only, or to all P-teams via a broadcast; (g) to conduct formal reviews, to provide feedback when needed, and to maintain synchronization between P-teams; (h) to gather and evaluate all required documentation, and to conduct acceptance tests for every version.

In the Six-Language Project, all communications between the C- team and the P-teams were allowed only in standard written format for possible post mortems about "fault leaks." Electronic mail between the C-team and the P-teams has proven to be the most effective medium for this purpose. Moreover, to reduce the unnecessary information exchange, answers to a particular question from a P-team were only sent to the corresponding P- team. A message was broadcast to all P-teams only when strictly necessary (e.g., a specification update). This protocol has avoided a possible bothersome overload of the message flood, as was observed in [6]. Ratio of P-team members to C-team members was 4:1 in the Six-Language Project, in which the C-team has performed a satisfactory job.

2.5   Exploit Presence of NVS in Testing Phase

An appealing and promising application of NVS is its reenforcement for current software verification and validation procedure during the testing phase, which is one of the hardest cores of any software system.

(1) Explore comprehensive verification procedures

For software verification, the NVS provides a thorough coverage for error detection since every discrepancy among versions needs to be resolved. Moreover, it is observed that consensus decision of the existing NVS may be more reliable than that of a "gold" model or version (usually provided by an application expert) [21].

(2) Enforce extensive validation efforts

NVP provides protective redundancy around requirement misinterpretations and specification ambiguities. Any single- version approach gets a single interpretation of the requirement, no matter how carefully a development procedure has been followed. Especially when the software development group is forces the system requirements and the software specifications to be assessed from independent observations and viewpoints, making the validation effort more effective and more extensive.

(3) Provide opportunities for "back-to-back" testing

There is a possibility that two or three versions can be executed "back-to-back" in a testing environment, completing verification and validation concurrently with productive execution. However, there is a risk here. If codes are brought together prematurely, the independent programming efforts would be violated, and "fault leaks" might be created among the program versions. In any case, should this scheme be applied in a project, it must be done by a testing team independent of the P-teams (e.g., the C-team), and the testing results should never be revealed to a P-team, if they contain information from other versions that would influence this P-team.

Some experiences in NVS testing were obtained from the Six-Language Project: (a) a golden reference model, derived by a Honeywell expert, was less reliable than the consensus of multiple program versions in defining correctness of a computation [2]; (b) multiple teams around testing explored erroneous test cases effectively; (c) the pace in testing phase was accurately tracked and controlled by monitoring and comparing the progress of the multiple teams.

2.6 Demonstrate Acceptance of NVS in Evaluation Phase

Evaluation of the software fault-tolerance attributes of an NVS system is performed by means of analytic modeling, simulation, experiments, or combinations of those techniques. Many open issues are yet to be investigated.

(1) Define NVS acceptance criteria

The acceptance criteria of the NVS system depend on the validity of the conjecture that residual software faults in separate versions will cause very few, if any, similar errors at the same cc-points. These criteria might depend on various applications and their developing procedures, thus need to be carefully elaborated case by case.

(2) Provide evidence of diversity

Diversity requirements support the objective of reducing common programming errors, since they provide more natural isolation against "fault leaks" between the teams of programmers. Furthermore, it is conjectured that the probability of a random, independent occurrence of faults that produce the same erroneous results in two or more versions is less when the versions are more diverse.

(3) Demonstrate effectiveness of diversity

Another conjecture is that even if related faults are introduced, the diversity of member versions may cause the erroneous results not to be similar at the NVX decision. Therefore, evidence and effectiveness of diversity should be carefully identified and assessed [22].

(4) Make NVS dependability prediction

For dependability prediction of NVS, there are two essential aspects: the choice of suitable software dependability models, and the definition of quantitative measures [23, 24, 25, 26]. These aspects should also characterize the level of fault-tolerance present in the NVS system [27, 28]. Usually, the dependability prediction of the NVS system is compared to that of the single- version baseline system.

In the Six-Language Project, nine flight simulations engaging various flight modes were imposed on the six program versions before they were finally accepted. This represented a total of 18440 program executions. Parallel to this testing, a structural analysis for the multiple programs was conducted in the evaluation phase. The efforts of finding more faults and the search for evidence of structural diversity among these programs were the major concerns. An additional benefit of this analysis was that it necessitated a thorough C-team inspection by *code comparisons,* in which seven additional faults that were not caught by any tests or any P-team inspections were detected [2].

2.7   Choose and Implement an Appropriate NVS Maintenance Policy in Operational Phase

The key point to remember regarding NVS maintenance policy in this phase is to follow a philosophy consistent to the overall design paradigm.

(1) Assure and monitor NVX basic functionality

The basic functionality of NVX should be properly assured and monitored during the operational phase. Critical parts of the NVS supervisory system could themselves be protected by the NVP technique. Operational status of the NVX running NVS should be carefully monitored to assure basic functionality. Any anomalies should be recorded for further investigation in order to improve this NVP paradigm, which is an on-going effort aiming at achieving ultra-reliability (e.g., 10^{-9} failures per hour) for safety-critical software systems. Such stringent requirements could not be achieved without a progressive evolution of the underlining design process [29].

(2) Keep the achieved diversity work in the maintenance phase

It is postulated that patching the software, as has been widely used in industry, might more easily reveal the existence of faults by exhibiting dissimilarities among the independently generated software versions. This would be a valuable feature of NVS units since such a patching technique could create the elegantly) designed single version software. An observation from the Six-Language Project was that it appeared extremely difficult to inject similar faults which were hard-to-detect in the six programs [30]. This was due to the achieved diversity among the programs.

(3) Follow the same paradigm for modification and maintenance

As for the modification and maintenance of the NVS unit, the same design paradigm should be followed, i.e., a common "specification" of the maintenance action should be "implemented" by independent maintenance teams. The potential cost for such a policy is by no means cheap, but it is hypothesized that the induced extra cost in maintenance phase, comparing with that for single software, is a factor relatively lower than the extra cost factor in development phase. This is due to two reasons: (a) the achieved NVS reliability is supposedly higher than that of a single version, leaving fewer costly operational failures to be experienced; (b) when adding new features to the operating software, the existence of multiple program versions should make the testing and certification tasks easier and more cost- effective. These tasks usually share a larger portion in maintenance phase than in development phase.

The configuration of the operational flight simulation system in the Six-Language Project consisted of single or multiple lanes of the control law

version	size (l.o.c.)	total executions	number of errors	error probability
Ada	2256	5127400	0	.0000000
C	1531	5127400	568	.0001108
Modula-2	1562	5127400	0	.0000000
Pascal	2331	5127400	0	.0000000
Prolog	2228	5127400	680	.0001326
T	1568	5127400	680	.0001326
average	1913	5127400	321	.00006267

Table 2: Errors in Individual Versions

computation, obtained from the six accepted program versions, and the pre-programmed Airplane Model. The Airplane Model computed the response of an airplane to each elevator command, with a landing geometry model describing the deviation relative to a glide slope beam. Outputs of this model was fed back to each lane for a subsequent round of execution. In order to provide a set of inputs to the Airplane Model that create larger variation magnitudes, and thereby force off-nominal software operating conditions, random turbulence in the form of vertical wind gusts was introduced. Moreover, these testing facilities could replaced by the Square Wave Model to form an open-loop testing configuration without feedback, for the purpose of boundary value analyses.

During the operational testing phase, 1000 flight simulations, or over five million program executions, were conducted. Table 2 shows the errors encountered in each single version, while Table 3 shows different error categories under all combinations of 3- version and 5-version configurations. Note that the discrepancies encountered in the operational testing were called "errors" rather than "failures" due to their non-criticality in the landing procedure, i.e., a proper touchdown was still achieved at their presence.

From Table 2 we can see that the average error probability for single version is .00006267. Table 3 shows that for all the 3-version combinations, the error probability concerning safety is .00002652 (category 4). This is a reduction of roughly 2.3. In all the combinations of 5-version configuration, the error probability for reliability is .000003413 (category 5; Two of the three errors are coincident, resulting in no-decision), a reduction by a factor of 18. This probability becomes zero in the safety measurement.

It is cautioned against interpreting these numbers as the expected depend-

category	3-version configuration		5 -version configuration	
	# of cases	probability	# of cases	probability
1.	102531685	.9998409	30757655	.9997807
2.	13385	.0001305	5890	.0001915
3.	210	.000002048	70	.000002275
4.	2720	.00002652	680	.00002210
5.	-	-	105	.000003413
Total	102548000	1.0000000	30764400	1.0000000

classifications of the category:
1 - no errors
2 - single errors in one version
3 - two distinct errors in multiple versions
4 - two coincident errors in multiple versions
5 - three errors in multiple versions

Table 3: Errors in 3-Version and 5-Version Execution Configurations

ability improvement of NVS over single-version software. The coincident errors produced by the Prolog and T programs were all caused by *one identical fault* in both versions, which was due to the ignorance of a slight specification update that was made very late in the programming process. This fault manifested itself right after these program versions were put together for the flight simulation. To eliminate causes for this type of faults in the future, the corresponding amendment to the NVP design paradigm is to deliberately request confirmation and validation for late specification changes in the C&D protocol, and to cautiously conduct multiple-version verification testing as part of the acceptance criteria. Had this fault been eliminated in the operational testing, categories 3, 4 and 5 for both 3-version and 5-version configurations in Table 3 would have been all zero, resulting in perfect dependability figures.

2.8 Refine by Iterations

Notice that some of the described stages occur at progressively later times, but backtracking from a given stage to its previous one may occur at any time. Alteration of requirements arising from use, revision of specification, change in environment, and erroneous implementation may interrupt the flow of the normal design paradigm or spawn sub-processes having their own life cycles.

This flexibility might allow the proposed paradigm to be tailored for other software engineering development models (e.g., the spiral model [31]).

3 Conclusions

Although at first considered as an impractical competitor of high-quality single-version programs, *N*-version software has gained some significant acceptance in academia and industry in the past few years. Since more and more critical systems are software-intensive or software-embedded, the trustworthiness of software is the principal prerequisite for the building of a trustworthy system. At present, *N*-version software might be an attractive alternative that can be expected to provide a higher level of trustworthiness and security for critical software units than test or proof techniques without fault tolerance. The ability to guarantee that any software fault, as long as it only affects minority members of an *N*-version unit, will be tolerated without service disruption may by itself be a convincing reason to adapt *N*-version software as a safety assurance technique for life-critical applications. The main focus of the proposed NVP design paradigm attempts to promote this ability.

In summary, this research has made the following contributions:

1. An NVP design paradigm has been formulated, applied, and evaluated. The proposed design paradigm, which integrates software engineering techniques and NVP design diversity guidelines and rules, could provide a fundamental model for the practical development of NVS.

2. The design paradigm has been used during the entire life cycle of the UCLA/ Honeywell Six-Language Project. This project served as an experimental means to executing and evaluating the proposed paradigm. In reviewing the objectives of the design paradigm, we believe that all of them were properly addressed, if not completely accomplished, by the experiment. All perceivable causes of related design faults were eliminated, and causes of the only two pairs of identical faults were identifiable and readily removable. The resulting amendment to the paradigm, as the lessons learned from this project, is to add extra guidelines in Section 2.3 for the production of graphical specification, in Section 2.4(2) for the confirmation of every specification update in the C&D protocol, and in Section 2.6(1) for the inclusion of multiple version verification testing as part of the NVS acceptance criteria.

3. The design paradigm tries to explore and support the idea of design diversity, and intends to prevent commonalities that could produce related software faults. The effectiveness of this design paradigm was shown by the experimental result that identical faults in two versions have occurred only twice in the Six-Language Project, comparing with a total of 93 faults found in the six software versions during the whole project life cycle. Identical faults involving more than two versions have never been observed. Moreover, in a mutation testing study which investigated all the 93 known faults [7], errors caused by every non-identical fault among program versions were *all* distinguishable and properly detected by the provided fault-tolerant mechanisms.

As the final concluding remark, it is obvious that coincident failures are indeed the Achilles' heel of NVS, and the main goal of the proposed design paradigm is to avoid them by two levels of treatment: (a) investigate design diversity to prevent possible *identical faults*, and (b) install error detection and recovery algorithms to handle potential *similar* errors. The advancement of the NVP technique could happen only when these two aspects are properly addressed and documented, and we hope that the proposed paradigm can serve as an evolving basis subject to public revision and amendment in order to achieve such an advancement.

Acknowledgement

The authors wish to thank the program committee member, Dr. J. Lala, for his valuable suggestions and observations for the revision of this paper. Comments from the program committee and the reviewers are also highly appreciated.

References

[1] A. Avižienis and L. Chen, "On the implementation of n- version programming for software fault-tolerance during program execution," in *Proceedings COMPSAC 77*, pp. 149–155, 1977.

[2] A. Avižienis, "The *n*-version approach to fault-tolerant software," *IEEE Transactions on Software Engineering*, vol. SE-11, pp. 1491–1501, December 1985.

[3] A. Avižienis, M. Lyu, and W. Schutz, "In search of effective diversity: A six-language study of fault-tolerant flight control software," in *Proceedings 18th Annual International Symposium on Fault Tolerant Computing*, IEEE, June 1988.

[4] L. Chen and A. Avižienis, "*N*-version programming: A fault-tolerance approach to reliability of software operation," in *Digest of 8th Annual International Symposium on Fault-Tolerant Computing*, pp. 3–9, June 1978.

[5] A. Avižienis and J. Kelly, "Fault-tolerance by design diversity: Concepts and experiments," *Computer*, vol. 17, pp. 67–80, August 1984.

[6] J. Kelly, A. Avižienis, B. Ulery, B. Swain, M. Lyu, A. Tai, and K. Tso, "Multi-version software development," in *Proceedings IFAC Workshop SAFECOMP'86*, pp. 43–49, October 1986.

[7] M. Lyu, *A Design Paradigm for Multi-Version Software*. PhD thesis, UCLA Computer Science Department, Los Angeles, California, May 1988.

[8] K. Kim, "Distributed execution of recovery blocks: An approach to uniform treatment of hardware and software faults," in *Proceedings IEEE 4th International Conference on Distributed Computing Systems*, pp. 526–532, May 1984.

[9] J.-C. Laprie, "Hardware-and-software dependability evaluation," in *Proceedings 11th World IFIP Congress*, pp. 109–114, September 1989.

[10] J. Lala, L. Alger, S. Friend, G. Greeley, S. Sacco, and S. Adams, "Study of a unified hardware and software fault tolerant architecture," Contract Number NAS1-18061 181759, NASA, January 1989.

[11] U. Voges, *Software Diversity in Computerized Control Systems*, ch. Use of Diversity in Experimental Reactor Safety Systems, pp. 29–49. Wien, Austria: Springer-Verlag, 1988.

[12] L. Gmeiner and U. Voges, "Software diversity in reactor protection systems: An experiment," in *Proceedings IFAC Workshop SAFECOMP'79*, pp. 75–79, May 1979.

[13] P. Bishop, D. Esp, M. Barnes, P. Humphreys, G. Dahll, and J. Lahti, "PODS- a project of diverse software," *IEEE Transactions on Software Engineering*, vol. SE-12, pp. 929–940, September 1986.

[14] M. Dyer, "Certifying the reliability of software," in *Proceedings Annual National Joint Conference on Software Quality and Reliability*, March 1988.

[15] J. Kelly and A. Avižienis, "A specification oriented multi-version software experiment," in *Digest of 13th Annual International Symposium on Fault-Tolerant Computing*, pp. 121–126, June 1983.

[16] T. Anderson, P. Barrett, D. Halliwell, and M. Moulding, "Software fault tolerance: An evaluation," *IEEE Transactions on Software Engineering*, vol. SE-11, pp. 1502–1510, December 1985.

[17] P. Traverse, *Software Diversity in Computerized Control Systems*, ch. AIRBUS and ATR System Architecture and Specification, pp. 95–104. Wien, Austria: Springer-Verlag, 1988.

[18] K. Tso and A. Avižienis, "Community error recovery in n-version software: A design study with experimentation," in *Digest of 17th Annual International Symposium on Fault-Tolerant Computing*, pp. 127–133, July 1987.

[19] S. S. Brilliant, J. C. Knight, and N. G. Leveson, "The consistent comparison problem in *n*-version software," *IEEE Transactions on Software Engineering*, vol. 15, pp. 1481–1485, November 1989.

[20] C. Ramamoorthy, Y. Mok, F. Bastani, G. Chin, and K. Suzuki, "Application of a methodology for the development and validation of reliable process control software," *IEEE Transactions on Software Engineering*, vol. SE-7, pp. 537–555, November 1981.

[21] J. P. J. Kelly, D. E. Eckhardt, A. Caglavan, J. C. Knight, D. F. McAllister, and M. A. Vouk, "A large scale second generation experiment in multi-version software: Description and early results," in *Proceedings The Eighteenth International Symposium on Fault-Tolerant Computing*, June 1988.

[22] J. J. Chen, *Software Diversity and Its Implications in the N-version Software Life Cycle*. PhD thesis, UCLA Computer Science Department, Los Angeles, California, 1990.

[23] J.-C. Laprie, "Dependability evaluation of software systems in operation," *IEEE Transactions on Software Engineering*, vol. SE-10, pp. 701–714, November 1984.

[24] A. Avizienis and J.-C. Laprie, "Dependable computing: From concepts to design diversity," in *Proceedings of the IEEE*, May 1986.

[25] J. D. Musa, A. Iannino, and K. Okumoto, *Software Reliability - Measurement, Prediction, Application*. New York, New York: McGraw-Hill Book Company, 1987.

[26] T.-F. P., "Software validation by means of statistical testing: Retrospect and future direction," in *Proceedings the First International DCCA Working Conference*, pp. 15–22, August 1989.

[27] D. Eckhardt and L. Lee, "A theoretical basis for the analysis of multiversion software subject to coincident errors," *IEEE Transaction on Software Engineering*, vol. SE-11, pp. 1511–1517, December 1985.

[28] B. Littlewood and D. Miller, "Conceptual modeling of coincident failures in multiversion software," *IEEE Transactions on Software Engineering*, vol. 15, pp. 1596–1614, December 1989.

[29] J.-C. Laprie and B. Littlewood, "Quantitative assessment of safety-critical software: Why and how?," in *Proceedings Probabilistic Safety Assessment and Management Conference*, February 1991.

[30] M. Joseph, *Architectural Issues in Fault-Tolerant, Secure Computing Systems*. PhD thesis, UCLA Computer Science Department, Los Angeles, California, May 1988.

[31] B. W. Boehm, "A spiral model of software development and enhancement," *IEEE Computer*, pp. 61–72, May 1988.

An Empirical Investigation of the Effect of Formal Specifications on Program Diversity

THOMAS I. MCVITTIE, JOHN P. J. KELLY, WAYNE I. YAMAMOTO

Department of Electrical & Computer Engineering

University of California at Santa Barbara

Santa Barbara, CA 93106, USA

Abstract

Formal specification languages are increasingly being employed as an aid in the design and implementation of highly reliable systems. Recent experimental evidence indicates that the syntax and semantics associated with a formal specification language can have a large effect on the subsequent program version. This paper analyses the effect formal specification languages have on program development by examining nine diverse versions of a communication protocol created using three different formal specification languages.

1 Introduction

For the past several years, we have been actively researching methods for developing and maintaining highly reliable software systems. Achieving highly reliable software is not a matter of measuring the effectiveness of fault avoidance or fault removal or fault tolerance methodologies, rather it is determining the best combination of fault avoidance and fault removal and fault tolerance techniques.

The design, modelling, verification and validation of highly reliable systems has become increasingly important in recent years. These systems are generally quite complex and the consequences of a failure are far reaching. For these systems, an unambiguous specification of the system behavior is necessary to ensure that the implementation will correctly provide the required services and interact smoothly with its environment. Formal specifications written using specially designed languages are increasingly employed to provide system designers with the tools needed to attain these highly reliable systems. Elements used by these languages to specify system behavior include: graphical symbols, psuedo-code instructions, petri-nets, and algebraic constructs. Unfortunately, no matter how diligently we apply formal design, verification, and validation techniques, it is still possible for undetected faults to exist in the final program version.

Back-to-Back testing and Multi-Version Software rely on Design Diversity to detect or to mask faults. Design Diversity involves the independent generation of multiple distinct versions from a common requirement. The advantage of design diversity is that reliable computing does not require the complete absence of design and implementation faults, but only that the versions should not produce similar errors which cause the versions to exhibit the same incorrect behavior.

Results of recent experiments indicate that many of the faults remaining in independently developed program versions are caused by errors in the common specification [1, 2, 3]. To address this issue, an experiment was performed in which the multiple program versions were independently developed from three diverse formal specifications [4, 5]. In this experiment, the versions developed from *different* formal specifications did not exhibit common mode failures when tested in a back-to-back configuration. However, independent versions created from the *same* formal specification did contain common mode faults and in some instances were alarmingly similar in structure and content. Clearly, formal specifications can have a substantial effect on programs, even those developed in isolation. The multiple diverse versions created during this experiment provide a unique opportunity to evaluate the effect of different formal specifications on program versions.

This paper examines nine diverse versions of a communication protocol created from specifications written in three formal languages. Section 2 introduces the application and the formal specification languages. Section 3 presents the program versions and briefly introduces the Ada tasking model. Section 4 compares versions created from the same specification with those created from diverse specifications. Section 5 presents conclusions.

2   The Application

The application is based on the Transport and Sliding Window (a data link) Layers of the Open Systems Interconnection (OSI) layered-model adopted by the International Standards Organization in 1979 [6]. The OSI model provides a reference for the design of *standardized* communication protocols, allowing different computing systems to exchange information.

In the OSI model, services needed to implement communication are partitioned into seven hierarchical layers. Each layer provides services to the layer directly above, and in turn uses the services provided by the layer below. Layers communicate by exchanging Access Service Primitives (ASPs). Identical layers at different sites, called peers layers, communicate by exchanging Protocol Data Units (PDUs). For example, a Transport Layer communicates with its peer by exchanging PDUs and with the Session layer above it by exchanging ASPs.

Selection of a formal language as the standard specification language for OSI protocol layers has caused considerable discussion within the protocol community. Three languages (LOTOS, Estelle, and SDL) have emerged as forerunners. Specifications for the Transport and the Sliding Window Layers are available in each of the three languages [7, 8]. The specifications were written by experts and, after some modification, have become standards. As far as practical, the authors checked their specifications for top-level consistency with the underlying informal description and against each of the other specifications.

2.1 Specification Languages

LOTOS [9], Estelle [10], and SDL [11] are industry standard languages which were developed by the ISO and CCITT for describing distributed or concurrent processing systems, in particular those which implement OSI services and protocols. Each language has mechanisms for describing both the static system representation and the dynamic behavior of the protocol. A brief introduction to each language follows.

As an example, consider the transmitter module which is one component of the Sliding Window protocol. The Sliding Window uses a positive handshake to confirm reliable delivery across a possibly unreliable underlying medium, and employs a transmit window for flow control and error recovery. The transmitter module receives messages from the layer above, encapsulates the message into a PDU by appending a unique ascending sequence number, and transmits the resulting PDU to the layer below. A copy of each PDU is stored by the transmitter until a suitable acknowledgement is received. To guard against the loss or corruption of a PDU by the underlying medium, a PDU is retransmitted if a corresponding acknowledgement is not received within a bounded time.

The capacity of the transmitter is limited to storing at most **tws** (the implementation specific size of the transmit window) PDUs. Thus the transmitter must have the ability to close, preventing additional message requests from being received from the layer above until one or more of the PDUs currently being processed is acknowledged. Generally, the Sliding Window determines whether it can accept additional messages by examining the highest sequence number yet assigned (hs), and the lowest sequence number associated with a stored message (lu). The window is open as long as $\mathbf{hs - lu + 1 < tws}$.

SDL

SDL models a system as one or more interconnected processes. Each process is composed of an extended finite state machine coupled with an input port. The behavior of a process is modeled in terms of states and possible state transitions. A transition from one state to another can occur in response to: receipt of a specified input, an internal event, or a change in the value of a shared variable or data structure. Data and operations performed upon data are described by means of abstract data types similar to those defined in ACT-ONE [12].

A process communicates with other processes or the environment via signals sent through bi-directional communication channels. These channels are predefined as FIFO queues of unbounded length. Sending a signal, the eventual delivery of the signal to the proper input port, and receipt of the signal by the appropriate process occur independently. Thus SDL uses an asynchronous communication model where the sender of the signal is unaware of *when* its signal is received.

The behavior of an SDL process is described by means of a flowchart like graphical language. Ovals depict the state of a process. Boxes indicate an internal function. Diamonds indicate decision points. Symbols indicating the transmission and receipt of signals are also provided. SDL provides only limited constructs for expressing non-deterministic behavior. Branches can be used to indicate that two or more specified signals can be received at a particular point. The signal that is received first determines the transition that will be taken.

Figure 1 depicts the SDL specification of the Transmitter process discussed previously. The Transmitter process can be in one of two states: *Data Transfer* (the initial state) or *Window Closed*. A transition from *Data Transfer* occurs when a UDTReq(data) signal is received on the proper channel. The appro-

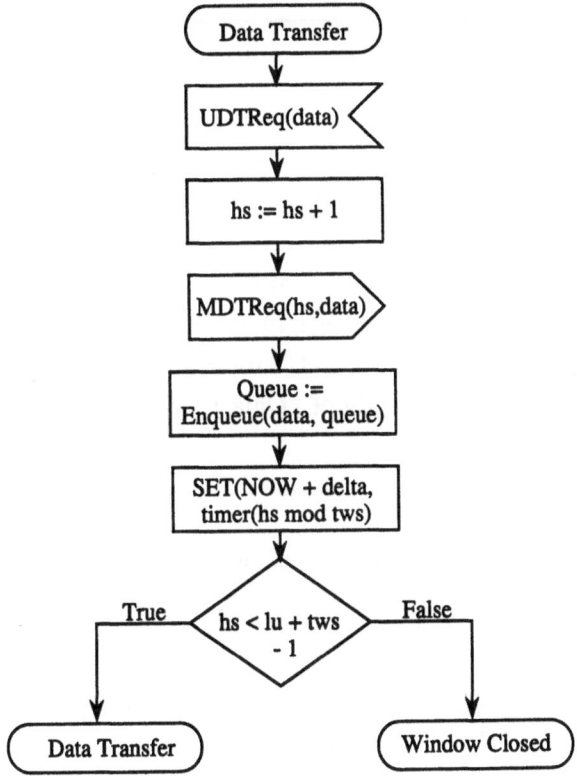

Figure 1: Transmitter Specification (SDL)

priate sequence number and original message are transmitted via the MDTReq channel. A copy of the message is enqueued and the appropriate timer is set. If the transmit window is full, the next state is *Window Closed*, otherwise it is*Data Transfer*.

Estelle

Estelle is based on the concept of communicating non-deterministic state machines. It models a system as a hierarchy of one or more modules which can communicate by exchanging messages. Each module is a finite state machine consisting of a set of primary states as well as a number of secondary states. Transitions between states are atomic and are specified using an extended Pascal -like grammar. A transition can occur in response to the receipt of a message, or some internal event. Transitions can be inhibited by the use of a *provided*

```
trans
  from  SENDING  to  same
    when  U.DataRequest
      provided  HighestSent - LowestUnacknowledged + 1 < TWS
        begin
          HighestSent := HighestSent + 1;
          output  T.TimerRequest( HighestSent );
          output  MT.DT( PDU_DT( HighestSent , Data ) );
          BufferSave( HighestSent , Data );
        end;
```

Figure 2: Transmitter Specification (Estelle)

clause or a *delay* statement. A *provided* clause specifies conditions, such as the value of certain variables, which must be satisfied before a transition is enabled. Delays inhibit a transition from being enabled for a specified period of time. Spontaneous state transitions can be modelled using a delay time of zero. Non-deterministic behavior is modelled by allowing multiple transitions from a specified state. Modules communicate by exchanging messages through bidirectional communication channels where messages can be queued at either end. Thus Estelle uses an asynchronous communication model. A Pascal like syntax is used to specify variables and allows for the creation of user defined types.

Figure 2 details the Estelle specification of the Transmitter process. The first four lines detail one of several possible transmissions from the SENDING state. Three conditions must be satisfied before this particular transition can occur. First, the transmitter must currently be in the SENDING state. Second, the transmitter must receive a message from the layer above. Finally, the transmitter must have the capacity to accept another message. All three of these conditions must be satisfied, otherwise the transition cannot occur. The next sequence number is determined, the appropriate timer is set, the resulting message is transmitted and a copy is stored in a buffer.

LOTOS

LOTOS is a mathematically-defined formal description technique based on the theories of CCS [13], CSP [14] and ACT-ONE [12]. It models the behavior of a system using constructs which indicate: sequencing, non-deterministic choice between events, process instantiation and termination, process synchronization, and conditional execution. Data and operations performed upon data are described by means of abstract data types similar to those defined in ACT-ONE.

A distributed system is modeled in the form of abstract processes which may be composed of several communicating sub-processes. A sub-process itself may be decomposed into several sub-processes. A process can perform actions which are internal to itself, and thus unobservable to any other process, or can interact with the other processes which comprise its environment. Processes may execute independently of each other (true interleaving) or may interact to perform some action. Interaction of two or more processes occurring at an interaction point (known as a *gate*) is referred to as an *event*. Events are atomic and synchronous. Upon arriving at a *gate*, a process will wait until the other required processes have arrived before the *event* occurs.

Figure 3 shows the LOTOS specification for the Transmitter process discussed previously. The Transmitter process is composed of three sub-processes, one of which is Sender. Sender and Transmitter synchronize at the **swt, mt**, and **t** gates (as indicated by [**swt, mt, t**]). The information contained in parenthesis **(tws, hs, lu, rq)** indicate that these variables are accessible to both processes. The accept statement indicates that the values associated with **hs** and **rq** will be updated by the Sender process.

The statement: **swt ? up : SP [hs lt (lu + tws - 1)]** indicates that the Sender will synchronize with one or more other processes at the swt gate only if the guard condition [**hs lt (lu + tws - 1)**], indicating that the Sliding Window is open, is satisfied. The **?** before **up** indicates that the value of the variable **up** (which holds messages received from the layer above) will be modified during the event. Sender creates a PDU containing the message it received and appends an appropriate sequence number. **!** in a statement indicates that the value of the associated variable will be exported to the other synchronized processes during the event. Thus the specification indicates that Sender will export the newly created message on gate **mt** and then set a timer on gate **t**. Upon exiting, Sender computes and exports the value of the next unique ascending sequence number **(hs + 1)** and stores a copy of the PDU **(pdu+- rq)**.

The three specification languages employ widely different methods to spec-

```
process Transmitter
    [swt, mt, t]
    ( tws : Nat, hs : Nat, lu : Nat, rq : PduQueue ) : noexit :=

    ( Sender [swt, mt, t] (tws, hs, lu, rq)
        >>
        accept hs : Natural, rq : PduQueue in
            Transmitter [swt, mt, t] (tws, hs, lu, rq)
    )
[]
    ( AckRec [mt, t] (hs, lu, rq) .................... )
[]
    ( TimeOut [mt, t] (hs, lu, rq) ................... )
endprocess (* Transmitter *)
```

```
process Sender [swt, mt, t]
    ( tws : Nat, hs : Nat, lu : Nat, rq : PduQueue );
        exit ( Natural, PduQueue ) :=

    swt ? up : SP [hs lt (lu + tws - 1)];
    (
      let pdu : Pdu = MakeDTPdu( Data(up), Succ(hs) ) in
        mt ! Mreq (pdu);
        t ! hs + 1 ! set;
        exit ( hs + 1, pdu +--rq )
    )
endprocess (* Sender *)
```

Figure 3: Transmitter Specification (LOTOS)

ify both the behavior of the system and data transformations. LOTOS uses algebraic constructs and abstract data types. Estelle uses a Pascal-like statements to represent both the behavior of the system and the encoding of data. SDL uses a graphical notation (with added textual statements) and abstract data types. In addition, LOTOS uses a synchronous communication model, while both SDL and Estelle use an asynchronous model.

3 Implementations

Specifications of an OSI Transport Protocol [15], and a Sliding Window Protocol [8] were available in each of the three specification languages [7] and served as the basis for the experiment. Nine independent versions of these protocols were developed in Ada from the formal specifications (three from each formal language) by graduate students in Computer Engineering or Computer Science at UCSB. All were experienced Pascal or C programmers, although none of the the programmers had extensive knowledge of communication protocols or of the formal specification languages.

Each of the programmers was provided with a formal specification of the Sliding Window and Transport Layers and appropriate tutorials covering the semantics and grammar of the specification language. The programmers were forbidden from referencing outside materials and from discussing the protocols or the project. Therefore the resulting programs accurately depict the programmers understanding of the protocols based solely on the descriptions provided by the formal specification language which they were assigned.

4 Effect of Specifications on Implementations

The LOTOS, SDL, and Estelle specification languages use widely different techniques to specify system behavior. The nine independently created versions of the protocols present a unique opportunity to evaluate the types of programs developed from each of the specification languages. We are particularly interested in features or constructs which are common in versions derived from the same formal specification.

In order to examine the effect that a specification has on program design and development, versions produced from identical specifications are compared with each other and with their source specification. In the following examples and figures, the program fragments have been passed through a source formatter

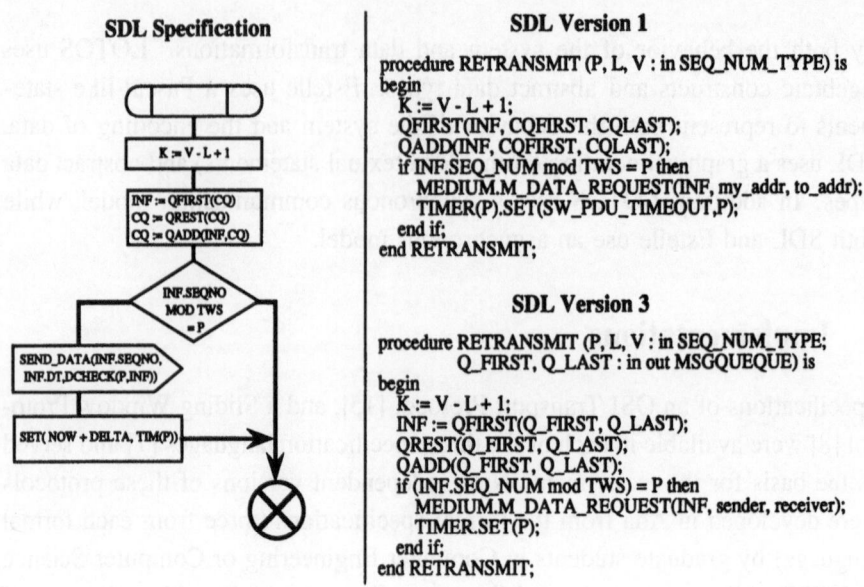

Figure 4: SDL implementations of the Retransmitter module.

and sections not covered in the analysis have been abbreviated. The ordering of statements and the choice of variable and state names have not been modified. An approach is adopted which examines both the large grain structure of the protocol version, as well as fine grain implementation details.

4.1   Effect on Fine Grain Implementation

As an example of how specifications can influence the lower level implementation, we arbitrarily focus on the retransmitter process in the Sliding Window protocol. The process is responsible for retransmitting messages whose timers have expired. Upon expiry of a timer, the retransmitter checks to determine if the timed out message is still stored and awaiting acknowledgment. If so, the appropriate message is retrieved, transmitted to the medium, and the timer is restarted.

Figure 4 compares the SDL specification of the appropriate transition with two of the resulting three implementations (Version 2 is not shown, but is similar to versions 1 and 3). With the exception of the QFIRST and TIMER routines, the versions are almost identical. It is interesting to note that neither QFIRST nor TIMER were explicitly defined in the SDL specification, thus their implementations would be expected to be somewhat different.

Estelle Specification

```
from SENDING to same
   when T.TimerResponse
      provided (Seq >= LowestUnacked) and
         (Seq <= HighestSent)
      begin
         output MT.DT(PDUDT(Seq, BufRetrieve(Seq)));
         output T.TimerRequest(Seq);
      end;

   provided otherwise
   begin
      {ignore timer response for sequence number
         outside the window}
   end;
```

Estelle Version 2

```
accept Timer_Response(S: in SEQ_NUM_TYPE) do
   S_TEMP := S;
end Timer_Response;
if S_TEMP >= LowestUnacked and S_TEMP <= HighestSent then
   medium.M_Data_Request(PDUDT(S_TEMP,BufRetrieve(S_TEMP)));
   Timer(S_TEMP mod TWS).START(S_TEMP);
else
   put_line("Timer Response for sequence outside of window");
end if;
```

Estelle Version 3

```
accept TimerResponse(timed_out_seq : in SEQ_NUM_TYPE) do
   lost_seq := timed_out_seq;
end TimerResponse;
if (lost_seq >= low_unack) and ( lost_seq <= highest_sent) then
   medium.M_DATA_REQUEST(trans_seq_queue(lost_seq), me, you);
   TIMER.TimerRequest(lost_seq);   .
end if;
```

Figure 5: Estelle Implementations of the Retransmitter Module.

Figure 5 compares the Estelle specification with two of its corresponding implementations (Version 1 is very similar to versions 2 and 3). The versions are very similar. The only substantial differences involve the calls to the Timer module, which was again unspecified.

The versions created from the LOTOS specification (figure 6) exhibit a larger amount of diversity, especially in the use of data structures. However, all three programmers implemented the timers as an array of timer tasks. The LOTOS specification fragment shown above does not indicate a structure of this type, and it is highly unlikely that three independent programmers would choose to use this complex structure when simpler methods were available. The cause of this structure does not appear in this specification section, but in the section detailing the creation of the timers (figure 7). Process AllTimers calls itself recursively and with each iteration it spawns a new **Timer** process with a unique identification number. This process continues until the number of timers spawned is equal to **tws**. Obviously creating and maintaining multiple timers creates unnecessary system overhead, yet all three programmers using the LOTOS specification chose to use this complex and expensive structure.

Overall, in instances where the specifications presented specific instructions concerning *how* an activity was to be performed, the subsequent versions tend to contain exactly the same method. However, in instances where the activity was only loosely described, or missing altogether, the versions tended to employ various different approaches.

LOTOS Specification

```
process retransmission
   [mt, t] ( sn : Nat , rq : PduQueue ) : exit :=

   ( choice newreq : PduQueue , pdu : Pdu []

      [ rq eq ( newrq –+ pdu )  and  ( sn eq SN(pdu) ) ] ->
         mt ! Mreq(pdu)
         t ! SN(pdu) ! set
   )
   []
      [ IsEmpty(rq) ] ->
         exit
end proc (* retransmission *)
```

LOTOS Version 1

{ section of code was not a separate procedure }

```
if ( lu <= sq ) and ( sq <= hs ) and ( Q_count > 0 )  then
   medium.m_data_request( Queue( sq mod tws ) , sen , rec );
   alltimers( sq mod maxis ).timer_set( sq );
end if;
```

LOTOS Version 2

```
procedure retransmit( sn : in Natural ) is
begin
   if ( sn >= lu )  and  ( sn <= hs )  then
      medium.m_data_request( tr_window( ( sn mod tws ) + 1 ) , 'A' , peer );
      time_a( ( sn  mod  tws ) + 1 ).set( sn );
   end if;
end;
```

LOTOS Version 3

```
procedure retransmitter ( sn : in seq_num_type ) is
   P : Pdu_list;
   I : seq_num_type;
begin
   I := sn;
   list_manager.in_list( I , P );
   if ( P /= null )  then
      alltimers( ( ( I-1 ) mod tws ) + 1 ).reset;
      alltimers( ( ( I-1 ) mod tws ) + 1 ).set;
      send_to_medium.send( P.pdu , tempreceiver );
   end if;
end;
```

Figure 6: LOTOS Implementations of the Retransmitter Module.

```
AllTimers: forking into all possible timers.  The number of
            timers needed is indicated by tws.

process AllTimers [t]  (MaxId : Nat, TimerId : Nat) : noexit :=

   [TimerId lt MaxId] ->
      (
         AllTimers [t] (MaxId, Succ(TimerId))
         |||
         Timer [t] (MaxId, TimerId)
      )
```

Figure 7: Creation of Timers in LOTOS.

4.2 Effect on Large Grain Program Structure

The influence which the specifications have on program development is not restricted solely to low level encoding of modules, the overall program structure can be dramatically influenced. As an example, we will examine the high level implementation of the transport protocol.

The Transport Layer is responsible for connection establishment and termination, data segmentation and reassembly, data transfer, and protocol error detection. It can receive six different ASPs (four from the layer above and two from the layer below) as well as signals from two internal timers. The Transport Layer has seven distinct states: IDLE, WFTCA (Wait for Transport Connect Accept), WFTRESP (Wait for Transport Response), OPEN, WAIT ,ERROR, and PRE-RELEASE. The response to a particular ASP depends on the current state of the Transport Layer. Given the large number of possible combinations of ASP and state, the overall layer structure can be complex and confusing. Therefore if the specifications do not exert undue influence on the implementations, we would expect the high level structures to vary greatly. Unfortunately this is not always the case.

The specification sections covering the relevant behaviors are too lengthy to be presented in full in this paper. Instead we will present the effect of the specifications by examining similarities in the resulting programs. Due to the length of the implementations, code segments have been abbreviated.

The SDL specification contains a separate diagram for every possible Transport Layer state. The diagrams list the signals that can be received in the particular state, the appropriate actions to be taken, and the associated next state. Thus a signal which can be received in any state will appear in each of the diagrams. Figure 8 shows the structure of two Transport Layers developed from the SDL specification (version 2 has been omitted but is similar to version 1). The ordering of statements and the choice of variable and state names have not been modified. Initially the versions appear to be fairly diverse, but upon closer examination it becomes apparent that each version employs a select statement for each possible state. All three SDL versions chose to implement the structure of the Transport Layer as a single loop and used multiple accept statements for each possible rendezvous. For example, version 3 has seven accept statements for Data_Indication, one for each possible state of the Transport Layer. Thus the implementations map the specification's high level design.

The similarities between the versions are further obscured by nested state assignments and rendezvous within the scope of an original accept statement. For

SDL Version 1

```
task body TPM is
  loop
    state := IDLE
    select
      accept Reset_Indication
    or
      accept Connect_Request
        state := WFTCA
        select
          accept TCR_Time
        or
          accept Disconnect_Request
        or
          accept Reset_Indication
        or
          accept Data_Indication
        end select;
    or
      accept Data_Indication
        state := CLOSED
        while state = CLOSED loop
          select
            accept Reset_Indication
          or
            accept Data_Indication
          end select
        end loop

      if state = WFTRESP then
          ...............
    end select
    while state = DATA_TRANSFER loop
      ...................
    while state = ERROR loop
      ..................
  end loop
end TPM
```

SDL Version 3

```
task body TPM is
  loop
    loop
      select
        accept Internal_Connect_Request
      or
        accept Internal_Data_Indication
      or
        accept Data_Indication
      or
        accept Reset_Indication
      end select;
      exit when state /= CLOSED;
    end loop;

    while state = WFTCA loop
      select
        accept Disconnect_Request
      or
        accept Reset_Indication
      or
        accept Data_Indication
      or
        delay TCR_Time
      end select
    end loop

    while state = WFTRESP loop
      ............
    while state = OPEN loop
      ............
    while (state = ERROR) or (state =
WAIT) loop
      ............
  end loop
end TPM
```

Figure 8: Transport Layer (SDL versions).

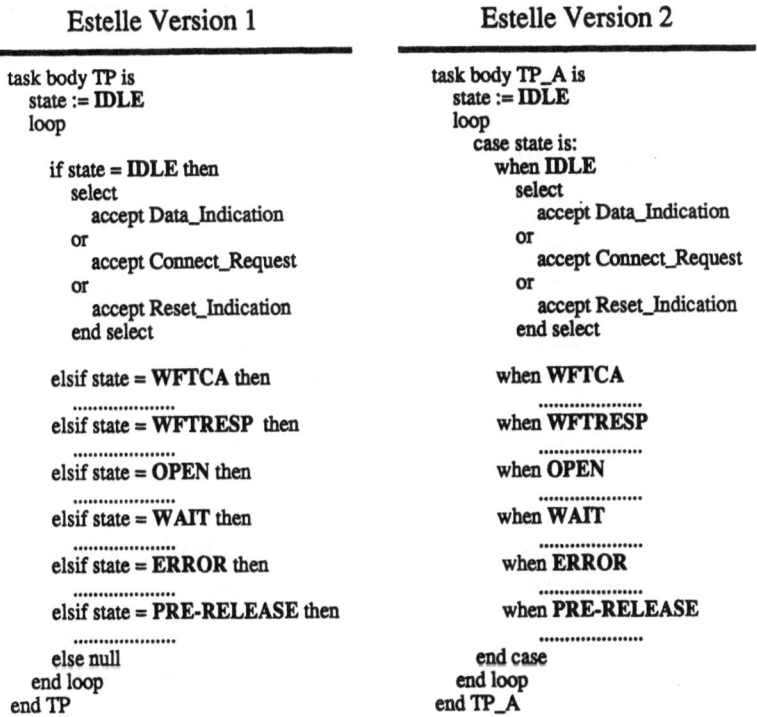

Estelle Version 1	Estelle Version 2

```
task body TP is                    task body TP_A is
   state := IDLE                      state := IDLE
   loop                               loop
                                        case state is:
      if state = IDLE then               when IDLE
         select                            select
            accept Data_Indication           accept Data_Indication
         or                                or
            accept Connect_Request           accept Connect_Request
         or                                or
            accept Reset_Indication          accept Reset_Indication
         end select                        end select

      elsif state = WFTCA then          when WFTCA
      ....................              ....................
      elsif state = WFTRESP then        when WFTRESP
      ....................              ....................
      elsif state = OPEN then           when OPEN
      ....................              ....................
      elsif state = WAIT then           when WAIT
      ....................              ....................
      elsif state = ERROR then          when ERROR
      ....................              ....................
      elsif state = PRE-RELEASE then    when PRE-RELEASE
      ....................              ....................
      else null                         end case
   end loop                            end loop
end TP                              end TP_A
```

Figure 9: Transport Layer (Estelle versions).

example, version 3 indicates that upon receipt of a Internal_Connect_Request while in CLOSED (or idle), the protocol should enter the WFTCA state and either delay TCR_Time or participate in one of the three possible rendezvous: Disconnect_Request, Reset_Indication, or Data_Indication. The same behavior is detailed in version 1. The initial state is IDLE, and the protocol is able to accept any of three different rendezvous (as indicated by the select loop). Upon accepting a Connect_Request, the protocol changes to the WFTCA state and can accept any of the four events detailed earlier.

Figure 9 shows two of the three Estelle implementations (version 3 is almost identical to 1). The versions created from this specification reflect Estelle's state-transition based style. Not only did the programmers adopt the basic structure of the layer, but also the exact ordering of states and signals were replicated verbatim.

A sample of the Estelle specification for possible transitions from the IDLE state is shown in figure 10. Note that the program versions use the exact order-

```
trans
    from IDLE
        when TC_IP.TPDU {a Data Indication}
            to WFTRESP  .....................
        when TCEP.TCON_REQ { a Connection Request}
            to WFTCA  .....................
        when TC_IP.SW_RST_IND { a Reset Indication}
            to PRE-DISCONNECT  .....................
```

Figure 10: Transport Specification (Estelle).

ing of events which are presented in the specification. Clearly, the specification had a large influence on the resulting program versions.

The LOTOS versions are shown in figure 11. The LOTOS specification divides the Transport Layer into a series of cooperating processes. Signals occur at gates shared between these processes, thus there is no clear concept of a process receiving a message from the outside, rather all processes sharing the specified gate receive the message.

The appropriate action taken by each process (possibly none) depends on code contained within the process itself. The resulting LOTOS versions depict two different approaches to implementing the specification. Versions 2 and 3 employed a single task to accept all possible rendezvous and then passed the received signal to an appropriate handling routine. For example, upon receipt of a Data_Indication, a call is made to a procedure which checks the current state and determines the subsequent sequence of events. The control structure is much cleaner than those derived from either the Estelle or SDL specification. The first version adopts the high level functional process decomposition shown in the specification. Thus the programmer uses six different tasks to implement the Transport Layer. Each task is well designed and handles exactly one of the primary functions of the Transport Layer. For example, the ESTABLISH task handles connection requests (both internal and external), and the RELEASE task handles the termination of the connection. Likewise DOWN_TASK and UP_TASK handle all of the ASPs which are received between establishment and termination of a connection.

LOTOS Version 1	LOTOS Version 2

```
task body ESTABLISH is           task body TP_A is
 loop                             loop
   select                           select
      accept Connect_Request            accept Data_Indication
   or accept Connect_Response        or accept Reset_Indication
   end select                       or accept Connect_Request
 end loop                           or accept Connect_Response
end ESTABLISH                       or accept Disconnect_Request
                                    or accept Data_Request
task body RELEASE is                or delay TCR_Time
 loop                               end select
   select                         end loop
      accept Disconnect_Request   end TP_A
   or accept Reset_Indication
   end select
 end loop
end RELEASE
```

	LOTOS Version 3

```
task body AFTER_TBR is ............

task body AFTER_TCA is .............

task body DOWN_TASK is ............

task body UP_TASK is ............
```

```
task body TRANSPORT is
 loop
   select
      accept Connect_Request
   or accept Data_Indication
   or accept Connect_Response
   or accept Data_Request
   or accept Reset_Indication
   or accept Disconnect_Request
   or accept Timer_Expired
   end select
 end loop
end TRANSPORT
```

Figure 11: Transport Layer (LOTOS).

4.3 Appropriate Level of Abstraction

The appropriate level of abstraction which should be incorporated in a specification is difficult to determine. A specification which is too abstract can result
in an implementation which does not meet the user requirements. Difficult
specifications may also encourage programmers to reference other possibly
less formal materials rather than work through the formal specification. On
the other hand, specifications which incorporate too much detail can strongly
influence the implementation whether this influence is intentional or not.

The specification must be complete enough so that two independently derived versions of the protocol can communicate, but not so detailed as to restrict
the choice of implementation mechanisms. *Design* specifications, as opposed
to a more high level specification, are much more implementation oriented, and
typically include descriptions of each process, the interfaces, and a description
of any data or data structures which are shared among the processes. A *Design*
specification would therefore be expected to contain more implementation specific detail than a high level specification, thus easing the job of translating the
specification into a program. The three specification languages each employed
various levels of abstraction. All were too implementation orientated to be
considered high level specifications, yet not detailed enough to be adequate
design specification.

In all three specifications the system was partitioned into processes, an activity which requires major design decisions to produce the best trade off between
parallelism and interprocess communication. However, none of the specifications addressed the problems of deadlock and concurrency control which can
occur in highly parallel systems. The detailed design lulled programmers into
erroneously assuming that the specifier had addressed these issues. Subsequently when deadlock was encountered, it often resulted in major redesigns
of processes and interprocess communication schemes.

As shown in the previous section, the syntax of the specification languages
had a large effect on the overall program structure. The versions created from
the SDL and Estelle languages reflect the state-transition oriented nature of the
language by implementing the Transport Layer as a sequence of states and state
transitions. Since a signal can be received in multiple states, the corresponding accept statement necessarily appeared in several locations. The resulting
Transport Layers tended to be monolithic and at times quite difficult to understand. The LOTOS specification tended to abstract away from implementation
details. The Transport Layer was divided into cooperating processes. Signals

are received at gates which are shared by all involved processes. The particular activity taken by each process depends upon the history of past events (i.e. a process may only be able to participate in an event if certain sequence of events have occurred) and the values associated with a signal. Thus there is no explicit concept of state nor does any one process receive the signal. The LOTOS versions depict two different approaches to implementing the specification. The first version adopted the functional decomposition of processes detailed by the specification. The other versions implemented the shared event concept by using a single task to accept all events and to distribute them to the appropriate handler depending on the previous sequence of events.

The semantics of the specification languages also had a large effect on the actual specifications and therefore the resulting program versions. For example, recall the retransmit example given in section 4.1. The SDL and Estelle specification languages both use a Pascal-like language to describe system activity. Thus in order to specify that the stored copy of the message should be retrieved, they both incorporated the concept of a queue or buffer into the specification. By defining the actual data structure, they are also constrained to identify operations on the data structure and the exact sequence of events required to retrieve the required message. For example the SDL specification maintains a FIFO queue in which the most recently transmitted message is at the tail of the queue and the oldest message is at the head. Thus retransmitting a message requires that it be removed from the head of the queue and then inserted at the tail. The SDL specifier assumed that unacknowledged messages would *always* time out in the order in which they were transmitted. Thus the appropriate message to be retransmitted would always be at the head of the queue. Unfortunately, the various interleavings of simultaneous events which can occur in a parallel system resulted in instances where the retransmitter modules handled the time out signals in a different order causing the actual implementations to fail. Clearly the exact data structure (be it linked list, queue, etc) used to store the data is unimportant, yet the semantics of a language may force implementation issues to be incorporated into the specifications. The LOTOS specification was the most abstract. Consequently, the programmers tended to use the specification as a guideline rather than an implementation blue print.

In addition, the informal protocol specifications indicate that three of the parameters in the TPDU blocks used to establish a connection may appear in any order. Both the LOTOS and the SDL specifications used abstract data types to specify the arbitrary ordering of these parameters and the resulting program versions correctly allowed various orderings of these parameters. However, the Estelle specification, which uses Pascal-like descriptions, showed the parameters occurring in only one fixed position. Thus these Estelle versions were capable of communicating amongst themselves, but often failed when they were tested with versions created from either LOTOS or SDL which (correctly) allowed these parameters to occur in a different order. In order to represent the arbitrary ordering in Estelle, it would be necessary to repeat the appropriate part of the specification for each possible ordering of parameters.

5   Conclusion

The use of formal specifications clearly has a large impact on program development. The language syntax, semantics, and representation style can influence not only the implementation details which are included in the specification itself, but also high level program structure, and low level implementation details. While specification writers often intend to merely illustrate a concept (i.e. *what* is to be created) with a somewhat arbitrary design as the vehicle of illustration, the detailed nature of a specification is often interpreted as actual program designs.

Program versions created from the same formal specification show an alarming level of similarity. Particularly when the specification language tends to focus on *how* a process should be implemented, rather than *what* the process should accomplish. It is reasonable to assume that versions which share similar structures and implementation details will have a higher probability of failing identically than versions which are markedly dissimilar.

The use of Multiple diverse formal specifications is an attractive alternative where fault tolerant strategies, such as Multi-Version Software, or fault detection techniques, such as Back-to-Back testing, are employed. While it is unlikely that specifications written in diverse specification languages would be routinely available, the additional expense and effort expended in their development may be balanced by a decrease in testing time [4, 5] and the ability to detect discrepancies in the specifications themselves before actual coding begins.

The results from this initial study support the conclusion that specifications do indeed have a strong influence on program versions. However, much work remains to be done to analyze the scope and ramifications of the influence.

We are actively researching *quantitative* measures, such as program metrics and symbolic representations, in order to more clearly portray the relationship between the specification and program versions.

References

[1] J. Kelly and A. Avižienis, "A specification-oriented multi-version software experiment," in *Digest of 13th Annual International Symposium on Fault Tolerant Computing*, pp. 120–126, June 1983.

[2] P. Bishop and *et al.*, "Project on diverse software - an experiment in software reliability," in *Proceedings IFAC Workshop Safecomp '85*, 1985.

[3] A. Avižienis and L. Chen, "On the implementation of n-version programming for software fault-tolerance during execution," in *Proceedings COMPSAC 77*, pp. 149–155, November 1977.

[4] J. Kelly and S. Murphy, "Achieving dependability throughout the development process : A distributed software experiment," *IEEE Transactions on Software Engineering*, vol. SE-16, pp. 153–165, February 1990.

[5] J. Kelly, T. McVittie, and S. Murphy, "Techniques for building dependable distributed systems - multi-version software testing," in *Proceedings of the 20th Annual Fault Tolerant Computing Symposium*, June 1990.

[6] ISO, "Basic reference model for open systems interconnection," Tech. Rep. ISO 7498, ISO, Geneva, 1984. also CCITT Recommendation X.200.

[7] K. Turner, ed., *Proceedings of the First International Conference on Formal Description Techniques, FORTE 88*, September 1988.

[8] ISO, *Guidelines for the Application of Estelle, LOTOS and SDL*. Stirling, January 1988.

[9] ISO, "Information processing systems - OSI - LOTOS - a formal description technique for the temporal ordering of observational behavior," Tech. Rep. ISO 8807, ISO, 1988.

[10] ISO, "Estelle: a formal description technique based on an extended state transition model," Tech. Rep. ISO 9074, ISO, 1988.

[11] CCITT, "SDL, specification and description language," (Blue Book) Z.100, International Consultative Committee for Telephony and Telegraphy, Geneva, March 1988.

[12] H. Ehrig and B. Mahr, *Fundamentals of Algebraic Specification*, vol. 1. Berlin: Springer-Verlag, 1985.

[13] R. Milner, *CCS, A Calculus of Communicating Systems*, vol. 92 of *LNCS*. Berlin: Springer-Verlag, 1980.

[14] C. Hoare, *Communicating Sequential Processes*. Prentice-Hall International, 1985.

[15] CCITT, *Recommendation T.70, Network Independent Basic Transport Service for the Telematic Services*. CCITT, (red book) ed., 1984.

Verification Techniques

The Proof of Correctness of a Fault-Tolerant Circuit Design

WILLIAM R. BEVIER, WILLIAM D. YOUNG*

Computational Logic, Inc.

1717 West Sixth Street, Suite 290

Austin, TX 78703-4776, USA

Abstract

We describe a formally verified implementation of the "Oral Messages" algorithm of Pease, Shostak, and Lamport. An abstract implementation of the algorithm has been verified to achieve interactive consistency in the presence of faults[1]. This abstract characterization is then mapped down to a hardware level design which inherits the fault-tolerant characteristics of the abstract version. The proof that the hardware level description is a correct implementation of the "Oral Messages" algorithm has been fully checked with a mechanical theorem prover. This design was then translated straightforwardly into a physical hardware implementation. A significant result of this work is the demonstration of a fault-tolerant device that is formally specified and whose implementation is proved correct with respect to this specification.

1 Introduction

A key problem facing the designers of systems which attempt to ensure fault tolerance by redundant processing is how to guarantee that the processors reach

*This work was sponsored in part at Computational Logic, Inc. by National Aeronautics and Space Administration Langley Research Center (NAS1-18878). The views and conclusions contained in this document are those of the authors and should not be interpreted as representing the official policies, either expressed or implied, of Computational Logic, Inc., NASA Langley Research Center or the U.S. Government. We with to thank our sponsors at NASA, particularly Ricky Butler, for invaluable guidance in formulating this problem and our colleagues at Computational Logic for building and fostering a marvelous research environment.

agreement, even when one or more processing units are faulty. Pease, Shostak and Lamport [2, 3] have devised the notion of *interactive consistency*, which formally characterizes what it means for non-faulty processors to reach agreement. They prove that, given certain assumptions about the type of inter-process communication, interactive consistency can be achieved if and only if the total number of processors exceeds three times the number of faulty processors. They also provide a clever algorithm which achieves interactive consistency.

Our goal was a verified hardware implementation of the "Oral Messages" (*OM*) algorithm of Pease, Shostak, and Lamport. Our approach to achieving this was to proceed in several phases. In the first phase, we defined an "abstract implementation" of the algorithm in the Boyer-Moore logic and proved that this high-level formalization achieved interactive consistency. We then defined a low-level characterization of the algorithm and proved that our low-level version is a correct implementation of our high-level version. As a consequence of this proof, we are guaranteed that our low-level implementation achieves interactive consistency.

Elsewhere [1] we report on our work to describe our abstract implementation of the algorithm in the Boyer-Moore logic. We stated the interactive consistency conditions in the logic, and used the Boyer-Moore theorem prover to check a proof that our formalization of *OM* satisfies these conditions. We also mechanically checked the result presented by Lamport, Shostak and Pease that *OM* provides an optimal solution: no algorithm exists which achieves interactive consistency via an exchange of oral messages if the number of faulty processors is at least one third of the total.

In this paper we present the design of a hardware implementation of *OM(1)*: the instance of *OM* which tolerates one faulty process when there are at least three non-faulty processes. We have mechanically checked the proof that the hardware design implements *OM(1)*, and therefore achieves interactive consistency.

The paper is organized as follows. The following section describes our formal specification of the Oral Messages algorithm and its correctness properties. Subsequent sections describe our implementation of the algorithm and sketch the proof that our implementation satisfies its specification. In the final section, we give some of our conclusions and observations on this development process.

2 The Specification

2.1 Interactive Consistency and The Function *OM*

The problem addressed by the interactive consistency algorithm is the following: given a number of communicating processors, how can they arrive at a consistent view of the system if there are faulty processors among them which potentially send conflicting information to different parts of the system. Lamport, Shostak, and Pease [3] describe the problem in terms of the metaphor of Byzantine Generals attempting to arrive at a common battle plan through an exchange of messages. One or more of the generals may be traitorous and attempt to thwart the loyal generals by preventing them from reaching agreement.

It is straightforward to state the problem in terms of a single commanding general communicating with a number of lieutenant generals. In this case we desire an algorithm which guarantees the following.

> A commanding general must send an order to his $n - 1$ lieutenant generals such that
>
> IC1. All loyal lieutenants obey the same order;
>
> IC2. If the commanding general is loyal, then every loyal lieutenant obeys the order he sends.
>
> Conditions IC1 and IC2 are called the *interactive consistency* conditions. [3]

The interactive consistency conditions can be formalized fairly straightforwardly. Let n be some number of processes. Let $\bar{v}$ be a vector of length n, where $\bar{v}[i]$ is the local value of process i, $i \in 0, \ldots, n - 1$. Let L be the set $0, \ldots, n - 1$ of indices into $\bar{v}$. This serves as the set of process names.

We assume a predicate $faulty : L \rightarrow \{T, F\}$ that identifies a potentially faulty process. A faulty process may or may not forward a message correctly. We also assume a function $faults : 2^L \rightarrow \mathbb{N}$ that counts the number of potentially faulty processes in a set of process names. The functions *faulty* and *faults* are used only in the specification of the problem; they are not computable.

Let g (the general) be a member of L and $\bar{w}_g$ be a vector of length n, where each entry $\bar{w}_g[i]$ is the value which process i concludes is process g's local value. $\bar{w}_g$ satisfies the interactive consistency conditions if for each $i, j \in L \subseteq g$ we have the following.

IC1. $\neg faulty(i) \land \neg faulty(j) \rightarrow \overline{w}_g[i] = \overline{w}_g[j]$

IC2. $\neg faulty(g) \land \neg faulty(i) \rightarrow \overline{w}_g[i] = \overline{v}[g]$

We have formalized in the Boyer-Moore logic a version of the Oral Messages algorithm of Pease, Shostak, and Lamport [2]. This algorithm, and our (abstract) implementation of it in the form of a function OM in the Boyer-Moore logic, produces a vector $\overline{w}$ which satisfies the interactive consistency conditions. The vector is computed after some number of rounds of *information exchange* among the processes.

Our Boyer-Moore function OM takes four parameters: n, the number of processes; g, the name of the general; $x = \overline{v}[g]$, the general's local value; and m, an integer which determines the number of rounds of information exchange to take place. Lamport, Shostak and Pease [3] prove that OM is guaranteed to achieve interactive consistency only if n is greater than three times the number of faulty processes. The number of rounds of information exchange m must be at least the number of faulty processes.

Elsewhere [1] we present our formal definition of OM, and describe our mechanically checked proof that our formalization of the algorithm produces a vector which satisfies the interactive consistency conditions. Our formal statements of the two theorems that OM satisfies IC1 and IC2, respectively, are given below. We believe that these are an intuitive and straightforward formalization of the interactive consistency conditions.

$$\neg faulty(i)$$
$$\land \neg faulty(j)$$
$$\land\, 3 \times faults(L) < n$$
$$\land\, faults(L) \le m$$
$$\rightarrow$$
$$OM(n, g, x, m)[i] = OM(n, g, x, m)[j],$$

$$\neg faulty(g)$$
$$\land \neg faulty(i)$$
$$\land\, 3 \times faults(L) < n$$
$$\land\, faults(L) \le m$$
$$\rightarrow$$
$$OM(n, g, x, m)[i] = x.$$

2.2 Multiple Applications of OM

To reach agreement, each process among a set of processes must act in turn as
the general in an application of *OM*. We define the function *OML* recursively
to apply *OM* to each member of a list of process names. [1]

$$OML(nil, \overline{v}, m) \equiv nil$$

$$OML(cons(g, l), \overline{v}, m) \equiv cons(OM(|\overline{v}|, g, \overline{v}[g], m)|_{\overline{v}[g]}^g), OML(l, \overline{v}, m))$$

The expression $OM(|\overline{v}|, g, \overline{v}[g], m)|_{\overline{v}[g]}^g)$ in this definition denotes the vector $\overline{w}$
in which, for each $i \neq g \in L$, $\overline{w}[i]$ records the value which process i concludes
is process g's local value, after m rounds of information exchange. The value
of $\overline{w}[g]$ is $\overline{v}[g]$, g's local value.

OML produces an $n \times n$ matrix in which the ith row is a vector $\overline{w}_i$ of
values such that $\overline{w}_i[j]$ is j's guess at i's local value. The ith column of the
matrix is the *interactive consistency vector* for process i. This vector contains the values which process i concludes is the local value for each process.
$OML(L, \overline{v}, m)[i, g]$ is the value which process i concludes is process g's local
value.

We can derive the following two facts about *OML* as a result of the interactive consistency conditions proved of *OM*.

1. In the matrix produced by *OML*, any two non-faulty processes agree on
   the local value of all other processes.

2. Each non-faulty process has the correct value for a non-faulty general.

These facts correspond to IC1 and IC2, respectively. The formal versions are

---

[1]The notation $|\overline{v}|$ denotes the length of vector $\overline{v}$, which in this case gives the number of
processes n. The notation $\overline{v}|_i^x$ denotes a vector with the following property.

$$\overline{v}|_i^x[j] \equiv \textit{ if } j = i \textit{ then } x \textit{ else } \overline{v}[j]$$

displayed below.

$$\neg faulty(i)$$
$$\land \neg faulty(j)$$
$$\land\; 3 \times faults(L) < n$$
$$\land\; faults(L) \leq m$$
$$\rightarrow$$
$$OML(L, \bar{v}, m)[i, g] = OML(L, \bar{v}, m)[j, g],$$

$$\neg faulty(g)$$
$$\land \neg faulty(i)$$
$$\land\; 3 \times faults(L) < n$$
$$\land\; faults(L) \leq m$$
$$\rightarrow$$
$$OML(L, \bar{v}, m)[i, g] = \bar{v}[g].$$

From these two properties, we can prove that two non-faulty processes have identical interactive consistency vectors. That is,

$$\neg faulty(i)$$
$$\land \neg faulty(j)$$
$$\land\; 3 \times faults(L) < n$$
$$\land\; faults(L) \leq m$$
$$\rightarrow$$
$$OML(L, \bar{v}, m)[i] = OML(L, \bar{v}, m)[j].$$

2.3   Traces of *OM* Applications

The function *OML* formally describes a single instance of n processes reaching agreement through m rounds of information interchange. This formalization is not conducive to mapping down to a lower-level implementation which executes the algorithm in a number of "steps." Therefore, we define a *trace function* O^* to model n processes attempting to reach agreement through time. The input to O^* is a sequence of n-tuples of sensed values where element $\bar{s}$ of the sequence is a vector in which the ith element represents the input to process i. O^* produces a sequence of output vectors. Each element $\bar{o}$ of the output sequence is a vector in which the ith element represents the output from process i.

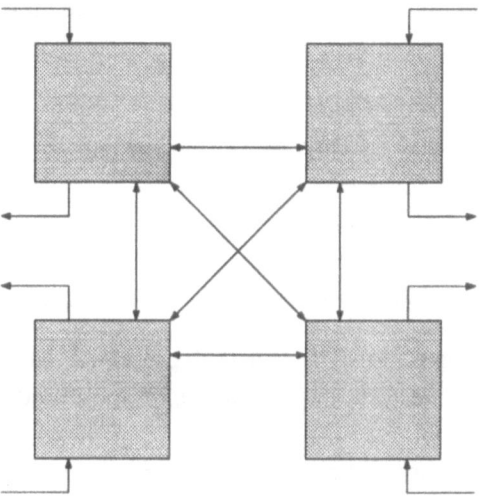

Figure 1: Four Redundant Processes

At each step, the trace function applies a *step function* O. The input to O is one of the input n-tuples $\bar{s}$, and its output is one of the vectors $\bar{o}$. The function O involves an application of *OML*, and of a *filter function* which computes an output value based on an interactive consistency vector. An example of such a filter function is the majority function *Maj*. O is defined by the following formula: for $j \in L = 0, \ldots, n - 1$,

$$O(\bar{s}, m)[j] \equiv \textit{filter}(\textit{OML}(L, \bar{s}, m)[j])$$

The trace function can be written as follows.

$$O^*(nil, m) \equiv nil$$

$$O^*(cons(\bar{s}, l), m) \equiv cons(O(\bar{s}, m), O^*(l, m))$$

An elementary theorem about O^* is that for an input sequence Σ,

$$O^*(\Sigma, m)[i] = O(\Sigma[i], m).$$

The Byzantine properties of *OML* are provably inherited by the trace function. In particular, we can prove that, given a sufficiently small number of faulty processes, two non-faulty processes always agree on their outputs. That is, for a trace of input n-tuples l, and for index k into that trace,

$$\neg faulty(i)$$
$$\wedge \neg faulty(j)$$
$$\wedge\ 3 \times faults(L) < n$$
$$\wedge\ faults(L) \leq m$$

$$\rightarrow$$

$$O^*(l, m)[k][i] = O^*(l, m)[k][j].$$

This conclusion follows from the fact that processes i and j have identical interactive consistency vectors, and therefore *filter* must produce the same value for both processes.

Instantiating this trace function with $n = 4$ and $m = 1$ gives us a specification for a system of four redundant processes that achieve Byzantine agreement, and which can tolerate up to one faulty process. The architecure of this system is illustrated in Figure 1.

3 The Implementation

Implementing our circuit entails describing the internal logic of each of the four processes represented by the boxes in Figure 1. These processes achieve agreement after exchanging messages. We desired a design in which the four processes behaved identically; this goal was achieved.

Each process has five inputs: a sensor value, clock, and data lines from each of the other three processes. Additionally, each process has four outputs: an actuator and data lines out to each of the other processes. These inputs and outputs are listed below. In our formal description of the circuit the widths of these data paths are not fixed. This leaves the implementor free to choose a data width.

- sense: A sensed value.

- clock: A clock waveform.

- data-in: Inputs from the three other processes.

- data-out: Outputs to the three other processes.

- actuator: Output to some actuator.

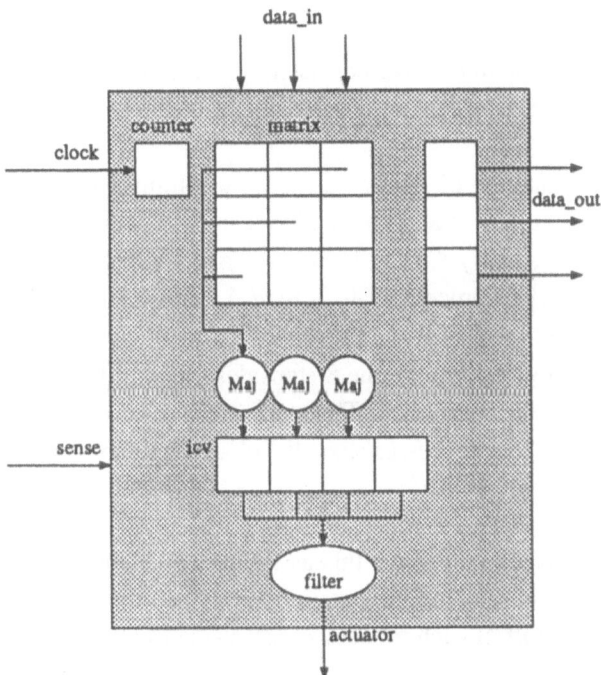

Figure 2: The Internal State of a Process

Figure 2 shows the *internal* state of a single process, along with some of the internal data paths. The internal state of a process contains the following components.

- counter: A 3-bit counter, used to cycle a process through 8 steps.

- matrix: A 3×3 matrix of data used to store values received during the information exchange.

- icv: The 1×4 interactive consistency vector for this process. icv[3] holds the process's local value, derived from the sense input.

The inter-connection of the processes to accomplish information exchange is depicted in Figure 3. Each arrow represents one-way communication. For each $i \in 0, 1, 2, 3$, and $j \in 0, 1, 2$, data-in[*j*] for process i is connected to data-out$[2 - j]$ of process $(i + j)$ *mod* 4. The interconnection scheme is designed to assure that all of the processes are identical.[2]

Each process cycles through the 8 steps described in Figure 4. The purpose of each step is described below. The steps are numbered by the value of the 3-bit counter. The four processes share the clock input and hence perform these steps synchronously.

0: Read the sensed input. Save this as the process's local value in icv[3]. Also, place this value on the output lines to the other three processes. This begins the report of each process's local value to all of the other processes.

1: Receive the local values of the other three processes, and store them in row 0 of the matrix matrix.

2,3: Fill the remaining rows of the matrix with the reports of each process's value. In steps 2 and 3 each process receives two values from each of the other three processes. At the end of step 3, the information exchange required for the four instances of $OM(1)$ is complete, as depicted in Figure 5.

---

[2]A result is that the interactive consistency vectors computed by two non-faulty processes are not actually identical, but are, in fact, rotations of one another. This implies that the filter function defined on the interactive consistency vector must be invariant under rotations of its vector argument.

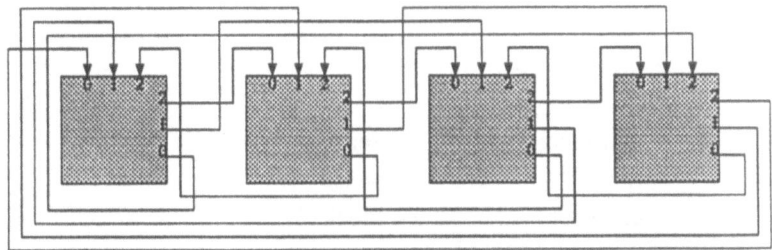

Figure 3: Process Interconnections

4: Compute the interactive consistency vector. This is accomplished by computing the majority of the three reported values for each of the other processes. (The circle labeled "Maj" in Figure 2 represents a 3-input majority circuit.)

5: Compute the actuator output based on the value of the interactive consistency vector. This is represented by a call to a function *filter*. In our specification *filter* is not defined, but is constrained to be invariant under rotation of its argument

6,7: No state change other than incrementing the counter.

The behavior of this circuit can be summarized as follows. Each process senses its input simultaneously, goes through three steps of information exchange, determines an interactive consistency vector, and then produces an actuator value 5 steps after the input was sensed. The actuator value is fixed until a new actuator value is computed on the next cycle. The clock has eight phases only to simplify its implementation as a three-bit counter

It may appear that our circuit design is still a bit removed from the gate level. However, our lowest level design was hand translated in a very straightforward fashion by L.M. Smith of Computational Logic, Inc. (CLI) into the PALASM formalism and implemented using Programmable Array Logic (PAL's) with approximately one day's effort. This translation was done mainly from the algorithm description in Figure 4 without any reference to the more abstract descriptions of the algorithm. The result was a physical circuit implementing the Byzantine General's algorithm for 1-bit wide data. A useful and (we believe) simple step would be to formalize the PAL and show the equivalence of our circuit design with the running implementation.

Case Counter: $i \in 0, 1, 2$

0:	data-out [i]	← sense
	icv [3]	← sense
	clock	← clock+1
1:	matrix [0,i]	← input [i]
	data-out [0]	← input [1]
	data-out [1]	← input [0]
	data-out [2]	← input [0]
	clock	← clock+1
2:	matrix [1,i]	← input [i]
	data-out [0]	← matrix [0,2]
	data-out [1]	← matrix [0,2]
	data-out [2]	← matrix [0,1]
	clock	← clock+1
3:	matrix [2,i]	← input [i]
	clock	← clock+1
4:	icv [0]	← maj (matrix [0,0], matrix [1,2], matrix [2,1])
	icv [1]	← maj (matrix [0,1], matrix [1,0], matrix [2,2])
	icv [2]	← maj (matrix [0,2], matrix [1,1], matrix [2,0])
	clock	← clock+1
5:	Actuator	← filter(icv)
	clock	← clock+1
6:	clock	← clock+1
7:	clock	← clock+1

Figure 4: Process Steps

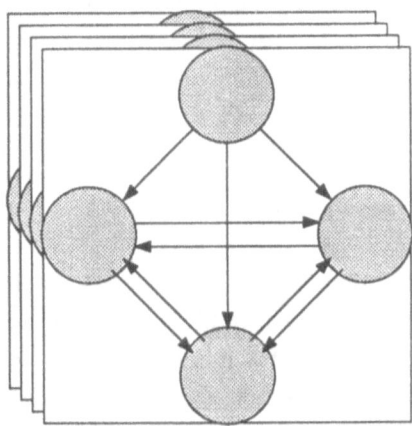

Figure 5: Rounds of Information Exchange for *OM(1)*

More recently, J. Moore at CLI has produced from our design a mechanically verified circuit description for an 8-bit wide version of the processor. Moore's work is based on a formalized hardware description language developed at CLI by Warren Hunt and Bishop Brock. Their formalism can be trivially transformed into LSI Logic Corporation's Network Description Language (NDL) for fabrication. Furthermore, properties of the described circuits (including input/output behavior) can be proved with the Boyer-Moore theorem prover. Using that formalism and working from this paper, Moore developed a circuit description generator that is parameterized by the width of the data paths. In his design the entire interactive consistency vector is output at step 5 since our filter function is unspecified. He used the Boyer-Moore theorem prover to prove that his generated circuits implement our design (modulo the interpretation of the output vector) for arbitrarily wide data. The generator was then used to produce a formal netlist describing an 8-bit wide implementation and that netlist was mechanically translated to NDL. The NDL description was analyzed without comment by the circuit analysis tools of LSI Logic Corporation. The resulting 8-bit processor could be built on an LSI LMA914C chip. The processor would require 738 cells (3254 gates) and 93 i/o signal pins. The processor would use only 23large chip because of the large number of pins (due the parallel i/o inherent in our design). Instantiating the design for larger data paths would produce logically correct but economically absurd designs due to the pin-out problem. Nevertheless, Moore's work supports our claim that our design does indeed supply the implementor with enough information to build the processor. It also serves to remind the reader that certain simplifying assumptions were made and that our work represents only a step toward the goal of mechanically verified fault-tolerance of devices of practical interest.

4   The Proof of Correctness

We want to be able to assert about our circuit design that the actuator values for all non-faulty processes agree, even in the presence of a single faulty process. How can we convince ourselves that this is true? Our design is fairly simple, but has some tricky details where mistakes can easily be made.

To convince ourselves of this assertion we have proved the design correct, and mechanically checked the proof with the Boyer-Moore theorem prover [4, 5]. The specification and proof of correctness consists of the following elements.

- **Correctness of the function** *OM*. The function *OM* is defined in the Boyer-Moore logic, and is proved to satisfy interactive consistency conditions.

- **The Circuit Specification Function.** The trace function O^* described earlier is defined in the logic. This function includes a call to *OM* to perform the information exchange. Because *OM* achieves interactive consistency, it is possible to prove that at any point in the trace all non-faulty processes agree if there are a sufficient number of non-faulty processes. The instance of this trace function with $n = 4$ and $m = 1$ serves as a specification function for our circuit design.

- **The Circuit Implementation Function.** The design of the circuit is formalized in the Boyer-Moore logic. A function named *local-step* is defined to formally express the state changes to a single process as described in Figure 4. A function *global-step* applies *local-step* on each clock tick to each of the four processes. *global-step* also formally describes the data flow among processes. A trace function C^* uses *global-step* as its step function. C^* is proved to "correspond" to the function O^* (defined in earlier). Describing the sense in which these functions correspond is the purpose of the remainder of this section.

The trace function C^* is defined as follows. Let $\overline{st}$ be a vector of four process states, and let G^* be the trace of these states. Each $G^*[k]$ contains a

4-tuple with the state of each process after k applications of *global-step*.

$$G^*(nil, \overline{st}) \equiv nil$$
$$G^*(cons(\overline{s}, l), \overline{st}) \equiv cons(\overline{st}, G^*(l, \textit{global-step}(\overline{s}, \overline{st})))$$

The trace function C^* is defined in terms of G^* by projecting out of the state of each process the value of its actuator at each step of the trace. The relationship between G^* and C^* is depicted in Figure 6. For $i \in 0, 1, 2, 3$ and for k an index into the trace, $C^*(l, \overline{st})[k][i] = actuator(G^*(l, \overline{st})[k])[i]$.

The time granularity of C^* is greater than that of O^*. It takes C^* five clock ticks to compute actuator values in response to a set of sense inputs. The intermediate steps of the trace are not of interest in the statement of interactive consistency. It is useful to define the notion of trace *selection*. The $n -$ *selection* of a trace t is the trace consisting of successive $(n - 1)^{st}$ elements of t. [3]

The proof of correctness of the circuit design requires the proof that some selection on C^* equals the trace O^*. We have chosen $n = 7$ as the selector value in our proof. The following theorem formally relates the behavior of the circuit design to the specification function. Figure 6 depicts this relationship proved between C^* and O^*; namely,

$$Select(7, C^*(l, \overline{st}))) = O^*(l, 1).$$

Recall that any two non-faulty processes agree on their outputs. That is, for a trace of input n-tuples l and for index k into that trace,

$$\neg faulty(i)$$
$$\wedge \neg faulty(j)$$
$$\wedge \, 3 \times faults(L) < n$$
$$\wedge \, faults(L) \leq m$$
$$\longrightarrow$$
$$O^*(l, m)[k][i] = O^*(l, m)[k][j].$$

---

[3]The function $NthCdr(t, n)$ returns the list of length $|l| \subseteq n$, where $NthCdr(l, n)[i] = l[i + n]$.

$Select(n, t) \equiv if \; |t| \geq n \; then \; cons(t[n - 1], Select(n, NthCdr(t, n))) \; else \; nil.$

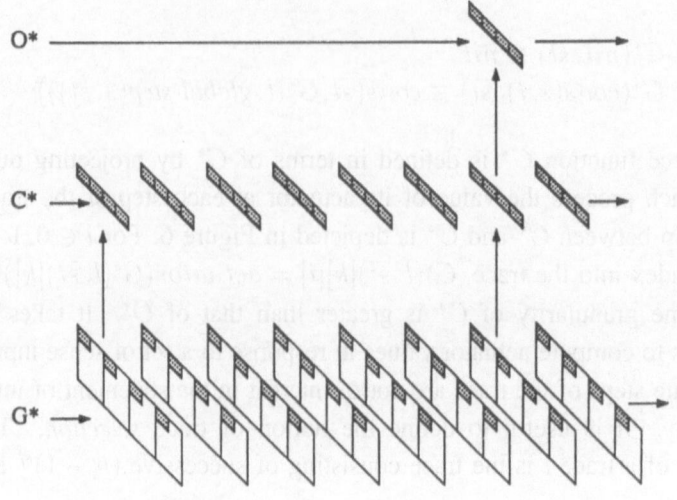

Figure 6: Correspondence among Trace Functions

Substituting $Select(7, C^*(l, \overline{st}))$ into this lemma, with $n = 4$, $m = 1$ and $L = \{0, 1, 2, 3\}$ gives a theorem which says that the circuit design, as defined by C^*, achieves agreement every 7th "tick" of the clock.

$$
\begin{aligned}
&\neg faulty(i) \\
\wedge\ &\neg faulty(j) \\
\wedge\ &3 \times faults(L) < n \\
\wedge\ &faults(L) \leq m \\
\rightarrow\ & \\
&Select(7, C^*(l, \overline{st}))[k][i] \\
=\ &Select(7, C^*(l, \overline{st})))[k][j].
\end{aligned}
$$

We take the proof of this theorem as a satisfactory formal demonstration of the correctness of the circuit design.

5 Conclusions

Much work has been done recently on verification of hardware and specific algorithms in hardware; see, for example, the papers in [6, 7, 8]. The work most similar to ours is a verification by Mark Bickford and Mandyam Srivas of Odyssey Research Associates of an implementation of the Oral Messages algorithm on four communicating Mini-Cayuga microprocessors[9]. Bickford and Srivas assumed the correctness of the algorithm and concentrated on the implementation.

We have verified a low-level hardware implementation of the Oral Messages algorithm of Pease, Shostak, and Lamport using a high-level abstract implementation as its specification. Because this abstract implementation has been formally proven to achieve interactive consistency, we are assured that our low-level implementation is fault-tolerant as well.

The main achievement of this work is the demonstration of a fault-tolerant device that can be formally specified, and whose implementation can be proved correct. We have shown how to formally relate an abstract algorithm like *OM* to a design which is implementable in hardware.

The main limitation of our device specification is that it does not explicitly account for distributed processes. Processes are described as operating synchronously. This simplifies the problem dramatically. Addressing this limitation is a future goal of our work.

All of the proofs, from the proof of correctness of the general Oral Messages Algorithm to the proof of the hardware implementation were fully machine checked. Proponents of the view that such fully formal and machine checked proofs do not contribute materially to mathematics or engineering may feel that our effort was superfluous.

From a mathematical perspective, we believe that two important goals of proof are to increase one's understanding and intuition about the content and significance of a theorem, and to provide a convincing argument that it is, in fact, valid. Our proof efforts led us to develop a very clean and unambiguous statement of the algorithm and its correctness properties. We believe that we understand this quite subtle algorithm and the reason it works much better for the effort. Moreover, our success in convincing a congenitally skeptical mechanical proof checker of the validity of this theorem practically guarantees that we have eliminated any errors which the much touted "social process" might overlook. Such confidence is particularly comforting in domains such as fault-tolerant and real-time computing where a well-developed intuition is difficult to cultivate; the theorem prover is not subject to being misled by the urgings of a misguided or ill-informed intuition.

From an engineering perspective, we feel that our approach has several benefits. By proving properties such as the interactive consistency conditions with respect to our high-level abstract implementation, we retain the clarity and abstractness of the published algorithm and benefit from the intuitions derived from the published proof. By then mapping down to a more concrete characterization, but one which provably retains the fault-tolerant characteristics of the abstract version, we are able to derive a hardware level characterization of the algorithm which is trivial to implement. We suspect that an attempt to implement the Oral Messages algorithm directly from the published abstract presentation would be extremely error-prone.

References

[1] W. Bevier and W. Young, "Machine checked proofs of a Byzantine agreement algorithm," Tech. Rep. 62, Computational Logic, Inc., June 1990.

[2] M. Pease, R. Shostak, and L. Lamport, "Reaching agreement in the presence of faults," *JACM*, vol. 27, pp. 228–234, April 1980.

[3] L. Lamport, R. Shostak, and M. Pease, "The Byzantine generals problem," *ACM TOPLAS*, vol. 4, pp. 382–401, July 1982.

[4] R. Boyer and J. Moore, *A Computational Logic*. New York: Academic Press, 1979.

[5] R. Boyer and J. Moore, *A Computational Logic Handbook*. Boston: Academic Press, 1988.

[6] L. Claesen, Ed., *Formal VLSI Specification and Synthesis*. Amsterdam: North-Holland, 1990.

[7] M. Lesser and G. Brown, Eds., *Hardware Specification, Verification and Synthesis: Mathematical Aspects*, vol. 408 of *Lecture Notes in Computer Science*. Berlin: Springer-Verlag, 1989.

[8] J. Staunstrup, Ed., *Formal Methods for VLSI Design*. Amsterdam: North-Holland, 1990.

[9] M. Bickford and M. Srivas, "Formal verification of a fault-tolerant microprocessor system design," contractor report, NASA, February 1991.

Verification of Memory Management Units*

E. THOMAS SCHUBERT, KARL N. LEVITT

Division of Computer Science

University of California at Davis

Davis, CA 95616, USA

Abstract

This paper describes the formal verification of two memory management units using the HOL theorem prover. The verification effort demonstrates the use of hierarchical decomposition and abstract theories. Both devices authorize memory requests and translate virtual addresses to real addresses. The first unit was designed and verified to the gate level. The second memory management unit is implemented with an abstract representation and provides some operating system support. Memory requests are validated based on a memory resident segment table. These units are being used as a basis for the verification of a composed chip set to form a trusted computing base.

Key Words: Hardware verification, abstraction, memory management unit, HOL.

1 Introduction

Life-critical systems are becoming increasingly dependent on computer systems. Though redundant components in fault-tolerant systems increase reliability, these systems do not exclude errors due to specification or implementation flaws. Building reliable systems out of unreliable components does not guarantee a safe and secure system. Faults resulting from design errors are especially difficult to protect against and can compromise critical functionality [1]. Simulation is often used to search for the presence of errors, but it cannot guarantee the absence of errors. Formal verification has attracted significant interest as

---

*This work was sponsored under Boeing Contract NAS1-18586, Task Assignment No. 3, with NASA-Langley Research Center

an alternative to simulation [2],[3],[4],[5]). Hunt suggests that it is faster to verify a microprocessor design than to exhaustively test one [6]. Hardware verification cannot reveal all specification flaws or errors in implementing the design, but will uncover all design errors.

Hardware verification requires that a system design is formally shown to satisfy its specification through a mathematical proof. Using theorem proving techniques, an expression describing the behavior of a device is proven to be equivalent in some sense to an expression describing the design of the device. Thus, the behavioral semantics are clearly defined and provide an accurate basis for building systems [7].

Research in the verification of computer systems has largely ignored the composition of chip sets which make up a computer [8]. The Computer Systems Verification Group at the University of California, Davis is verifying a chip set which will comprise a realistic hardware base for a secure operating system. The initial phase of our efforts is to verify a set of hardware devices including a central processing unit [9], a floating point coprocessor [10], an interrupt controller, a direct memory access controller and a memory management unit. Future work will compose these devices to form a complete system.

The heart of this system is AVM-1, to our knowledge, one of the largest microprocessor yet verified [11]. AVM-1 supports interrupts as well a supervisor mode. The verified MMU described here provides a mechanism to ensure process isolation as well as virtual memory support.

To demonstrate the usefulness of hierarchical decomposition and abstract representation, we verified a number of simple memory management units that form a complexity hierarchy [12]. By developing a sophisticated MMU in steps, the construction of the final proof is more tractable. The two most complex memory management verification efforts are described here. Both devices authorize memory requests and translate virtual addresses to real addresses. The first unit was designed and verified to the gate level. The second memory management unit demonstrates the use of abstract theories and provides greater operating system support. Memory requests are validated based on a memory resident segment table.

1.1 Related Work

Neumann proposes a unified hierarchy that accomodates all critical requirements [13]. Responsibility to satisfy each requirement can then be delegated to an appropriate layer of the design. The layers remain interdependent; the more abstract layers rely on the correctness of the lower levels. Formal proofs about the hardware level discharge some of the assumptions made by higher, software levels. Similarly, hardware level proofs often make assumptions about the behavior of the software which are discharged when the levels are composed [14].

Hardware verification efforts thus far have focused primarily on a microprocessor as the base for computer systems [15], [16], [6], [17]. Perhaps the best known verification effort is that of the VIPER microprocessor ([18], [15]). VIPER is the first microprocessor intended for commercial distribution where a formal verification has been attempted. These processors are however, quite limited. Only Joyce's microprocessor Tamarack-3, provides interrupts and none provide memory management functions necessary to support a secure operating system.

Previous efforts to verify systems have attempted to construct vertically verified systems with a microprocessor/memory as the system's base. Joyce has specified and verified a compiler for the verified Tamarack-3 microprocessor [19].

Computational Logic Inc. has verified a "stack" of interpreters where the implementation of a level is the specification of the next lower level [7]. In this way, higher levels of the stack define new functionality by collecting the next lower level's functionality. The "stack" consists of a compiler (Micro-Gypsy), an assembler and linking loader, and a microprocessor.

Bevier has verified a simple operating system (KIT) which ensures that tasks are isolated from one another. An implementation of the hardware base has not been verified [8] [20]. He assumes extensions to the FM8502 microprocessor to provide interrupts, asynchronous I/O, memory management, and supervisor-mode instructions.

1.2 HOL

The specification and verification were done using the HOL theorem proving system [21]. The object language of HOL is a formulation of higher-order logic. Universally quantified variables are used to specify input and output device lines while internal device lines are existentially quantified. Conditional expressions have the form:

    a →b | c

which means "if a then b else c."

The following HOL expression defines an and gate implementation using an inverter and a nand gate. The existentially quantified variable p, represents an internal line which links the output of the nand gate with the input of the inverter.

$$\vdash_{def} \quad \text{and2\_imp a b out} = \exists p.\text{nand2 a b p} \ \wedge \ \text{inv p out}$$

HOL provides the human verifier with a selection of tactics for use in goal-directed proofs. The tactics are very similar to the kinds of steps a human theorem prover would take in solving a goal. New tactics can be written that allow the theorem prover to be extended and customized for a particular task.

New theorems can only be created in a controlled manner. All proofs can be reduced to one containing only the 8 primitive inference rules and 5 primitive axioms. High-level inference rules and tactics derived from some combination of primitive inference rules.

In the text, various fonts will be used to denote constants, definition names and *object types*. The turnstile symbol ⊢, is used to indicate the term is a theorem which has been formally proven in the logic. When the subscript "def" is present (e.g. $\vdash_{def}$), the theorem is simply a definition.

2   Virtual Address MMU

The memory management unit is programmed through two memory mapped control registers:

1. A protection register governs the range of valid virtual memory addresses a process may access.

2. A translate address register designates the base real address accessible in memory.

Processes cannot be trusted on their own to leave the unit's registers un-modified. Only when the supervisor line is high will the unit permit a register write. This ensures that the security protection scheme intended by the operating system kernel cannot be bypassed intentionally or unintentionally by user processes. This scheme assumes that non-kernel tasks execute in user state. The supervisor bit can be extended into a process identifier field or a security ring field.

The protection register and virtual addresses are partitioned into a segment and an offset[1]. A request is validated if the segment address matches the stored

---

[1]Here a page is a contiguous block of memory words; each block being a fixed length. Segments are blocks of words but all segments need not be of the same length.

segment component and the offset is less than or equal to the stored bounds component. When a request is validated, the MMU constructs a real address using the offset of the requested address and the translate address register. When the supervisor line is asserted, all accesses are authorized and address translation is not performed.

2.1 Specification

The specification (virtBBCk_spec) is defined as a state transition system. The state and output environment at time *t+1* is a function of the state and input environment at time *t*. The state is maintained in variables (bbReg, vaReg). The input environment consists of the address bus value, data bus value, and control bus signals (addr, data, super, rw). The output environment consists of a request validation line and a real address (ack, outAddr). The functions vSUPERV and vCOMP define the supervisor and user mode behaviors respectively. The parameters n, s and ADDR serve as constants defining the most significant bitVector bit, the most significant address offset bit and the base address of the MMU registers. The size of the bitVectors must be greater than the segment offset for the specification to be meaningful. When a register write operation occurs, the least significant bit (addr 0) determines which register is updated.

The function definition VtoR, creates a real address by replacing the segment identifier with the real base offset; the bottom *s* bits of the virtual address remaining unchanged.

```
⊢_def virtBBCk_spec n s ADDR bbReg vaReg addr data super rw ack outAddr =
  (s < n) ⇒  ∀. t (bbReg(t+1),vaReg(t+1),outAddr(t+1),ack(t+1)) =
  super t →  vSUPERV n (bbReg t) (vaReg t) (addr t) (data t) ADDR (rw t)
          | vCOMP n s (bbReg t) (vaReg t) (addr t)

⊢_def vSUPERV n bbReg vaReg addr data ADDR rw =
  ((rw ∧ (bvEQUAL n (bvPART n 1 addr) (bvPART n 1 ADDR)))
  →  (addr 0)  →  (data,  vaReg,  addr,  T:bool)
            | (bbReg, data,  addr,  T:bool)
  | (bbReg, vaReg, addr, T)   )

⊢_def vCOMP n s bbReg vaReg addr =
  (bvEQUAL n (bvPART n s bbReg)(bvPART n s addr) ∧ ¬(bvGREATER s addr bbReg) )
    →  (bbReg, vaReg, (VtoR vaReg addr s), T)
    | (bbReg, vaReg, addr, F)
```

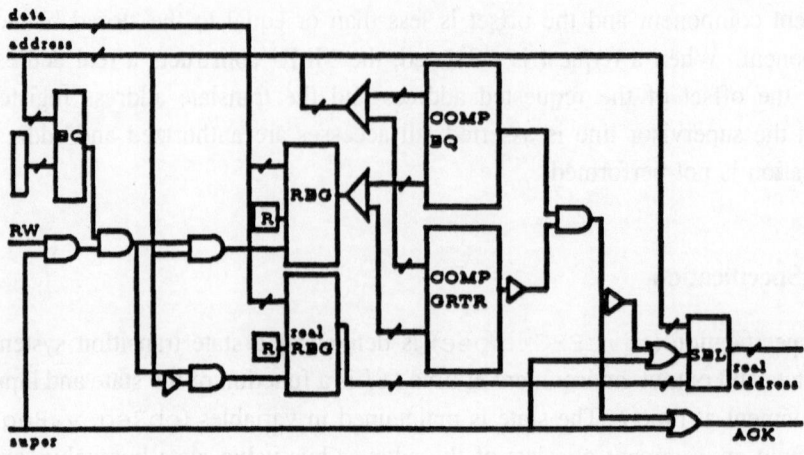

Figure 1: Base and Bounds MMU with Virtual Address Translation

2.2 Implementation

The implementation (virtBBCk_imp) is defined using primitive gates and previously verified registers and bitVector comparison units. The function pick_imp defines a bitVector MUX. The datapath can be seen in figure 1. The abstraction functions PRT and PRTA are used to split off a subsection of a bitVector.

The verification served as a learning experience and required approximately 3 man weeks. Several simple intermediate lemmas were proven with the final theorem requiring 1209 seconds of CPU time executing on a Sun SparcStation. The final proof generated 64185 primitive inferences.

A set of these units could be combined to provide sufficient hardware support for a segmented and paged memory. This design also supports multiple process requirements assuming the top bits of an address specify a process identifier.

3 Abstract MMU

The Abstract MMU extends the function of the Virtual Address MMU. The segment identifier is used to fetch the segment descriptor. Segment descriptors are located in a memory resident table and consist of two words. The first word specifies the segment size and read, write, and execute permissions. The second word acts as a base address for the segment's real location in memory.

```
⊢def pick_imp (wordA wordB :num→bool)(pick:bool) res
    = (pick=T) → (res=wordA) | (res=wordB)

⊢def virtBBCk_imp n s ADDR bbReg vaReg addr data super rw ack outAddr=
    (s < n) ⇒  ∀ t.
    (∃ wBB wVA select x y z aM goodSeg goodOfs ok nok nxlat g l e.
      (and2_imp (rw t) (super t) (x t))                              ∧
      (compEq_imp n (PRT n n 1 addr t)  (PRTA n n 1 ADDR) (y t))     ∧
      (and2_imp (y t) (x t) (z t))                                   ∧
      (inv (addr 0 t) (aM t) )                                       ∧
      (and2_imp (z t) (aM  t)    (wBB t))                            ∧
      (and2_imp (z t) (aM  t)    (wVA t))                            ∧
      (reg_imp n data wBB bitFalse bbReg)                            ∧
      (reg_imp n data wVA bitFalse vaReg)                            ∧
      (compEq_imp n (PRT n n s bbReg t) (PRT n n s addr t) goodSeg   ∧
      (comp_imp s  (ABS n addr t) (ABS n bbReg t)  g l e)            ∧
      (inv g goodOfs)                                                ∧
      (and2_imp goodOfs goodSeg ok)                                  ∧
      (or2_imp ok (super t) (ack (t+1)) )                            ∧
      (inv ok nok )                                                  ∧
      (or2_imp nok (super t) nxlat)                                  ∧
      (pick_imp (ABS n addr t) (ABS n vaReg t) nxlat (select t))     ∧
      ( (outAddr (t+1))= (VtoR (select t) (ABS n addr t) s ) )
```

```
⊢   virtBBCk_imp n s ADDR bbReg vaReg addr data super rw ack outAddr ⇒
    virtBBCk_spec n s ADDR (ABS n bbReg) (ABS n vaReg) (ABS n addr)
                    (ABS n data) super rw ack outAddr
```

To translate from a virtual address to a real address, the MMU adds the segment offset to the segment base address. To support segment paging, the first word also contains a bit indicating whether the segment is presently in memory. If this bit is F, the operating system is free to use the second word as a disk offset or in any other fashion.

The location of the table is determined by the MMU's segment table pointer register. This register is accessible only in supervisor mode. The MMU assumes the table provides an entry for all possible segment descriptors. The MMU described here must fetch a descriptor from memory for each access. A cache to speed up performance is being specified and verified separately.

We take a top-down approach to the verification of this more complicated device. The implementation level here is the electronic block level. We construct a generic theory describing an MMU where several functions are left abstract. For example, the security unit is abstract. Note that the verification of a concrete implementation of the security unit is an simple extension to the base and bounds unit verification described above.

Instead of dealing with concrete data types such as bitVectors with a specific length, the abstract MMU works with data values of abstract types *wordn*, *address* and *memory*.

Support for generic or abstract theories is not directly provided by HOL. However, a theory about abstract representations can be defined in the object language [11]. An *abstract representation* contains a set of uninterpreted constants, types, abstract operations and a set of abstract objects. The semantics of the abstract representation is unspecified. Inside the theory, we do not know what the objects and operations mean.

Using the abstract theory package, a set of selector functions [19] can be created. When applied to an abstract representation, a selector function extracts the desired function. For example, given an abstract representation *r*, a selector function *foo*, the term "*foo r*" extracts the abstract function *foo*.

Previous device theories have considered the size of the segment identifier and segment offset fields within a bitVector. The abstract representation ignores these details by providing functions which return the segment identifier or segment offset fields from an address (segId and segOfs, respectively). There is also a function segIdshf which returns the offset of a segment descriptor within the memory resident segment table for a given address. Since descriptors require two words, the implementation of this function simply shifts the segment identifier to the left one bit position (e.g. it adds a trailing zero bit).

The abstract functions selected by availBit, readBit, writeBit and execBit extract a bit value from an argument of type *wordn*. These functions are applied to the first word of a segment descriptor.

Several functions which operate on two-tuples are available. Given a pair of *wordn* values, add returns a value of *wordn*. Functions addrEq, ofsLEq and validAccess replace the concrete comparison units used in previous units.

Additional abstract coercion functions are available to convert values between types. If the theory were instantiated, the abstract types would likely be implemented with bitVectors; leaving these functions unnecessary.

Memory is also treated abstractly. The abstract representation provides a fetch function fetch.

A type abbreviation *RWE* is also defined to be a three tuple of bit values. Selector functions rBIT, wBIT and eBIT access the first, second and third bits, respectively.

```
let mmu_abs = new_abstract_representation
[
('segId',         ":(*address -> *wordn)"           );
('segOfs',        ":(*address -> *wordn)"           );
('segIdshf',      ":(*address -> *wordn)"           );
('availBit',      ":(*wordn -> bool)"               );
('readBit',       ":(*wordn -> bool)"               );
('writeBit',      ":(*wordn -> bool)"               );
('execBit',       ":(*wordn -> bool)"               );
('add',           ":(*wordn # *wordn ->*wordn)"     );
('addrEq',        ":(*address # *address -> bool)"  );
('ofsLEq',        ":(*address # *wordn   -> bool)"  );
('validAccess',   ":(*address # *wordn # RWE -> bool)" );
('val',           ":(*wordn -> num)"                );
('wordn',         ":(num-> *wordn)"                 );
('address',       ":(*wordn -> *address)"           );
('fetch',         ":(*memory # *address) -> *wordn" );
];;
```

3.1 Specification

The specification is decomposed into several rules and ignores timing details. The state and output environment of the MMU specification is a three-tuple consisting of a boolean acknowledgment, a memory address and the table pointer register value. Note the variable *r* in the definitions below is the abstract representation.

Functions `superMode` and `userMode` describe the behavior of the MMU when operating in their respective modes. `legalAccess` uses many of the abstract functions to fetch from memory the appropriate segment descriptor and compare it with the request's access parameters. `vToR` constructs a real address from a virtual address.

3.2 Implementation

The implementation is constructed from electronic block model components. These are defined as specifications for the behavior of a gate level implementation. Many of the devices specify their timing behavior as well. The building blocks consist of a security comparison unit, an address match unit, a memory fetch unit, an adder, registers, latches, muxes, and a control unit. Most of the device definitions are straight forward with the exception of the memory and the control unit. These two units will be described in greater detail. Internal and External block diagrams are presented in figures 2 and 3.

```
⊢_def legalAccess r vAddr tblPtr rwe mem =
  let a = (fetch r)(mem,(address r)((add r) (segIdshf r vAddr,tblPtr))) in
  ((validAccess r) (vAddr,a,rwe)      ∧
  (ofsLEq r) (vAddr,a))

⊢_def vToR r vAddr tblPtr mem =
  let a = (fetch r) (mem, (address r) ((add r)( (wordn r 1),
    (add r)(segIdshf r vAddr,tblPtr) ))) in
  (address r) ((add r) (segOfs r vAddr, a))

⊢_def superMode r vAddr rwe tblPtrADDR tblPtr data mem =
  ((wBIT rwe)  ∧  (addrEq r (vAddr,tblPtrADDR)))
    → ( T, vAddr, data )
    | ( T, vAddr, tblPtr )

⊢_def userMode r vAddr rwe tblPtrADDR tblPtr data mem =
  legalAccess r vAddr tblPtr rwe mem
    → ( T, (vToR r vAddr tblPtr mem), tblPtr )
    | ( F, vAddr,                     tblPtr )

⊢_def mmu_spec r vAddr rwe tblPtrADDR  tblPtr data mem superv =
  superv → superMode r vAddr rwe tblPtrADDR  tblPtr data mem
         | userMode  r vAddr rwe tblPtrADDR  tblPtr data mem
```

Memory

The memory unit specification defines a synchronous interface to memory. If the request line *req* is high at *t*, then at *t+1*, *data* will contain the requested memory value and the *done* line will be T. If there is no request at time *t*, then *done* at *t+1* will be F. To construct an asynchronous version, this specification could be modified to state that given a request at time *t*, the next time *done* is T, *data* will hold the requested value from memory for at least one time unit.

When composing the MMU with a cache, the synchronous specification will also change. If there is a cache hit, a value would be returned much sooner

```
⊢_def addUnit_spec r a b c =      ∀ t:num. c (t+1) = (add r ( (a t),(b t) ))
⊢_def muxUnit_spec r a b out w =  ∀ t:num. out (t+1) =(w(t+1))→ address r(b(t+1)) | (a t)
⊢_def mux3Unit_spec a b c out w = ∀ t:num. out t =(w t = 0)→ a t | (w t = 1)→ b t | c t
⊢_def latchUnit_spec r i out ctrl = ∀ t:num. out (t+1) = ctrl (t+1) → out t | (i (t+1))
⊢_def matchUnit_spec r a b m =    ∀ t:num. m (t+1) = ( addrEq r (a t, b t) ) → T:bool | F
⊢_def oneUnit_spec r t = (wordn r) 1
⊢_def bitFalse t = F

⊢_def secUnit_spec r a b rwe ok = ∀ t. ok (t+1) =
  ((validAccess r) ((a t),(b t),(rwe t))  ∧   (ofsLEq r) ((a t),(b t)))

⊢_def splitUnit_spec r virt id ofs = ∀ t:num.
  ((id t) = (segIdshf r) (virt t))  ∧  ((ofs t) = (segOfs r) (virt t))

⊢_def regUnit_spec r i ld clr out =
  (∀ t:num. out(t+1)=(clr t→ (wordn r 0) | ld t → i t| out t))
    ∧ (out 0 = (wordn r 0) )
```

```
⊢_def memoryUnit_spec r req addr data done mem =
    ( (data 0 = wordn r 0)  ∧  (done 0 = F) )  ∧
    ∀ t. ( (req t)
        → ( (data (t+1) = fetch r (mem t, addr t) )  ∧  (done (t+1) = T) )
        | ((data(t+1)=wordn r 0)  ∧  (done(t+1)=F) ))
```

(perhaps an order of magnitude) than if main memory were to be accessed.

The control unit and the final correctness statement do not rely on a synchronous memory unit specification. The proof could be easily modified to fit these other models.

Control Unit

To process each memory request, the control unit will pass through several clocked phases. At each clock tick the control unit may change its phase depending on the results computed by the other internal units and the MMU input from the system bus. The control unit state is maintained by the variable phase. The control unit inputs (environment) include: the request line reqIn, supervisor line super, the request type (read/write/execute) lines rwe, the address compare result line match, the security unit result line secOk, and the memory fetch result line fdone. The control unit output lines include: the MUXes which control the adder's inputs muxC, the adder output latch lC, the MUX which controls the bus memory address lines xlat, the register update lines tmpC, tblC, the memory request line rReq, the MMU done line done, and the MMU access acknowledgment line ack.

There are six distinct phases, however, not all phases are executed for each request. Which phases are executed depends on the validity of the memory request. Request evaluation begins with the control unit in phase 0 and completes when phase 0 is again reached. A valid request will require five phases with a delay of at least one time unit before each phase change. Most phases require one clock cycle, however, memory requests for a segment descriptor may take several. The control unit will busy-wait until a memory fetch completes.

The dataPath definition describes the interconnection between all the units other than the control unit. The mmu_imp definition joins the control unit with the data path. The state consists of the table pointer register value, the security data register and the control unit phase (tblPtr, secData, phase). The input environment is provided by the system bus and the memory (vAddr, vData, rwe, superv, reqIn, mem). The output environment includes a

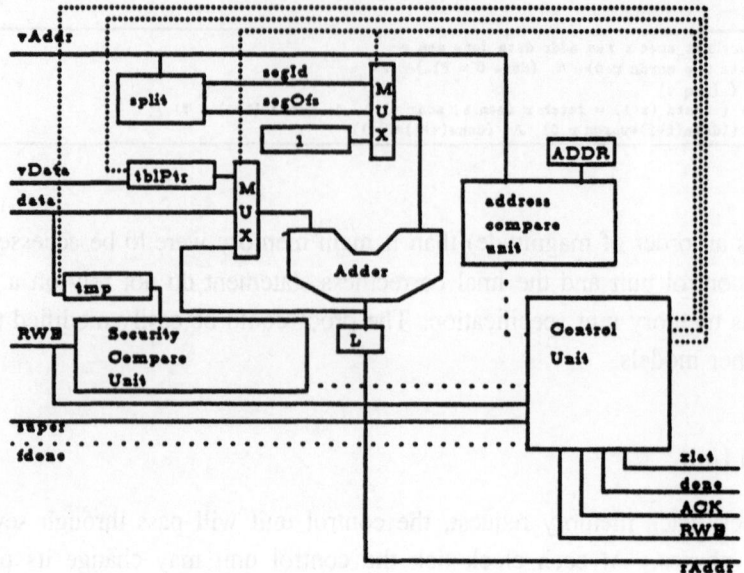

Figure 2: Abstract MMU Internal Block Diagram

real address and several control unit outputs (rAddr, done, ack, xlat). The memory address of the table pointer register is specified by the constant tblPtrADDR.

Memory Management Execution Cycle

When in phase 0, the control unit will busy-wait for a request and then proceeds to phase 1. During phase 0, the address comparison unit (matchUnit_spec) can determine whether the bus address matches the MMU's table pointer ad-

```
⊢def dataPath  r vAddr vData rwe mem tblPtrADDR tblPtr rAddr muxC tmpC
               tblC lC rReq xlat match secOK fdone =
∃ (mux1 mux2 id ofs addOut data latOut:num→*wordn)(secData:num→*wordn).
(regUnit_spec    r vData tblC bitFalse tblPtr)          ∧
(regUnit_spec    r data  tmpC bitFalse secData)         ∧
(secUnit_spec    r vAddr secData rwe secOK)             ∧
(splitUnit_spec  r vAddr id ofs)                        ∧
(mux3Unit_spec   id ofs (oneUnit_spec r) mux1 muxC)     ∧
(mux3Unit_spec   tblPtr data  latOut     mux2 muxC)     ∧
(addUnit_spec    r mux1 mux2 addOut)                    ∧
(latchUnit_spec  r addOut latOut lC)                    ∧
(matchUnit_spec  r vAddr tblPtrADDR match)              ∧
(muxUnit_spec    r vAddr latOut rAddr xlat)             ∧
(memoryUnit_spec r rReq rAddr data fdone mem)
```

⊢*def* mmu_imp r vAddr vData rwe superv tblPtr tblPtrADDR reqIn rAddr done
        ack xlat mem phase =
∃ (muxC :num→num)(tmpC tblC lC rReq match secOK fdone :num→bool) .
(controlUnit_spec reqIn superv rwe match secOK fdone muxC tmpC tblC lC rReq
        xlat done ack phase)   ∧
(dataPath r vAddr vData rwe mem tblPtrADDR tblPtr rAddr muxC tmpC tblC lC
        rReq xlat match secOK fdone)

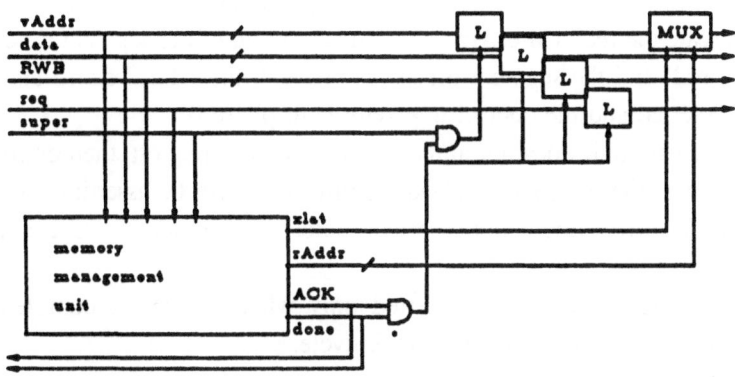

Figure 3: Abstract MMU External Block Diagram

dress. The split unit `splitUnit_spec` divides the address into its segment table offset and segment offset components.

In phase 1 if the supervisor line is high, two results are possible. When the request is a write and the `match` line is T, the control unit will direct the table pointer register to store the value on the data bus and set the next phase to 5. After one clock tick in phase 5, the acknowledge and done lines are asserted and the control unit returns to phase 0. This ensures the data bus value will remain constant while the register updates its store. If the request is not directed to the segment table pointer register, the done and acknowledge lines are asserted and the phase is set to 0. The `xlat` line remains F, the original request is passed on to memory without modification.

During this time, the adder will compute the memory address of the segment descriptor using the shifted segment identifier and the segment table pointer (MUX outputs). When the supervisor line is not high and the control unit is in phase 1, a memory fetch will be initiated using the adder output. The adder output latch control line is asserted to keep this value constant. The temporary register write control line (`tmpC`) will be asserted to capture the first word of the fetched segment descriptor. The control unit will move on to phase 2.

The control unit will remain in phase 2 until the `fdone` line is asserted indicating the memory fetch has completed. During this time, the adder will have incremented the address so that the second word of the segment descriptor can be fetched. The control unit will then move on to phase 3.

The control unit will also remain in phase 3 until the `fdone` line is asserted indicating the memory fetch has completed. If the security unit has asserted the `secOK` line, phase 4 is entered. The delay provides sufficient time for the adder to create the real address from the second word of the segment descriptor (fetched word) and the segment offset. In phase 4, the `xlat`, `done` and `ack` lines are asserted and the control unit returns to phase 0.

If the security unit does not authorize the memory request, the control unit does not enter phase 4, but, instead, returns to phase 0, asserting only the `done` line. The `ack` line is not asserted, indicating the memory request is not authorized.

Note that the `done` line is asserted only when the MMU completes its execution cycle—and only for one clock cycle.

```
⊢def controlUnit_spec reqIn super rwe match secOK fdone
                muxC tmpC tblC lC rReq xlat done ack phase =
    ((muxC 0,tmpC 0,tblC 0,lC 0,rReq 0,xlat 0,done 0,ack 0, phase 0) = (0, F,F,F, F,F,F,F, 0) )
 ∧
    (∀ t .(muxC(t+1),tmpC(t+1),tblC(t+1),lC(t+1),rReq(t+1),xlat(t+1),done(t+1),
       ack(t+1),phase(t+1) ) =                     %  M  t t l  r x d a  P  %
                                                    %  U  m b a  e l o c  H  %
                                                    %  X  p l t  q t n k  A  %
       (phase t = 0) → (reqIn t →                   ( 0, F,F,F,  F,F,F,F, 1)  |
                                                    ( 0, F,F,F,  F,F,F,F, 0))  |
       (phase t = 1) → (super t →
            ((wBIT (rwe t)) ∧ match t) →            ( 0, F,T,F,  F,F,F,F, 5)  |
                                                    ( 0, F,F,F,  F,F,T,T ,0)  |
                                                    ( 2, T,F,T,  T,T,F,F, 2))  |
       ((phase t = 2) ∧ fdone t) →                  ( 1, F,F,F,  T,T,F,F, 3)  |
       ((phase t = 3) ∧ fdone t) → (secOK t →       ( 0, F,F,F,  F,T,F,F, 4)  |
                                                    ( 0, F,F,F,  F,F,T,F, 0))  |
       (phase t = 4) →                              ( 0, F,F,T,  F,T,T,T, 0)  |
       (phase t = 5) →                              ( 0, F,F,F,  F,F,T,T ,0)  |
       (muxC t,tmpC t,tblC t,lC t, F ,xlat t,done t,ack t,phase t))
```

3.3 Verification

Several auxiliary definitions are used to express the final correctness statement. To relate the implementation to the specification, a temporal abstraction is constructed using the two predicates `Next` and `First`[17]. The predicate `First` is true when its argument *t* is the first time that *g* is true. The predicate

Next is true when *t2* is the next time after *t1* that *g* is true. The predicate stable_sigs states that between *t1* and *t2* the MMU inputs will remain constant.

```
⊢_def First g t = (∀ p:time. p<t ⇒ ¬(g p)) ∧ (g t)
⊢_def Next g (t1,t2) = (t1<t2) ∧ (∀ t:time . t1<t ∧ t<t2 ⇒ ¬ (g t)) ∧ (g t2)
⊢_def stable_sigs t1 t2 vAddr rwe tblPtrADDR data mem super = ∀ t'. t1 < t' ∧ t' < t2 ⇒
        (super t' = super t1) ∧ (vAddr t' = vAddr t1) ∧ (rwe t' = rwe t1) ∧
        (data t' = data t1) ∧ (mem t' = mem t1) ∧ (tblPtrADDR t' = tblPtrADDR t1)
```

The correctness theorem states that if the implementation is in phase 0 and a memory request is made, the implementation will eventually respond (*c* time steps later), when the state of the implementation matches the state defined by the specification for a set of given MMU inputs. The inputs must remain stable until the MMU responds to a request. If a memory request is not made, the acknowledgment line remains F, the phase remains 0 and the MMU table pointer register remains unchanged.

```
⊢ mmu_imp r vAddr vData rwe super tblPtr tblPtrADDR reqIn rAddr done ack xlat mem phase ⇒
(∀ t. (phase t = 0) ⇒
  (reqIn t →
   (∃ c. Next done(t,t + c) ∧ (phase(t + c)=0) ∧
       (stable_sigs t (t + c) vAddr rwe tblPtrADDR vData mem super ⇒
       (mmu_spec r (vAddr t) (rwe t) (tblPtrADDR t) (tblPtr t) (vData t) (mem t) (super t)
       = ack(t + c),rAddr(t + c),tblPtr(t + c))))
   | ( (ack(t + 1) = F) ∧
       (phase(t + 1) = 0) ∧
       (tblPtr(t + 1) = tblPtr t) ) ))
```

In the process of proving the correctness result several control unit lemmas were useful:

- Each phase was shown to be distinct.

- The control unit phase state can be only one of six possible values.

- Phase 0 can never follow phase 2.

- During phase 0 the state of the MMU does not change

The verification of this device required six man weeks. The correctness theorem required 2635.2 seconds of CPU time running on a SPARCStation with 16 Mbytes of memory. HOL generated 121858 primitive inferences to prove the theorem.

4 Conclusion

Abstract theories provide a mechanism to ignore many details that can be handled at lower levels of a design. For example, the abstract memory management unit focuses attention on the correctness of the control unit. Using the abstract theory package, abstract devices can be instantiated with verified gate level implementations of the abstracted functions.

The abstraction mechanism also permits design changes without the need for a complete reverification effort. The correctness theorem for the abstract MMU is not dependent on the layout of the segment protection descriptor or on the specific protection requirements.

The basis for a secure hardware platform is a fully functional memory management unit. The memory management unit presented here serves as a model to verify a more sophisticated device, such as the hardware reference monitor SIDEARM [22].

The memory management units verified provide sufficient hardware support for an operating system kernel to ensure process isolation and virtual memory. The device designs can be simplified to define a paging unit. Future work will investigate the composition of segmentation and paging units.

A register stack that implements a FIFO replacement strategy has also been verified. This is being enhanced to construct a MMU cache with either an LRU or LFU replacement strategy. Future work will investigate composing the memory management unit with the CPU and other chips to form a complete hardware base.

References

[1] V. P. Nelson, "Fault-Tolerant Computing: Fundamental Concepts," *Computer*, July 1990.

[2] H. G. Barrow, "VERIFY: A Program for Proving Correctness of Digital Hardware Designs," *Artificial Intelligence*, vol. 24, 1984.

[3] J. J. Joyce, "Formal Specification and Verification of Microprocessor Systems," *Microprocessing and Microprogramming, North-Holland*, vol. 24, 1988.

[4] G. Milne and P. Subrahmanyam, *Formal Aspect of VLSI Design*. Publishers B.V., 1986.

[5] D. Weise, "Functional Verification of MOS Circuits," *24th ACM/IEEE Design Automation Conference*, 1987.

[6] W. A. Hunt, "Microprocessor Design Verification," *Journal of Automated Reasoning*, vol. 5, 1989.

[7] W. R. Bevier, W. A. Hunt, and W. D. Young, "Toward Verified Execution Environments," *IEEE Symposium on Security and Privacy*, 1987.

[8] W. R. Bevier, "Kit and the Short Stack," *Journal of Automated Reasoning*, vol. 5, 1989.

[9] P. J. Windley, "A Hierarchical Methodology for Verifying Micropro-grammed Microprocessors," *IEEE Symposium on Research in Security and Privacy*, 1990.

[10] J. Pan and K. N. Levitt, "A Formal Specification of the IEEE Floating-Point Standard with Application to the Verification of Floating-Point Co-processors and Numerical Arithmetic," *Proceedings of 24th Asilomar Conference on Signals, Systems and Computers*, 1990.

[11] P. J. Windley, *The Formal Verification of Generic Interpreters*. PhD thesis, University of California, Davis, 1990.

[12] E. T. Schubert, "Verification of Memory Management Units using HOL," technical report CSE-90-27, University of California, Davis, August 1990.

[13] P. G. Neumann, "On Hierarchical Design of Computer Systems for Critical Applications," *IEEE Transaction on Software Engineering*, vol. SE-12, No. 9, September 1986.

[14] J. D. Guttman and H. Ko, "Verifying A Hardware Security Architecture," *IEEE Symposium on Research in Security and Privacy*, 1990.

[15] A. Cohn, "A Proof of Correctness of the VIPER Microprocessor: the Second Level," in *VLSI Specification,Verification, and Synthesis*, (G. Birtwhistle and P. Subrahmanyam, eds.), Springer-Verlag, 1989.

[16] W. A. Hunt, "A Verified Microprocessor," technical report 47, The University of Texas at Austin, Dec. 1985.

[17] J. J. Joyce, *Multi–Level Verification of Microprocessor–Based Systems*. PhD thesis, Cambridge University, December 1989.

[18] W. J. Cullyer, "Implementing Safety Critical Systems: The VIPER Microprocessor," in *VLSI Specification,Verification, and Synthesis*, (G. Birtwhistle and P. Subrahmanyam, eds.), Kluwer Academic Press, 1988.

[19] J. J. Joyce, "Totally Verified Systems: Linking Verified Software to Verified Hardware," *Hardware Specification, Verification and Synthesis: Mathematical Aspects*, July 1989.

[20] W. R. Bevier, "A Verified Operating System Kernel," technical report 11, Computational Logic, Inc., October 1989.

[21] M. Gordon, "HOL: A Proof Generating System for Higher-Order Logic," in *VLSI Specification, Verification, and Synthesis*, (G. Birtwhistle and P. Subrahmanyam, eds.), Kluwer Academic Press, 1988.

[22] W. Boebert, "The LOCK Demonstration," *11th National Computer Security Conference*, 1988.

High Level Design Proof of a Reliable Computing Platform

BEN L. DIVITO

Vigyan, Inc., 30 Research Drive
Hampton, VA 23666-1325, USA

RICKY W. BUTLER, JAMES L. CALDWELL

NASA Langley Research Center
Hampton, VA 23665-5225, USA

Abstract

An architecture for fault-tolerant computing is formalized and shown to satisfy a key correctness property. The *reliable computing platform* uses replicated processors and majority voting to achieve fault tolerance. Under the assumption of a majority of processors working in each frame, we show that the replicated system computes the same results as a single processor system not subject to failures. Sufficient conditions are obtained to establish that the replicated system recovers from transient faults within a bounded amount of time. Three different voting schemes are examined and proved to satisfy the bounded recovery time conditions.

Key Words: Fault tolerance, formal methods, correctness proofs, majority voting, modular redundancy.

1 Introduction

NASA has initiated a major research effort towards the development of a practical validation and verification methodology for digital fly-by-wire control systems. Researchers at NASA Langley Research Center (LaRC) are exploring formal verification as a candidate technology for the elimination of design errors in such systems. In a detailed technical report [1], we put forward a high level architecture for a *reliable computing platform* (RCP) based on fault-tolerant computing principles. This paper presents initial results of applying formal methods to the verification of a fault-tolerant operating system that schedules and executes the application tasks of a digital flight control system.

The major goal of this work is to produce a verified real-time computing platform, both hardware and operating system software, which is useful for a wide variety of control-system applications. Toward this goal, the operating system provides a user interface that "hides" the implementation details of the system such as the redundant processors, voting, clock synchronization, etc. We describe an abstract model of the architecture, a first level decomposition of the model towards a physical realization, and a proof sketch that the decomposition is an implementation of the model.

2   Design of the Reliable Computing Platform

Traditionally, the operating system function in flight control systems has been implemented as an executive (or main program) that invokes subroutines implementing the application tasks. For ultra-reliable systems, the additional responsibility of providing fault tolerance makes this approach untenable. We propose an operating system that provides the applications software developer a reliable mechanism for dispatching periodic tasks on a fault-tolerant computing base that *appears* to him as a single ultra-reliable processor.

Our system design objective is to minimize the amount of experimental testing required and maximize our ability to reason mathematically about correctness. The following design decisions have been made toward that end:

- the system is non-reconfigurable
- the system is frame-synchronous
- the scheduling is static, non-preemptive
- internal voting is used to recover the state of a processor affected by a transient fault

A four-level hierarchical decomposition of the reliable computing platform is shown in figure 1.

The top level of the hierarchy describes the operating system as a function that sequentially invokes application tasks. This view of the operating system will be referred to as the *uniprocessor model*, which is formalized as a state transition system in section 5 and forms the basis of the specification for the RCP.

Fault tolerance is achieved by voting results computed by the replicated processors operating on the same inputs. Interactive consistency checks on sensor inputs and voting actuator outputs requires synchronization of the replicated processors. The second level in the hierarchy describes the operating system as a synchronous system where each replicated processor executes the

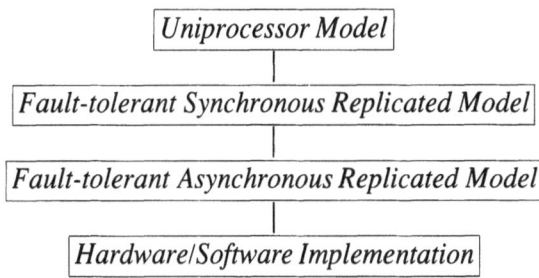

Figure 1: Hierarchical Specification of the Reliable Computing Platform.

same application tasks. The existence of a global time base, an interactive consistency mechanism and a reliable voting mechanism are assumed at this level. The formal details of the model, specified as a state transition system, are described in section 6.

At the third level, the assumptions of the synchronous model must be discharged. Rushby and von Henke [2] report on the formal verification of Lamport and Melliar-Smith's [3] interactive-convergence clock synchronization algorithm. This algorithm can serve as a foundation for the implementation of the replicated system as a collection of asynchronously operating processors. Elaboration of the asynchronous layer design will be carried out in Phase 2 of our research effort.

Final realization of the reliable computing platform is the subject of the Phase 3 effort. The research activity will culminate in a detailed design and prototype implementation. Figure 2 depicts the generic hardware architecture assumed for implementing the replicated system. Single-source sensor inputs are distributed by special purpose hardware executing a Byzantine agreement algorithm. Replicated actuator outputs are all delivered in parallel to the actuators, where force-sum voting occurs. Interprocessor communication links allow replicated processors to exchange and vote on the results of task computations.

3 Previous Efforts

Many techniques for implementing fault-tolerance through redundancy have been developed over the past decade, e.g. SIFT [4], FTMP [5], FTP [6], MAFT [7], and MARS [8]. An often overlooked but significant factor in the

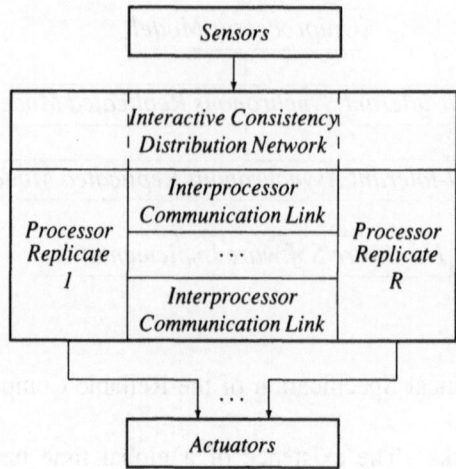

Figure 2: Generic hardware architecture.

development process is the approach to system verification. In SIFT and MAFT, serious consideration was given to the need to mathematically reason about the system. In FTMP and FTP, the verification concept was almost exclusively testing.

Among previous efforts, only the SIFT project attempted to use formal methods [9]. Although the SIFT operating system was never completely verified [10], the concept of Byzantine Generals algorithms was developed [11] as was the first fault-tolerant clock synchronization algorithm with a mathematical performance proof [3]. Other theoretical investigations have also addressed the problems of replicated systems [12].

4 Application Definition

We present a method for specifying an operating system workload that characterizes the interface between the application software and the operating system. The specification consists of a generic set of mathematical definitions serving as a schema. For an actual application, these definitions would be instantiated with appropriate values.

4.1 Tasks

Let $T_1, \ldots, T_K$ be the application tasks. Assume each task produces either actuator output or data values drawn from some domain. These data values may be provided as inputs to other tasks or serve as long term state variables. Tasks have no persistent state variables; the effect of persistent state is achieved by recirculating task outputs.

Let $S_1, \ldots, S_p$ be the sensors and $A_1, \ldots, A_q$ be the actuators. Let these symbols also stand for the sets of values received from the sensors and sent to the actuators. Also let D_i be the set of data values produced as output by task T_i. These values may be structured objects such as arrays, records, etc. Thus, if T_i is an actuator task, $D_i = A_j$ for some j. Note that this precludes an actuator task from producing non-actuator data in addition to actuator output. Let $D = \bigcup_i D_i$.

Task T_i computes a function f_i on a set of input values. Inputs may be taken from sensor data or the outputs of other tasks. Tasks are prohibited from having side effects; their only effects are their explicit outputs.

4.2 Schedules

Application tasks are scheduled via a fixed, deterministic sequence of task executions. A complete, repeating task schedule comprises a *cycle*.

...	$Cycle_{i-1}$	$Cycle_i$	$Cycle_{i+1}$	...

Cycles are repeated indefinitely and the task execution sequence of one cycle is identical to the others. A cycle is divided into M *frames* of equal duration.

$Frame_0$	...	$Frame_{M-1}$

$$|\longleftarrow - - - - \quad Cycle \quad - - - \longrightarrow|$$

The frame length is a *fundamental unit of time* for the application (typically ~ 50 ms). The sensors are read at most once per frame and actuators are written at most once per frame. Each frame is divided into *subframes* of variable length. The number of subframes is variable also.

$Subframe_1$	...	$Subframe_{M_i}$	////

$$|\longleftarrow - - - - - - \quad Frame_i \quad - - - - - \longrightarrow|$$

The number of subframes for the ith frame is given by M_i (distinct from M, the number of frames). The time from the end of the last subframe until

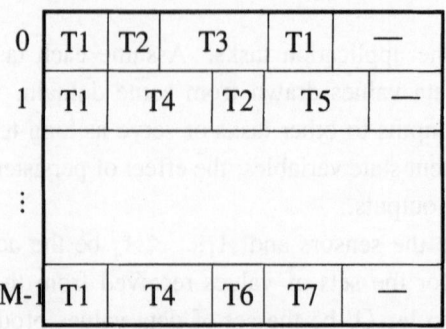

Figure 3: Structure of a task schedule.

the end of the frame is slack time for performing OS overhead functions and dispatching pre-emptable, non-critical tasks.

The schedule for an entire cycle would assign task executions to each subframe (figure 3). We refer to each site in a task schedule as a *cell*. A cell is denoted by the pair (i, j) for the ith frame and j^{th} subframe. A schedule is then given by a mapping from cells into the scheduled task:

$$ST : \{0..M - 1\} \times nat \to \{0..K\}$$

$ST(i, j)$ gives the task index of the scheduled task for cell (i, j), and 0 for $j > M_i$ (*nat* denotes the natural numbers).

Now consider the binding of input values for task execution. For task T_i, we must supply inputs for the arguments of f_i. Each input must come from a prior task execution or be taken as sensor input. So the designation of a task input will be a triple (i_type, i, j) where $i\_type \in \{sensor, cell\}$ with the meaning:

sensor	value from sensor i in current frame
cell	value from task output in cell (i, j) of current or previous cycle

A task may get input from a prior task output up to one cycle length in the past (M frames). By convention, if the task in cell (k, l) receives input from the task in cell (i, j) where

$$i > k \vee (i = k \wedge j \geq l)$$

then the input comes from (i, j)'s task execution during the previous cycle.

A mapping from cells into sequences of triples defines the assignment of input values to task executions.

$$TI : \{0..M - 1\} \times nat \rightarrow sequence(triple)$$

$$TI(i, j) = [(t_1, i_1, j_1), ..., (t_n, i_n, j_n)]$$

for a task with n inputs. Let $TI(i, j) = [\,]$ when $j > M_i$ or the task at (i, j) has no inputs.

The functions ST and TI need to be supplemented by a binding of task outputs to actuators for "actuator" tasks:

$$AO : \{0..M - 1\} \times nat \rightarrow \{0..q\}$$

$AO(i, j) - a$ to designate that the output of the task at cell (i, j) should go to actuator a. As before, $AO(i, j) = 0$ if $j > M_i$ or the task at (i, j) does not produce actuator output.

Provided the functions TI and AO satisfy certain well-formedness constraints, they suffice to uniquely characterize a task schedule. AO may not allow multiple outputs to the same actuator within a single frame.

Since task results may be carried forward from one cycle to the next, it is necessary to account for the "previous" cycle at system initialization. The application must define what these previous-cycle task outputs should be for the first cycle to use as suitable task inputs. A function

$$IR : \{0..M - 1\} \times nat \rightarrow D$$

is used to characterize the initial task results values.

5 Top Level Specification

The top level OS machine specification captures the behavior of the application tasks running on an ideal computer. It defines the net effect of task execution as seen by the control application. All details of the replicated system implementation are hidden.

5.1   OS State and I/O Types

The state of the ideal OS consists of a frame counter and task outputs produced in the current and previous cycle. Thus an OS_state is a pair:

$$OS\_state = (\ frame : \{0..M-1\},$$
$$results : cycle\_state\ )$$
$$\text{where } cycle\_state : \{0..M-1\} \times nat \rightarrow D.$$

$OS.frame$ denotes the frame counter while $OS.results(i,j)$ denotes the task output at cell (i,j) during the current or previous cycle.

The application definition needs to provide the initial state values for the results portion of the state. The initial OS state is given by the pair $(0, IR)$, where IR defines the initial results state values.

The following data types represent vectors of sensor inputs and actuator outputs.

$$Sin \quad = \quad vector([1..p])\ of\ \cup_i S_i$$
$$Aout \quad = \quad vector([1..q])\ of\ \cup_i A_i$$

5.2   State Transition Definitions

Transitions correspond to the execution of tasks for a single frame. The state variable $OS.frame$ gives the number of the frame to be executed by the next transition. After the Mth state transition of the current cycle, $OS.frame$ is reset to 0. After the ith state transition of the current cycle, $OS.results(i,j)$ contain the results of the latest task executions. Later cells of $OS.results$ still contain the results of the prior cycle's task executions.

Since the frame number is incremented by one, with a wrap-around when it reaches M, we use the shorthand notation defined as follows.

$$x \oplus y = (x + y) \bmod M$$

$$x \ominus y = (x + M - y) \bmod M$$

The function OS defines the state transition.

$$OS : Sin \times OS\_state \rightarrow OS\_state$$

$$OS(s, u) = (u.frame \oplus 1, \lambda i, j.\ new\_results(s, u, i, j))$$

The result of the function is a pair (f, r) containing the new frame counter and results state. The subordinate function $new\_results$ is defined below.

$$new\_results(s, u, i, j) = \text{if } i = u.frame$$
$$\text{then } exec(s, u, i, j)$$
$$\text{else } u.results(i, j)$$

To refer to the execution of tasks within the current frame, the function $exec(s, u, i, j)$ gives the result of executing the task in the ith frame and j^{th} subframe, i.e., at cell (i, j) in the schedule. Because the tasks in a frame may use the outputs of prior tasks within the same frame, which are computed in this frame rather than found in the result state, the definition involves recursion through the task schedule. Details can be found in [1].

5.3 Actuator Output

Since actuator outputs are always taken from task outputs, which are recorded as part of the OS state, we find it convenient to define actuator outputs as a function only of the OS state, as in a "Moore" style state machine. To cast actuator outputs into a functional framework, we must account for the case of an actuator not being sent an output value in a given frame. We assume an actuator may be sent commands as needed by the application, which may choose not to sent output during some frames. Let us denote by the symbol ϕ the null actuator output, i.e., an output value ϕ indicates the absence of anything to send to the actuator. Then we define actuator outputs as a function of the OS state using the function UA.

$$UA(u) = [\,_{k=1}^{q} Act(u, k)\,]$$

We use the notation $[\,_{i=1}^{m} a_i\,]$ to mean $[a_1, \ldots, a_m]$.

The function Act is used to define the output for each individual actuator.

$$Act(u, k) = \begin{cases} u.results(u.frame \ominus 1, j) \\ \quad \text{if } \exists j : AO(u.frame \ominus 1, j) = k \\ \phi \quad \text{otherwise} \end{cases}$$

Because of the application restriction that at most one task output may be assigned to an actuator, the axiom above leads to a well-defined result. The frame count is decremented by one because UA is applied to the new state after a transition, where the frame count has already been incremented.

6   Second Level Specification

The replicated OS machine specification represents the behavior of the OS and application tasks running on a redundant system of synchronized, independent processors with a mechanism for voting on intermediate states. Let R be the number of redundant processors. We use $\{1, \ldots, R\}$ as processor IDs. Each processor runs a copy of the OS and the application tasks. The uniprocessor OS state is replicated R times and this composite state forms the replicated OS state. Transitions for the replicated OS cause each individual OS state to be updated, although not in exactly the same way because some processors may be faulty.

6.1   Faulty Processors

The possibility of processors becoming faulty requires a means of modeling the condition for specification purposes. We adopt a worst case fault model. In each frame, a processor and its associated hardware is either faulty or not. A *fault status vector* is introduced to condition specification expressions on the possibility of faulty processors.

Voting intermediate results is the way a previously faulty processor recovers valid state information. The voting pattern determines which portions of the state should be voted on each frame. A state variable that is voted will be replaced with the voted value regardless of what its current value is in memory. We will vote the frame counter on every frame and hence, will not include it in the voting pattern definition.

Let the predicate VP represent the voting pattern.

$$VP : \{0..M - 1\} \times nat \times \{0..M - 1\} \to \{T, F\}$$

$VP(i, j, n) = T$ iff we are to vote $OS.results(i, j)$ during frame n.

Since processors may be faulty and the values of their state variables may be indeterminate, we introduce a special *bottom* data object to denote questionable or unknown data values. The symbol "$\perp$" is used for this purpose. We regard it as a special data object distinct from known "good" objects. This usage is intended to model the presence of potentially erroneous data.

Voting is the primary application for $\perp$. We use the function

$$maj : sequence(D \cup \{\perp\}) \to D \cup \{\perp\}$$

to denote the majority computation. It takes a sequence of data objects of type D and produces a result of type D. If a majority does not exist, then $maj(S) = \bot$; otherwise, $maj(S)$ returns the value within S that occurs more than $|S|/2$ times.

6.2 The Replicated State

The replicated OS state is formed as a vector of uniprocessor OS states:

$$Repl\_state = vector([1..R]) \ of \ OS\_state$$

Thus, if r is a Repl_state value, then $r[k]$ refers to the OS_state for the kth processor. The OS_state definition is identical to that of the top level OS specification. The state variable $r[k].frame$ gives the number of the frame to be executed by the next transition within processor k. To refer to a results element of a replicated OS state we use the notation $r[k].results(i,j)$.

The initial state of the replicated OS is formed by merely copying the uniprocessor initial state R times. Thus, we have:

$$Initial\_Repl\_state = [\textstyle{R \atop k=1} \ (0. IR)\,]$$

where IR denotes the initial results state values as provided in the application task definitions.

Inputs to the replicated processors come from the same sensors as in the uniprocessor case. The act of distributing sensor values via some kind of interactive consistency algorithm is assumed to produce R values to present to the replicated system. Therefore, we introduce a vectorized data type to use for input variables in the functions below.

$$ICin = vector([1..R]) \ of \ Sin$$

Thus, if c is an $ICin$ value, then $c[k]$ refers to the sensor inputs for the kth processor.

6.3 Replicated System Transitions

Transitions correspond to the execution of all tasks in a single frame for all replicates. Since the replicated OS state is a vector of uniprocessor OS states, we can first decompose the Repl_state transition into R separate cases.

$$Repl : ICin \times Repl\_state \times fault\_status \rightarrow Repl\_state$$

$$Repl(c, r, \Phi) = [^R_{k=1} \, RT(c, r, k, \Phi)]$$

RT is the function used to define the OS state transition for each replicate.

The additional argument Φ is used to supply assumptions about the current fault status of the replicated processors.

$$fault\_status \;=\; vector([1..R]) \; of \; \{T, F\}$$

$\Phi[k]$ is true when processor k is faulty during the current frame. Various specification functions take Φ arguments as a way to model assumptions about fault behavior and show what the system response is under those assumptions.

To define *RT* we must take into account whether the processor is faulty and apply voting at the appropriate points. Because voting incorporates values from all the processors, the entire Repl state is required as an argument to *RT* even though it only returns the OS state for the kth processor.

$$RT(c, r, k, \Phi) = \text{if } \Phi[k]$$
$$\text{then } \bot$$
$$\text{else } (\; frame\_vote(r, \Phi),$$
$$Repl\_results(c, r, k, \Phi)\,)$$

If processor k is faulty, we regard its entire OS state as suspect and therefore assign it the value $\bot$.

RT requires the frame counter be voted on every transition. All processor frame counters are input to a majority operation. Voting for a frame is based on values computed during that frame. Consequently, the incremented frame counter values are used in the specification.

$$frame\_vote(r, \Phi) = maj([^R_{l=1} \, FV_l])$$

where $FV_l = \text{if } \Phi[l] \text{ then } \bot \text{ else } r[l].frame \oplus 1$

Because some of the $r[l]$ may be faulty, we assume their frame counters are questionable and produce $\bot$ as their votes.

For the results state variables, we need to incorporate selective voting. The VP predicate determines when and where to vote.

$$Repl\_results(c, r, k, \Phi) =$$
$$\lambda i, j. \; \text{if } VP(i, j, r[k].frame)$$
$$\text{then } results\_vote(c, r, i, j, \Phi)$$
$$\text{else } new\_results(c[k], r[k], i, j)$$

The function $new\_results$ is defined in the uniprocessor OS specification. It gives the value of the task results part of the state after a state transition.

Defining the vote of task results is similar to that for the frame counter.

$$results\_vote(c, r, i, j, \Phi) = maj([^R_{l=1} RV_l])$$
where $RV_l = $ if $\Phi[l]$ then $\perp$ else $new\_results(c[l], r[l], i, j)$.

As before, some of the processors may be faulty so some $r[l]$ may have value $\perp$. We assume task execution on faulty processors produces $\perp$ as well.

Note that voting within a frame occurs after all computation has taken place. In particular, the voted value of a task's output is not immediately available to a later task within the same frame.

6.4 Replicated Actuator Output

As in the uniprocessor case, outputs from the replicated processors go to the actuators. Each processor sends its own actuator outputs separately. Therefore, we introduce a vectorized data type to describe the replicated system outputs.

$$RAout \;=\; vector([1..R]) \; of \; Aout$$

Thus, if b is an $RAout$ value, then $b[k]$ refers to the actuator outputs for the kth processor.

The actuator output variables are updated according to the application function AO in the same manner as the uniprocessor OS. We use the OS function UA to extract the actuator outputs for each processor in the replicated system.

$$RA : Repl\_state \times fault\_status \rightarrow RAout$$

$$RA(r, \Phi) = [^R_{k=1} RA_k]$$
where $RA_k = $ if $\Phi[k]$ then $\perp$ else $UA(r[k])$

RA produces a vector of actuator outputs, one for each processor. Faulty processors are assumed to produce indeterminate output ($\perp$).

7 Replicated System Proofs

We develop a methodology for showing that the replicated OS is a correct implementation of the uniprocessor OS. Previously presented concepts are put together with a framework for the replicated and uniprocessor state machines. Sufficient conditions based on commutative diagram techniques are derived for showing correctness. Issues stemming from real-time considerations are not included in the following. In subsequent work we will address requirements such as having adequate real time to execute the task schedule and OS overhead functions.

7.1 Fault Model

In each frame, a processor is either faulty or not. A function

$$\mathcal{F} : \{1..R\} \times nat \rightarrow \{T, F\}$$

represents a possible fault history for a given set of redundant processors. $\mathcal{F}(k, n) = T$ when processor k is faulty in frame n, where n is the global frame index ($n \in \{0, 1, \ldots\}$). Let $fault\_fn$ be the type representing the signature of $\mathcal{F}$.

Faults are often distinguished as being either *permanent* or *transient*. A permanent fault would appear in $\mathcal{F}$ as an entry that becomes true for a processor k in frame n and remains true for all subsequent frames. A transient fault would appear as an entry that becomes true for several frames and then returns to false, indicating a return to nonfaulty status.

Application task configurations and voting patterns determine the number of frames required to recover from a transient fault. Let N_R represent this number ($N_R > 0$). We define a processor as *working* in frame n if it is nonfaulty in frame n and nonfaulty for the previous N_R frames. We use a function $\mathcal{W}$ to represent this concept.

$$\mathcal{W} : \{1..R\} \times nat \times fault\_fn \rightarrow \{T, F\}$$
$$\mathcal{W}(k, n, \mathcal{F}) =$$
$$(\forall j : 0 \leq j \leq min(n, N_R) \supset \sim \mathcal{F}(k, n - j))$$

The number of working processors is also of interest:

$$\omega(n, \mathcal{F}) = |\{k \mid \mathcal{W}(k, n, \mathcal{F})\}|$$

A processor that is nonfaulty, but not yet working, is considered to be *recovering*.

Finally, the key assumption upon which correct state machine implementation rests is given below.

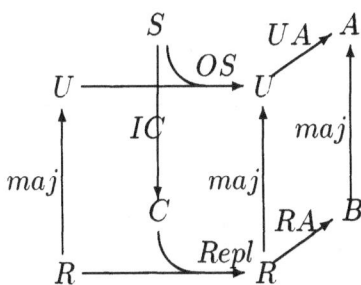

Figure 4: Commutative diagram for UM and RM.

Definition 1 *The* Maximum Fault Assumption *for a given fault function $\mathcal{F}$ is that $\omega(n, \mathcal{F}) > R/2$ for every frame n.*

All theorems about state machine correctness are predicated on this assumption that there is a majority of working processors in each frame.

7.2 Framework For State Machine Correctness

Mappings are needed to bridge the gap between the two state machines. Let us refer to the uniprocessor state machine as UM and the replicated state machine as RM. We map from RM to UM by applying the majority function. We map from UM to RM by distributing data objects R ways.

For sensor inputs, we assume an interactive consistency process is used in the system, so the net effect is that sensor data is merely copied and distributed.

$$IC : Sin \rightarrow ICin \qquad IC(s) = [\,_{i=1}^{R}\, s\,]$$

The majority mapping on replicated states and actuator outputs captures the notion that a majority of the processors should be computing the right values.

Relationships among the various entities for the two state machines are characterized by the commutative diagram in figure 4. Table 1 summarizes the sets involved.

Assume the inputs to UM are drawn from an infinite sequence of sensor values $S = [s_1, s_2, \ldots]$. Further assume UM will have states $[u_0, u_1, u_2, \ldots]$ and outputs $[a_1, a_2, \ldots]$. Similarly, the inputs to RM are drawn from the infinite sequence $C = [c_1, c_2, \ldots]$, and RM will have states $[r_0, r_1, r_2, \ldots]$ and outputs $[b_1, b_2, \ldots]$.

Set	Type	Description
S	Sin	Uniprocessor sensor inputs
A	$Aout$	Uniprocessor actuator outputs
U	OS_state	Uniprocessor OS states
C	$ICin$	Replicated sensor inputs
B	$RAout$	Replicated actuator outputs
R	$Repl\_state$	Replicated OS states

Table 1: Sets of inputs, outputs, and states.

Definition 2 *The state machines UM and RM are defined by initial states u_0 and r_0, and state transitions that obey the following relations for $n > 0$:*

$$\begin{aligned} u_n &= OS(s_n, u_{n-1}) \\ a_n &= UA(u_n) \\ r_n &= Repl(c_n, r_{n-1}, \mathcal{F}_n^R) \\ b_n &= RA(r_n, \mathcal{F}_n^R) \end{aligned}$$

where $\mathcal{F}_n^R = [\,_{k=1}^{R}\, \mathcal{F}(k, n-1)\,]$.

Not shown in figure 4 is the fault status vector argument to the functions *Repl* and *RA*.

7.3 The Correctness Concept

Our approach to the correctness criteria is based on state machine concepts of behavioral equivalence, specialized for this application. In essence, what we want to show is that the I/O behavior of RM is the same as that of UM when interpreted by the mapping functions IC and maj. We say that the machine RM *correctly implements* UM iff they exhibit matching output sequences when applied to matching inputs sequences and the Maximum Fault Assumption holds.

Definition 3 RM correctly implements UM under assumption $\mathcal{P}$ *iff the following formula holds:*

$$\forall \mathcal{F}, \mathcal{P}(\mathcal{F}) \supset \forall S, \forall n > 0 : a_n = maj(b_n)$$

where a_n and b_n can be characterized as functions of an initial state and all prior inputs.

We parameterize the concept of necessary assumptions using the predicate $\mathcal{P}$. For the replicated system, it will be instantiated by the Maximum Fault Assumption:

$$\mathcal{P}(\mathcal{F}) = (\forall m : w(m, \mathcal{F}) > R/2).$$

Definition 3 provides the formal means of comparing the effects of the two machines and reasoning about their collective, intertwined behavior. It focuses on the correctness of the actuator outputs as a function of the sensor inputs; this is what matters to the system under control.

We now introduce the usual sufficient conditions for correctness based on commutative diagram techniques. The following criteria can be understood as showing that two subdiagrams of figure 4 commute: one for the state transition paths and another for the output function paths. Although the second subdiagram is a nonstandard form for commutative diagrams, since it does not depict a homomorphism, it is nevertheless useful for characterizing the relationship between the two machines' output values.

Definition 4 (RM Correctness Criteria) *RM correctly implements UM if the following conditions hold:*

(1) $u_0 = maj(r_0)$

(2) $\forall \mathcal{F}, (\forall m : w(m, \mathcal{F}) > R/2) \supset$
$\quad \forall S, \forall n > 0,$
$\quad\quad OS(s_n, maj(r_{n-1})) =$
$\quad\quad\quad maj(Repl(IC(s_n), r_{n-1}, \mathcal{F}_n^R))$

(3) $\forall \mathcal{F}, (\forall m : w(m, \mathcal{F}) > R/2) \supset$
$\quad \forall S, \forall n > 0,$
$\quad\quad UA(maj(r_n)) = maj(RA(r_n, \mathcal{F}_n^R))$

The conditions of Definition 4 are shown to imply the correctness notion of Definition 3 in [1].

8   Design Proofs

Proving replicated system correctness for a particular voting pattern can be simplified by first establishing some intermediate sufficient conditions. The following treatment is based on the formulation of a Consensus Property, which relates the state of working processors to the majority of the replicated states. We use this property to prove the RM Correctness Criteria. This proof is independent of a particular voting pattern; it need be done only once. Similarly, the Consensus Property can be established by introducing a Replicated State Invariant. Then we construct a proof of the invariant based on the Full Recovery Property, whose statement is generic, but whose proof is different for each voting pattern.

Adopting this methodology creates the following general proof structure.

RM Correctness Criteria

⇑

Consensus Property

⇑

Replicated State Invariant

⇑

Full Recovery Property

⇑

Voting Pattern

8.1   Consensus Property

The Consensus Property relates certain elements of the replicated OS state to the majority of those elements. It asserts that if the pth processor is working during a frame, i.e., not faulty and not recovering, then its element of the replicated OS state equals that of the majority, both before *and* after the transition. This reflects our intuition that if a processor is to be considered productive, it must have established a state value that matches the consensus and will continue to do so after the computations of the current frame.

Definition 5 (Consensus Property) *For $\mathcal{F}$ satisfying the Maximum Fault Assumption,*

$$\mathcal{W}(p, n-1, \mathcal{F}) \supset$$
$$r_{n-1}[p] = maj(r_{n-1}) \wedge r_n[p] = maj(r_n)$$

holds for all p and all $n > 0$.

Having stated a generic Consensus Property, we assume its truth to prove the RM Correctness Criteria hold. See [1] for a detailed proof of the following result.

Theorem 1 *The RM Correctness Criteria follow from the Consensus Property.*

8.2 Full Recovery Property

We introduce a predicate, rec, that captures the concept of a state element having been recovered through voting. It is a function of the last faulty frame, f, and the number of frames, h, a processor has been nonfaulty.

$$rec(i, j, f, h, e) =$$
$$\text{if } h \leq 1 \text{ then } F$$
$$\text{else } (VP(i, j, f \oplus h) \wedge e) \vee$$
$$\quad \text{if } i = f \oplus h$$
$$\quad\quad \text{then } \bigwedge_{l=1}^{|TI(i,j)|} RI(TI(i,j)[l], i, j, f, h)$$
$$\quad\quad \text{else } rec(i, j, f, h - 1, T)$$

$$RI(t, i, j, f, h) =$$
$$(t.type = sensor) \vee$$
$$\text{if } t.i = f \oplus h \wedge t.j < j$$
$$\quad \text{then } rec(t.i, t.j, f, h, F)$$
$$\quad \text{else } rec(t.i, t.j, f, h - 1, T)$$

By recursively following the inputs for the scheduled task at cell (i, j), $rec(i, j, f, h, e)$ is true iff $results(i, j)$ should have been restored in frame $f \oplus h$, provided the processor has been nonfaulty for h frames and f was the last faulty frame. The boolean argument, e, indicates whether the recovery status applies at the end of the frame or sometime before computation is complete. This is necessary to account for the block voting that occurs at the end of a frame.

The conditions for rec can obtain if (i, j) is voted in frame $f \oplus h$, or it is computed in frame $f \oplus h$ and all inputs have been recovered, or it is not computed in frame $f \oplus h$ and was recovered by frame $f \oplus (h - 1)$. Thus, cell (i, j) is not recovered if it results from computations involving unrecovered data, or it has not been voted since the last faulty frame f.

Definition 6 (Full Recovery Property) *The predicate* $rec(i, j, f, N_R, T)$ *holds for all* i, j, f.

This definition equates full recovery with the predicate rec becoming true for all state elements (i, j) after N_R frames have passed since the last fault.

8.3 Replicated State Invariant

As a practical matter, it is necessary to prove the Consensus Property by first establishing an invariant of the replicated OS state. Such an invariant relates the values of the nonfaulty processor states to the majority value of replicated OS states. To do so, it is necessary to identify the partially recovered values of OS states for recovering processors.

Expressing the invariant below requires a means of determining how many consecutive frames a processor has been healthy (without fault). Let $\mathcal{H}(k, n, \mathcal{F})$ give the number of healthy frames for processor k prior to the nth frame. In an analogous way, let $\mathcal{L}(k, n, \mathcal{F})$ give the last faulty frame for processor k prior to the nth frame.

The Replicated State Invariant states that if the pth processor is nonfaulty during a frame, i.e., working or recovering, then its frame counter after the transition equals that of the majority. It also relates this processor's results state values to the majority if they have been recovered, as determined by the function rec.

Definition 7 (Replicated State Invariant) *For fault function* $\mathcal{F}$ *satisfying the Maximum Fault Assumption, the following assertion is true for every frame* n:

$$(n = 0 \vee \sim \mathcal{F}(p, n - 1)) \supset$$
$$r_n[p].frame = maj(r_n).frame = n \bmod M \wedge$$
$$(\forall i, j : rec(i, j, \mathcal{L}(p, n, \mathcal{F}), \mathcal{H}(p, n, \mathcal{F}), T) \supset$$
$$r_n[p].results(i, j) = maj(r_n).results(i, j)).$$

Theorem 2 *The Replicated State Invariant follows from the Full Recovery Property.*

Theorem 3 *The Consensus Property follows from the Replicated State Invariant and the Full Recovery Property.*

Again, complete proofs of these theorems can be found in [1] along with definitions for $\mathcal{H}$ and $\mathcal{L}$.

9 Specific Voting Patterns

With the general framework established thus far, the replicated system design is verified on the premise that the Full Recovery Property holds. This property depends on the details of each voting pattern and must be established separately for each. Following are three voting schemes and their proofs. The last one is the most general and constitutes the goal of this work; the other two can be seen as special cases whose proofs are simpler and instructive.

9.1 Continuous Voting

We begin with the simplest case, namely when the voting pattern calls for voting all the data on every frame. Clearly, this leads to transient fault recovery in a single frame. Although the entire state of a recovering processor is restored in one frame, our formalization of rec assumes one frame is used to recover the frame counter, so the conservative assignment $N_R = 2$ is used.

Definition 8 *The* continuous voting *version of the replicated OS uses the assignments* $VP(i, j, k) = T$ *for all* i, j, k, *and* $N_R - 2$.

Theorem 4 *The* continuous voting *pattern satisfies the Full Recovery Property.*

Proof. Since $VP(i, j, k)$ holds for all i, j, k, and $N_R = 2$, expanding the definition of rec shows that $rec(i, j, f, N_R, T)$ reduces to T for all i, j, f. ∎

9.2 Cyclic Voting

Next we consider a more sparse voting pattern, namely voting only the data computed in the current frame. Only the portion of $r.results(i, j)$ where $i = r.frame$ is voted; the other $M - 1$ portions are voted in later frames. This leads to voting each part of the results state exactly once per cycle and therefore leads to transient fault recovery in $M + 1$ frames. (One frame is required to recover the frame counter.) The proof in this case is only slightly more difficult.

Definition 9 *The* cyclic voting *version of the replicated OS uses the assignments* $VP(i, j, k) = (i = k)$ *for all* i, j, k, *and* $N_R = M + 1$.

Theorem 5 *The* cyclic voting *pattern satisfies the Full Recovery Property.*

Proof. Since $VP(i, j, f \oplus h)$ reduces to $i = f \oplus h$, the definition of rec becomes

$$rec(i, j, f, h, T) =$$
$$\text{if } h \leq 1 \text{ then } F$$
$$\text{else } i = f \oplus h \vee rec(i, j, f, h - 1, T).$$

Thus, it follows that

$$rec(i, j, f, N_R, T)$$
$$= rec(i, j, f, M + 1, T)$$
$$= (i = f \oplus 2) \vee \ldots \vee (i = f \oplus (M + 1))$$

Because the modulus of $\oplus$ is M this expression evaluates to T. ∎

9.3   Minimal Voting

The last case is concerned with the most general characterization of voting requirements. *Minimal voting* is the name used to describe these requirements because they represent conditions necessary to recover from transient faults via the most sparse voting possible.

Central to the approach is the use of task I/O graphs, constructed from the application task specifications embodied in the function TI. Nodes in the graph denote cells in the task schedule and directed edges correspond to the flow of data from a producer task to a consumer task. Sensor inputs and actuator outputs have no edges in these graphs. Associated with edges of the graph are voting sites that indicate where task output data should be voted before being supplied as input to the receiving task.

The essence of the Minimal Voting scheme is that every cycle[1] of the task I/O graph should be covered by at least one voting site. It is possible to place more than one vote along a cycle or place votes along noncyclic paths, but they are unnecessary to recover from transient faults. Such superfluous votes may be desirable, however, to improve the transient fault recovery rate.

Definition 10 *A task I/O graph $G=(V,E)$ contains nodes $v_i \in V$ that correspond to the cells (i, j) of a task schedule. Edges consist of ordered pairs (v_1, v_2) where $((i_1, j_1), (i_2, j_2)) \in E$ iff output from cell (i_1, j_1) is used as input to (i_2, j_2).*

---

[1] We are using the graph theoretic concept of cycle here, as opposed to the terminology introduced earlier of a frame cycle consisting of M contiguous frames in a schedule.

Definition 11 *A path through the task I/O graph* $G = (V, E)$ *consists of a sequence of nodes* $P = [v_1, \ldots, v_n]$ *such that* $(v_i, v_{i+1}) \in E$. *A cycle is a path* $C = [v_1, \ldots, v_n, v_1]$. *The frame length of an edge* $e = ((i_1, j_1), (i_2, j_2))$ *is given by:*

$$fl(e) = \begin{cases} M & \text{if } i_1 = i_2 \wedge j_1 \geq j_2 \\ i_2 \ominus i_1 & \text{otherwise} \end{cases}$$

The frame length of a path, $FL(P)$, *is the sum of the frame lengths of its edges.*

Definition 12 *Let* $C_1, \ldots, C_m$ *be the (simple) cycles of graph* G, *and* $P_1, \ldots, P_n$ *be the noncyclic paths of* G. *Define the following maximum frame length values for cycles and noncyclic paths:*

$$\begin{aligned} L_C &= max(\{FL(C_i)\}) \\ L_N &= max(\{FL(P_i)\}) + 1 \end{aligned}$$

Note that noncyclic paths may share edges with cycles in the graph, but may not contain a complete cycle. L_N is increased by one to account for the frame at the beginning of the path.

Definition 13 *The* minimal voting *condition is specified by the following constraint on* VP:

$\forall C \in cycles(G):$
$\quad \exists((a,b),(c,d)) \in C, \exists f:$
$\quad\quad VP(a,b,f) \wedge (a = c \wedge b \geq d \vee 0 \leq f \ominus a < c \ominus a)$

and the assignment $N_R = L_C + L_N + M$.

The condition requires at least one vote along each cycle. There is a caveat, however, on where the votes may be placed. Because voting occurs at the end of a frame, a vote site may not be specified on an edge between two cells of the same frame. Such placements are ruled out by the condition above. The bound N_R includes a worst case length to restore a state element, $L_C + L_N$, plus an additional M frames to account for maximum latency due to when the last fault occurred within the schedule. Note that all cycles must have frame lengths that are multiples of M.

Figure 5 illustrates the definitions above for a graph embedded in a four frame schedule. The graph shown has one cycle with frame length four $(L_C = 4)$ and a single vote site. The voting pattern would be specified by

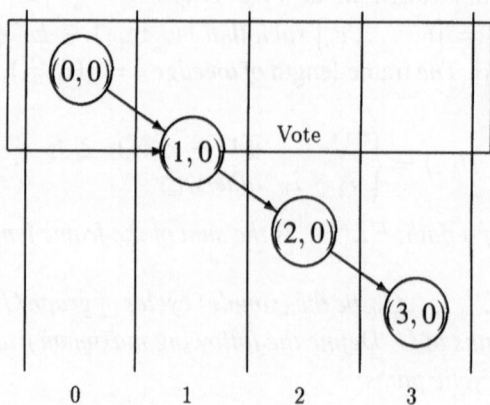

Figure 5: Example of task I/O graph.

$VP(1, 0, 2) = T$ to indicate that results cell $(1, 0)$ is voted in frame 2. The longest noncyclic path has frame length three ($L_N = 4$). Thus, the voting pattern meets the Minimal Voting condition and we assign it $N_R = 12$.

Definition 14 *A recovery tree is derived from the expansion of the recursive function, rec applied to specific arguments. Nodes of the tree are associated with terms of the form $rec(i, j, f, h, e)$. The tree is constructed as follows. Associate the root with the original term $rec(i, j, f, h, e)$. At each node, expand the rec function. If $VP(i, j, f \oplus h) \wedge e$ is true, mark the node with a T. Otherwise, evaluate the conditional term of the rec definition. Create a child node for each recursive call associated with the appropriate term and repeat the process. If evaluation shows only sensor inputs are used at a node, mark it with a T. If evaluation terminates with $h \leq 1$, mark the node with an F. After building the tree out to all its leaves, work back toward the root by marking each parent node with the conjunction of its child node markings.*

Thus, construction of the recovery tree for a term $rec(i, j, f, h, e)$ corresponds to building a complete recursive expansion of the boolean term. The marking at the root after the construction process is the value of the term.

Definition 15 *The frame length of an edge (v_1, v_2) in a recovery tree, where $v_1 = (i_1, j_1, f_1, h_1, e_1)$ and $v_2 = (i_2, j_2, f_2, h_2, e_2)$, is given by $|h_2 - h_1| \in \{0, 1\}$. The frame length of a path $[v_1, \ldots, v_n]$ in the tree is the sum of the frame lengths of the edges in the path, which is given by $|h_n - h_1|$.*

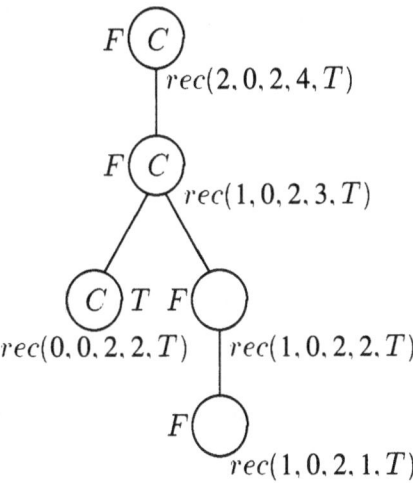

Figure 6: Recovery tree for the term $rec(2, 0, 2, 4, T)$.

Figure 6 shows the recovery tree for term $rec(2, 0, 2, 4, T)$ applied to the graph in figure 5. Nodes labeled with a "C" are computation nodes, i.e., they correspond to state elements in frames where $i = f \oplus h$. In this case, the four healthy frames are insufficient to recover the value of cell $(2, 0)$; eight frames are required.

Lemma 1 *If all leaves of a recovery tree are marked* T, *then the root must be marked* T.

Proof. Follows readily by induction on the height of the tree. ∎

Definition 16 *Let* $GP(P)$ *map a path* $P = [u_1, \ldots, u_m]$ *from a recovery tree into the analogous path in the corresponding task I/O graph. Form* $P' = [v_1, \ldots, v_n]$ *by retaining only those nodes from P arising from a computation frame* $(i = f \oplus h)$. *Then let* $GP(P) = [(i_1, j_1), \ldots, (i_n, j_n)]$ *where* (i_k, j_k) *is taken from the* rec *term of* v_k.

Lemma 2 *If a path P from a recovery tree begins and ends with a computation node, then* $FL(GP(P)) = FL(P)$.

Proof. Along the path P, between every pair of computation nodes there will be $fl(e) - 1$ noncomputation nodes one frame apart, where e is the edge in the task graph corresponding to this pair. Summing them all makes $FL(GP(P)) = FL(P)$. ∎

Theorem 6 *The* minimal voting *condition satisfies the Full Recovery Property.*

Proof. To show $rec(i, j, f, N_R, T)$, construct the recovery tree for this term. Consider each leaf node v_i and its path P_i to the root w. Let P_i be the concatenation of three subpaths X, Y, Z, where Y is the maximal subpath beginning and ending with a computation node. Let u be the first node of Y and let G denote the task graph. By Lemma 2 it follows that $FL(GP(Y)) = FL(Y)$ and because the maximum frame separation between computation nodes is M, $FL(Z) < M$.

We show that all leaves are marked with T. The only way for a v_i to be marked F is for $FL(P_i) \geq N_R - 1$, causing v_i's $h \leq 1$.

Case 1. P_i maps to an acyclic path in G. Since $GP(Y)$ is acyclic $FL(Y) = FL(GP(Y)) < L_N$. Moreover, $N_R = L_C + L_N + M$ so $FL(YZ) < N_R - 1$. In the worst case, u represents a task with sensor inputs only, X is empty and $u = v_i$. Otherwise, Y is shorter than the worst case length and $FL(XY) < L_N$. In either case, $FL(P_i) < N_R - 1$.

Case 2. P_i covers part of a cyclic path in G. P_i cannot map to a complete cycle because it would contain a vote site, terminating the recursion of rec. The worst case is that X and part of Y follow a partial cyclic path in G and the rest of Y is acyclic. Thus, we have $FL(P_i) < L_C + L_N + M - 1 = N_R - 1$.

By Lemma 1, it follows that the root is marked with T and therefore $rec(i, j, f, N_R, T)$ holds. ∎

The results presented above are conservative, being based on a loose upper bound for N_R. The actual N_R for most graphs will be somewhat smaller. The worst case for the graph of figure 5 is actually 10 frames versus the estimated value of $N_R = 12$. In addition, for more dense and highly regular voting patterns such as Continuous Voting and Cyclic Voting, we can obtain more accurate values and it would be inadvisable to apply the Minimal Voting bound to these cases.

An important consequence of the Minimal Voting result is that if a graph has no cycles, then no voting is required! In this case the recovery time bound would be given exactly by $N_R = L_N + M$. Although such a task graph is untypical for real control systems, there may be applications that could be based on this kind of design.

10 Summary

We have presented a method for specifying and verifying architectures for fault-tolerant, real-time control systems. The paper develops a uniprocessor top-level specification that models the system as a single (ultra-reliable) processor and a second-level specification that models the system in terms of redundant computational units. The paper then develops an approach to proving that the second-level specification is an implementation of the top-level. We have explored different strategies for voting and presented a correctness proof for three voting strategies. The Minimal Voting results offer real promise for building fault-tolerant systems with low voting overhead.

Acknowledgements

Comments received from John Rushby, Paul Miner, and Chuck Meissner during the course of this work are gratefully acknowledged.

References

[1] B. L. Di Vito, R. W. Butler, and J. L. Caldwell, "Formal design and verification of a reliable computing platform for real-time control," Technical Memorandum 102716, NASA, Oct. 1990.

[2] J. Rushby and F. von Henke, "Formal verification of a fault tolerant clock synchronization algorithm," Contractor Report 4239, NASA, June 1989.

[3] L. Lamport and P. M. Melliar-Smith, "Synchronizing clocks in the presence of faults," *Journal of the ACM*, vol. 32, pp. 52–78, Jan. 1987.

[4] J. Goldberg *et al.*, "Development and analysis of the software implemented fault-tolerance (SIFT) computer," Contractor Report 172146, NASA, 1984.

[5] A. L. Hopkins, Jr., T. B. Smith, III, and J. H. Lala, "FTMP — A highly reliable fault-tolerant multiprocessor for aircraft," *Proceedings of the IEEE*, vol. 66, pp. 1221–1239, Oct. 1978.

[6] J. H. Lala, L. S. Alger, R. J. Gauthier, and M. J. Dzwonczyk, "A Fault-Tolerant Processor to meet rigorous failure requirements," Tech. Rep. CSDL-P-2705, Charles Stark Draper Lab., Inc., July 1986.

[7] C. J. Walter, R. M. Kieckhafer, and A. M. Finn, "MAFT: A multicomputer architecture for fault-tolerance in real-time control systems," in *IEEE Real-Time Systems Symposium*, Dec. 1985.

[8] H. Kopetz *et al.*, "Distributed fault-tolerant real-time systems: The Mars approach," *IEEE Micro*, vol. 9, pp. 25–40, Feb. 1989.

[9] L. E. Moser and P. M. Melliar-Smith, "Formal verification of safety-critical systems," *Software–Practice and Experience*, vol. 20, pp. 799–821, Aug. 1990.

[10] "Peer review of a formal verification/design proof methodology," Conference Publication 2377, NASA, July 1983.

[11] L. Lamport, R. Shostak, and M. Pease, "The Byzantine Generals problem," *ACM Transactions on Programming Languages and Systems*, vol. 4, pp. 382–401, July 1982.

[12] L. V. Mancini and G. Pappalardo, "Towards a theory of replicated processing," in *Lecture Notes in Computer Science*, vol. 331, Springer Verlag, 1988.

Distributed Systems II

Distributed Systems II

A Membership Protocol Based on Partial Order*

SHIVAKANT MISHRA, LARRY L. PETERSON, RICHARD D. SCHLICHTING

Department of Computer Science

The University of Arizona

Tucson, AZ 85721, USA

Abstract

Membership information is used to provide a consistent, system-wide view of which processes are currently functioning or failed in a distributed computation. This paper describes a *membership protocol* that is used to maintain this information. Our protocol is novel because it is based on a multicast facility that preserves only the partial order of messages exchanged among the communicating processes. Because it depends only on a partial ordering of messages rather than a total ordering, our protocol requires less synchronization overhead. The advantages of our approach are especially pronounced if multiple failures occur concurrently.

1 Introduction

The construction of highly dependable applications is often simplified by the use of suitable system abstractions. For distributed systems, many of the most important abstractions are related in some way to providing consistent information to a collection of cooperating processes despite failures. One example is a global (virtual) clock that will return consistent values to all processes despite failures [1, 2, 3]. Another example is a reliable multicast or atomic broadcast, which typically guarantees reliable dissemination of messages and that all processes receive messages in the same sequence. In both cases, it is the consistency of the information that is the key; as has been amply demonstrated,

---

*This work supported in part by the National Science Foundation under grants CCR-8811923 and CCR-9003161, and the Office of Naval Research under grant N00014-91-J-1015.

such a property facilitates construction of dependable systems by providing all processes with identical information upon which to make decisions [4, 5, 6, 7].

This paper focuses on a third type of consistent information in a distributed system: membership information [8, 9, 10]. This information consists of a consistent, system-wide view of which processes are functioning and which have failed at any given moment in time. Implementation is typically by a *membership protocol*, an agreement protocol that is executed by the cooperating processes when the failure or recovery of another process is suspected.

There are actually two different types of membership protocols, each serving different purposes [11]. The first can be viewed as a user-level service that typically translates the failure or recovery confirmation into an event that is then ordered with respect to other events in the system. This ordering is then made available to the application to use in making decisions. Examples of this kind of protocol include [12, 13, 8, 10].

The other type of membership protocol—of which the one presented here is an instance—is sometimes called a *monitor protocol* [11]. In contrast to the user-level orientation of the first type, these protocols are used by the system itself to maintain a consistent view of which processes are functioning and hence participating in system decisions. For example, such information is used in reliable multicast protocols to determine when a message has been received and acknowledged by every functioning process so that it can be committed to the application. The processor failure or recovery event must again be consistently ordered with respect to other events such as interprocess communication to guarantee that messages are committed consistently, but the failure notification is not necessarily passed on to the application. Examples of this kind of protocol include [11].

The specific purpose of the protocol we describe is to maintain the consistency of the *membership list* that is used for message committment decisions in the Psync multicast mechanism [14]. Like other multicast primitives, Psync is targeted at the problem of maintaining a consistent ordering of messages exchanged among a collection of processes in the presence of failures. Unlike many others, however, Psync maintains a *partial* ordering of messages rather than a total ordering. Our membership protocol, which also uses Psync for its communication, takes advantage of this property to provide a more efficient implementation. In particular, it allows different processes to conclude that a failure has occurred at different times relative to the sequence of messages received, thereby reducing the amount of synchronization required. This increases the concurrency of the system, while still preserving the semantic correctness of the application. The advantages of our approach are especially pronounced if multiple processor failures occur concurrently.

This paper is organized as follows. Section 2 provides a brief overview of the Psync multicast mechanism. The basic membership protocol is then presented in Section 3; this protocol handles a single failure or recovery event at a time. Section 4 extends the protocol to handle multiple concurrent events. It also argues the correctness of the algorithm. Finally, Section 5 offers some conclusions.

2 Overview of Psync

Psync supports a *conversation* abstraction through which a collection of processes exchange messages. The key idea of the abstraction is that it explicitly preserves the partial order of the exchanged messages. This section gives a brief overview of Psync, introducing only those terms and concepts critical to understanding the membership protocol. A more complete description of Psync can be found in [14].

A conversation is explicitly opened by specifying a set of participating processes. This set is also called the *membership list*, ML. A message sent to the conversation is received by all the processes in ML. Fundamentally, each process sends a message in the *context* of those messages it has already sent or received. Informally, "in the context of" defines a relation among the messages exchanged through the conversation. This relation is represented in the form on a directed acyclic graph, called a *context graph*. Psync provides a set of operations for sending and receiving messages, as well as for inspecting the context graph. For example, Figure 1 depicts the context graph associated with a conversation that has three participants, denoted a, b and c, where $a_1, a_2, \ldots$ denotes the sequence of messages sent by process a, and so on.

Several important terms can be defined relative to the context graph. First, if there is a path in the context graph from one message to another, then the former message is said to *precede* the latter message; for example, a_1 precedes b_3. Second, a pair of messages that are not in the context of each other are said to have been sent at the *same logical time*; for example, c_1 and a_2. Third, a message sent by a process becomes *stable* once it is followed by a message sent by all other participants in ML; for example, messages a_1, b_1, and c_1 are the only stable messages in Figure 1. Finally, a *wave* is a set of messages sent

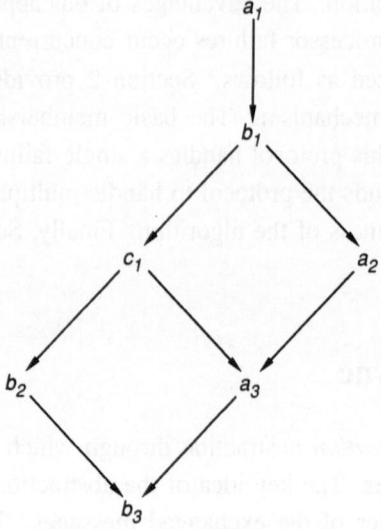

Figure 1: Example Context Graph

at the same logical time, that is, the context relation does not hold between any pair of messages in a wave. A wave is known to be *complete*—i.e., a process inspecting the context can be certain that no future messages will arrive that belong to the wave—as soon as a single message in the wave becomes stable. This is because for a message to be stable implies that all processes other than the sender have received it, which in turn implies that all future messages sent to the conversation must be in the context of the stable message; they cannot precede or be at the same logical time as the stable message. Stable messages and complete waves are used to decide when to commit messages to the application [15].

Psync maintains a copy of a conversation's context graph at each processor on which a participant in the conversation resides. A distinct copy of the membership list ML is also maintained at each such processor. For simplicity, we assume only a single participating process is running on each of these processors. Each time a process sends a message, Psync propagates a copy of the message to each of these other processors. This propagated message contains the ids of all the messages upon which the new message depends; i.e., it identifies the nodes to which the new message is to be attached to the context graph. Psync recovers from network failures by recognizing when a new message is to be attached to a message that is not present in the local copy of the graph. In this case, Psync asks the processor that sent the new message to retransmit the missing message. That processor is guaranteed to have a copy of the missing message because it just sent a message in the context of it.

Psync provides only minimal support for recovering from a process failure, where by process failure we mean that the processor on which the process is running has crashed without undergoing any erroneous state transitions. Specifically, Psync supports three operations that affect the local definition of ML. Note that these operations are purely local; they do not affect the definition of ML at other processors. First, the local process can instruct Psync to ignore a particular process; i.e., to stop accepting messages from that process. This action does not affect the definition of ML, but rather makes it look as though messages sent by the ignored process have been dropped by the network while being delivered to the local processor. Second, the local process can tell Psync to maskout a certain process. This operation actually removes the process from the local definition of ML; it also has the same affect as ignoring a message. Third, the local process can tell Psync to maskin a certain process. This has the effect of returning the process to the local definition of ML and accepting future messages sent by that process.

In addition, a process can instruct Psync to restart a conversation. This operation is used to initiate recovery actions following a failure. Restarting a conversation has two effects: to inform other processes that the invoking process has restarted, and to initiate reconstruction of the local copy of the context graph. Psync accomplishes this by broadcasting a special restart message. When this message is received at a processor, the local instance of Psync performs two actions. First, it transmits its current set of leaf nodes to the participant that generated the restart message; this allows reconstruction of the lost context graph image using the standard lost message protocol. Second, it generates a local notification of the restart event; this is implemented as an out-of-band control message that is delivered to the local process.

3 Basic Protocol

As outlined in the Introduction, the purpose of the membership protocol is to reach agreement about the failure or recovery of a process. The protocol is initiated when one of these two events is suspected. In our scheme, the task of detecting these events is assigned to another protocol called the *detection proto-col*. Specifically, this protocol monitors the messages exchanged in the system and on suspecting a change of state of a process, initiates the membership protocol by submitting a distinguished message to the conversation. A failure is typically suspected when no message has been received from a process in some interval of time, while recovery is based on the asynchronous notification generated when the recovering process executes the Psync `restart` primitive. However, the detection protocol is independent of the membership protocol and may employ any strategy to detect process status changes without affecting the membership protocol.

3.1   Correctness Criteria

To ensure the correctness of the application implemented over Psync in the presence of failures, the membership protocol must guarantee two properties: that all functioning processes receive the same set of messages in partial order, and that the decisions taken by the application based on these messages are correct even in the presence of failures. We call these aspects *external consistency* and *internal consistency*, respectively. These are defined more precisely as follows:

External Consistency: The conversation graph is same at all the processes. This has two aspects. First, all functioning processes reach the same decision about a failed (or suspected failed) process. Second, every functioning process starts accepting messages from a recovering participant at the same time.

Internal Consistency: Decisions made by the application based on the process' view of the context graph are correct. This has two aspects. First, stability and completeness decisions are made correctly; i.e., a message is considered stable only if it is followed by a message from all other functioning processes, and a wave is considered complete only when it has all of its messages. Second, processes receive all messages in the conversation.

We prove the correctness of the membership protocol by demonstrating that it guarantees both internal and external consistency.

3.2   Single Failures

Consider the case where at most one process fails at a time. Assume that ML initially contains n processes. The membership protocol is based on the effect that the failure has on the context graph. In particular, since a process obviously sends no messages once it has failed, it can be guaranteed there is no message from the failed participant at the same logical time as the membership protocol's initiation message sent by the detection protocol. If, on the other hand, there *is* a message from the suspect process at the same logical time as the initiation message, then it can be viewed as evidence that the process has in fact not failed. In this case, it is likely that the original suspicion of process failure was caused by the process or network being "slow" rather than an actual failure. The membership protocol uses this heuristic to establish the failure of a process.

The goal of the protocol is to establish an agreement among the $n - 1$ alive process about the failure or recovery of the nth process. As outlined above, the basic strategy is to agree on the failure of the process if and only if none of the $n - 1$ participants have received a message from the suspect process at the same logical time as the protocol initiation message. In case of recovery, the process is incorporated in the membership list once all the remaining $n - 1$ participants have acknowledged its recovery.

The actual details of the protocol are illustrated in Figure 2. Upon suspecting the failure of a process p, the detection protocol submits a $\langle$p is down$\rangle$ message to the conversation. p is subsequently considered to have failed if the $\langle$p is down$\rangle$ message is later followed in the context graph by a $\langle$Ack, p is down$\rangle$ message from every other participant.

Internal and external consistency are easily demonstrated for this algorithm. Every process stops accepting messages from the failed process at the same logical time in the conversation—on receipt of the $\langle$p is down$\rangle$ message. Similarly, a process is incorporated at the same logical time—the wave containing the $\langle$p is up$\rangle$ message—at all the processes. As a result, every process starts accepting messages from the recovered participant at the same time—as soon as it is incorporated. Since Psync guarantees the delivery of messages, every participant receives the same set of messages, and as result, every participant reaches the same conclusion about the failure of a process. The failed process is removed when every process has sent an $\langle$Ack, p is down$\rangle$ message. Since a process stops accepting messages from p before sending $\langle$Ack, p is down$\rangle$ message, all messages from failed process have been

Message ⟨p is down⟩:

If a message from *p* at the same logical time as ⟨p is down⟩ has been received, then send ⟨Nack, p is down⟩; otherwise send ⟨Ack, p is down⟩ and stop accepting messages from *p*.

Message ⟨Nack, p is down⟩:

Start accepting messages from *p* and terminate protocol.

Message ⟨Ack, p is down⟩:

If message ⟨p is down⟩ is stable then remove *p* and terminate protocol.

Message ⟨p is up⟩:

Send ⟨Ack, p is up⟩

Message ⟨Ack, p is up⟩:

If ⟨p is up⟩ is stable, then add *p* to the membership list and terminate protocol.

Figure 2: Membership Protocol Assuming Single Failure

received at the time of its removal. Thus, a process receives all the messages in the conversation. Stability decisions are correct because the failed process is removed at the same time and a recovering process is incorporated at the same time at all the processes. This means that every process determines the stability of a message with respect to the same set of processes. Thus, both internal and external consistency are satisfied.

Finally, notice that the events associated with the failure of a process, i.e., the halt in accepting messages and its removal from the membership list, are only partially ordered with respect to other messages in the system. Compared with other protocols in which these events are totally ordered with respect to other messages, this approach enhances the concurrency and efficiency of the application.

4 Multiple Failures

This section extends the membership protocol to handle concurrent failures and recoveries. In the presence of such concurrent events, the protocol becomes much more complex. Perhaps the predominant reason for this is the inherent lack of knowledge about the set of processes that participate in the membership agreement process itself. That is, processes may fail or recover at any time and, in particular, they may fail or recover while the membership protocol is in progress. Another source of complexity stems from the requirement that a consistent order of removal or incorporation of processes in the membership list be maintained. This order must be the same at all the processes to ensure correctness of the application, however, it is not at all clear what this order should be, or even the correct interpretation of "the same."

In this section, we first address this latter question by deriving an order in which these list modification events must be performed. We show that the semantics of *remove*—removing a failed process from the membership list ML—and *join*—incorporating a recovering process in the membership list ML—put a restriction on the order in which the modifications of the membership list take place. We then describe the actual membership protocol.

4.1 Ordering List Modification Events

Suppose that two processes p and q fail at approximately the same time. If the last message sent by p is at the same logical time as the last message sent by q (that is, neither is in the context of the other in the context graph), then p and q can obviously not participate in each other's failure agreement protocol. Since establishing agreement about the failure of a process requires concurrence of all functioning processes, agreement for processes that fail in this way must be done simultaneously. On the other hand, if the last message sent by q is in the context of the last message sent by p, then q may contribute messages to the agreement about p having failed, i.e., q may participate in the failure agreement of p.

Now, expand this scenario to include a third failing process r. Suppose that the last message sent by r is at the same logical time as the last message sent by q, but follows the last message sent by p. By the argument made above, this implies that the failure agreement of q and r must also be done simultaneously, leading to the conclusion that all three processes must be treated as a group. In general, then, the failure agreement of a set of processes must be done simultaneously whenever the last message sent by any process in the set is at the same logical time as the last message sent by at least one other process in the set.

We formalize this notion by defining a *simultaneous failure group (sf-group)* S as follows:

S is an equivalence class of failed processes under the relation $\leftrightarrow^*$, where $p \leftrightarrow q$ if the last message sent by p is at the same logical time as the last message sent by q and $\leftrightarrow^*$ is a reflexive transitive closure of $\leftrightarrow$.

From the point of view of the membership protocol at a given process, all processes in an sf-group are treated as a unit: one failure agreement algorithm is used for the entire group and they are eventually removed from the membership list simultaneously. Thus, as execution of a process proceeds, there are a series of sf-groups totally ordered with respect to one another. Specifically, an sf-group S_2 is said to *follow* another sf-group S_1 if the last message sent by any process in S_2 follows the last message sent by all processes in S_1. In this situation, the failure agreement for S_2 is typically performed after the failure agreement for S_1.

Notice, however, that it is also correct to perform the agreement for S_1 and S_2 simultaneously; i.e., as if they were a single sf-group. This type of merging may be necessary in certain situations, such as if one or more processes in S_2 fail before they can participate in the failure agreement associated with S_1, or if an alive process receives the protocol initiation message for a process in S_2 before receiving all messages associated with the protocol for S_1. Interestingly, since messages are received in partial order, this latter situation can result in different sf-groups being formed at different functioning processes. As discussed more fully in Section 4.3, our protocol exploits this fact to allow some processes to remove processes from the membership list earlier than others, while still preserving the semantic correctness of the application.

The other type of membership list modification is the addition of recovering processes. In this case, it is sufficient to add a process to the membership list at every process sometime before the recovered process sends its first message. Once this incorporation is complete, the process participates normally in system activity, including execution of future membership protocol agreement algorithms. Since the set of alive processes must be same at all the processes while executing—for example, a failure agreement algorithm—the recovering process must be incorporated at the same logical time with respect to all other membership protocol events at all processes.

In summary, the order in which membership list modification events are handled is as follows:

1. All processes in the same sf-group are removed simultaneously. The order

of removal of processes in different sf-groups follow the relative order of the sf-groups.

2. A recovering process is incorporated in the membership list at the same logical time at all the processes.

4.2 Protocol Preliminaries

For simplicity, we first define the terms and data structures used by the protocol. We say a message m_2 *immediately follows* m_1 if there is a direct edge from m_1 to m_2 in the context graph. We say m_2 *follows* m_1 if there is a path from m_1 to m_2 in the context graph. A process p is *suspected down* if it is in the membership list and a $\langle p$ is down$\rangle$ message has been received. Similarly, a process p is *suspected up* if it is not in the membership list and $\langle p$ is up$\rangle$ has been received.

The membership protocol maintains two lists: SuspectDownList and SuspectUpList. SuspectDownList contains the list of $\langle p$ is down$\rangle$ messages that have been received and SuspectUpList contains the list of $\langle p$ is up$\rangle$ messages that have been received. As described below, messages are removed from these two lists once the process can reach a conclusion about the status of each p. The protocol also maintains an integer variable count that contains the total number of messages in SuspectUpList plus the number of unstable messages in SuspectDownList. Initially the value of count is zero and the two lists are empty.

We define the following logical times related to process failures and recoveries, where by logical time we are referring to a wave in the context graph.

Suspected Down Time (SDT): The SDT of a failed process p is the logical time containing the $\langle p$ is down$\rangle$ message.

Actual Down Time (ADT): The ADT of a failed process p is the earliest logical time such that there are no messages from p at or after ADT.

Realized Down Time (RDT): The RDT of a failed process p is the logical time when p is masked out of the membership list.

Suspected Up Time (SUT): The SUT of a process p is the logical time containing the $\langle p$ is up$\rangle$ message.

Realized Up Time (RUT) The RUT of a process ML is the logical time when
p is masked back into the membership list.

Furthermore, define a *membership check state* as a state where `Suspect-`
`DownList` or `SuspectUpList` is non-empty. In a similar manner, let a
membership check period be the time interval over which the system is in a
membership check state. A membership check period always starts at the SDT
or SUT of some process and ends when the two suspect lists are empty. In
other words, a membership check period starts when `count` becomes non-
zero and continues until count becomes zero. The end of a membership check
period is always signified by the RDT of some process.

In the membership protocol, a message is considered stable if it is followed
by messages from all the processes that are not in `SuspectDownList`. Note
that this definition of stability is applicable only in the membership protocol;
all other protocols, including those that run concurrently with the membership
protocol, use the standard definition of stability given in Section 2.

4.3 Membership Protocol

The main idea of the membership protocol is to establish agreement among all
the alive processes about the membership list at the end of the corresponding
membership check period. Thus, if there are n processes of which k are
suspected to have failed, agreement on the failure of the k processes is reached
if none of the other $n - k$ processes have a message from any of the suspected
processes at the same logical time as the first protocol initiation message sent
by the detection protocol. To maintain external consistency, the membership
protocol also synchronizes the RUT of rejoining processes and the SDT of
failed processes among all the alive participants.

Informally, the membership protocol may be described as follows. Upon
receiving a ⟨p is down⟩ message, the message is added to `Suspect-`
`DownList`. Upon receiving a ⟨p is up⟩ message, the message is added to
`SuspectUpList`. The message associated with a suspected down process
is subsequently removed from `SuspectDownList` if there is any process
that has evidence to contradict the hypothesis that it has failed; that is, if a
⟨Nack, p is down⟩ message is received immediately following the ⟨p is
down⟩ message. A suspected up process is removed from `SuspectUpList`
and added to the membership list as soon as the appropriate ⟨p is up⟩

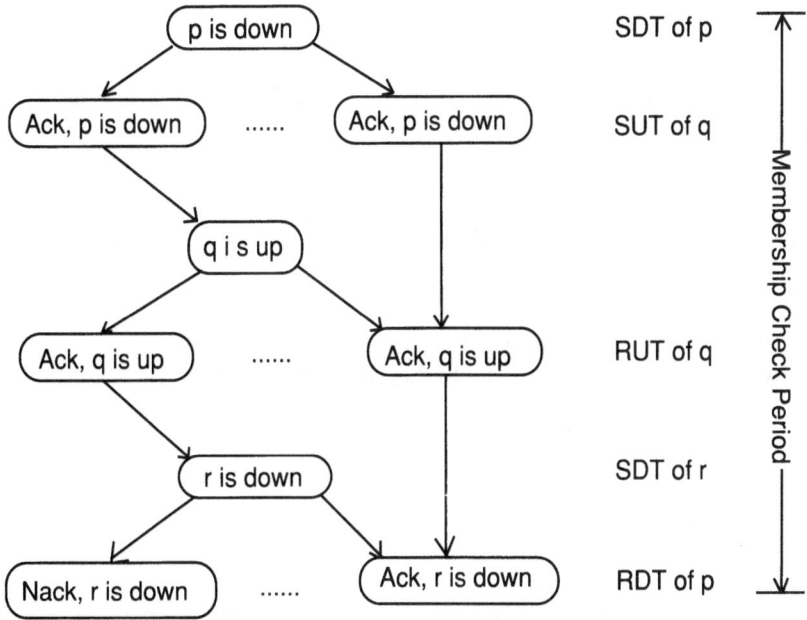

Figure 3: View Representing a Membership Check Period

message becomes stable. The membership check period ends when all messages in SuspectDownList become stable and SuspectUpList becomes empty. At this point, all of the suspect down processes are removed from the membership list, and SuspectDownList is reinitialized to empty.

Figure 3 illustrates one possible scenario, in which processes p and r are checked for a possible failure and process q rejoins the membership list. The membership check period starts with the arrival of the ⟨p is down⟩ message. At the end of this period—that is, the RDT of p—process p is removed from the membership list. However, r remains since ⟨r is down⟩ is immediately followed by ⟨Nack, r is down⟩.

Figure 4 gives a more formal description of the protocol based on actions taken by a given process upon receipt of each type of message. The phrase "count is adjusted" means that the variable count is assigned the number of messages in SuspectUpList plus the number of unstable messages in SuspectDownList. In addition, if count goes to zero as the result of processing a message, then SuspectDownList is reinitialized to empty and

Message ⟨p is down⟩:

If a message from p at the same logical time as ⟨p is down⟩ has been received, then send ⟨Nack, p is down⟩; otherwise send ⟨Ack, p is down⟩, stop accepting messages from p, insert ⟨p is down⟩ in SuspectDownList, and adjust count.

Message ⟨Nack, p is down⟩:

Remove ⟨p is down⟩ message from SuspectDownList and start accepting messages from p.

Message ⟨Ack, p is down⟩:

If message ⟨p is down⟩ is stable then decrement count.

Message ⟨p is up⟩:

Send ⟨Ack, p is up⟩, insert ⟨p is up⟩ in the SuspectUpList, and increment count.

Message ⟨Ack, p is up⟩:

If ⟨p is up⟩ is stable, then incorporate p into the membership list, remove ⟨p is up⟩ from SuspectUpList, and adjust count.

Figure 4: Membership Protocol

the corresponding processes are removed from the membership list.

The key observation about the algorithm is that different processes can form different sf-groups, or equivalently, different processes can have different membership check periods. The reason is that the membership protocol includes a process p for consideration as soon as it receives ⟨p is up⟩ or ⟨p is down⟩ protocol initiation message. To see the effect of this, consider the scenario outlined in Figure 5; here the numbers above the nodes represent the order in which messages are received. The process on the left receives ⟨q is down⟩ before the final ⟨Ack, p is down⟩ message.[1] This is possible because there is no path from the final ⟨Ack, p is down⟩ to the ⟨q is down⟩ message. Thus, both p and q are considered in one membership check period. In contrast, the process on the right receives all ⟨Ack, p is down⟩ messages before receiving ⟨q is down⟩. As a result, the right-most process

---

[1]Keep in mind that any pair of messages in the context graph for which there is no path leading from one to the other could have been received by the process in either order, and in fact, in different orders at different processes.

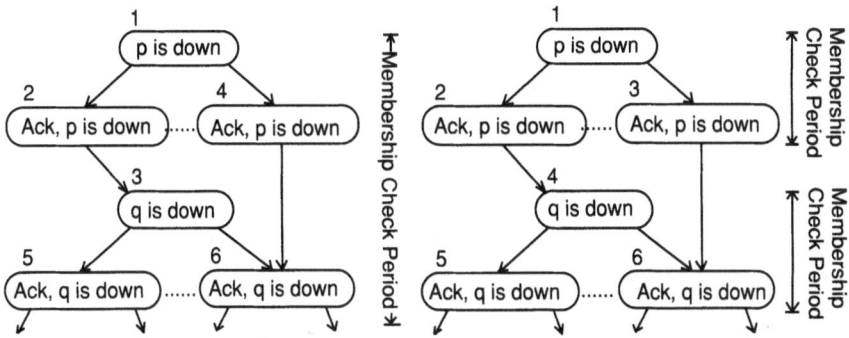

Figure 5: Two Different Processes' View

uses two membership check periods. The implication of this situation is that p is removed from the membership list earlier at the right process than in at the left process, thus allowing its execution to proceed without unnecessary delay.

4.4   Correctness Arguments

We now argue that internal consistency and external consistency are maintained by the membership protocol.

Theorem 1 *All messages from process p have been received by a process q before q decides the failure of p.*

Proof: The set of messages from process p is the union of all messages from p received at all alive processes. At the RDT of p, the message ⟨p is down⟩ is immediately followed by message ⟨Ack, p is down⟩ from every alive process. Thus, all messages sent by p that have been received by any alive participants prior to sending ⟨Ack, p is down⟩ have been received by q. Since a process ignores any further incoming messages from p as soon as it sends ⟨Ack, p is down⟩ q receives no other messages from p. Thus, q has received all messages from p at its RDT; i.e., when it decides that p is down. □

Theorem 2 *An alive process receives all the messages in the conversation.*

Proof: The membership check period ends at the RDT or RUT of any participant. From Theorem 1, we know that p has received all messages from failed participants. Since the membership check period terminates at a wave that contains messages from all alive processes (ACK or NACK), all messages from all alive processes have also been received. Thus, a processes receives all messages in the conversation in the membership check period. Since Psync guarantees the delivery of all the messages in the normal state (i.e., when a process is not in a membership check state), an alive process receives all the messages in the conversation. □

Theorem 3 *A recovering process is incorporated in the conversation at the same logical time at all alive processes.*

Proof: A recovering process p is incorporated into the membership list when the ⟨p is up⟩ message is immediately followed by a ⟨Ack, p is up⟩ message from every alive process, i.e., when the wave containing the message ⟨p is up⟩ is complete. Since this wave is the same at all processes, p is added to the membership list at the same time by all processes. □

Theorem 4 *Stability and completeness decisions taken in the application during the execution of membership protocol are valid.*

Proof: Stability decisions are correct if a message is decided stable when it is followed by messages from every alive process. The stability check of a message m can go wrong if some process that has a message m' following m is not considered for the stability check and m' has not been received. If a process p fails, then it is removed from the membership list at its RDT. Thus, the failed participant is removed from consideration for stability checks only after its RDT. Since all the messages from a failed participant have been received by its RDT (Theorem 1), all of its messages have been received before it is removed from consideration for stability checks. Thus, removal of a failed participant does not introduce any inconsistency in the stability checks. When a participant p recovers, it starts sending messages only after its RUT. Since the recovering participant is incorporated into the membership list at its RUT, it is considered for stability checks for all the messages at or following RUT. Thus, stability checks remain consistent during the recovery of a process.

Completeness decisions are correct if a wave is decided to be complete only when it contains all of its messages. As noted in Section 2, this is typically done by testing the stability of a message in the wave. In the presence of failures, however, a failed process is not considered to determine stability, and a wave may contain messages from processes that later fail. This does not cause a problem because a failed process is not considered when determining the completeness of a wave only after it has been removed from the membership list; i.e., after its RDT. But by Theorem 1, all messages from failed processes have been received by its RDT. Therefore, a wave that is determined to be complete has all messages from failed processes and all the alive processes (since all the alive processes are considered to determine stability), and as a consequence, removing a failed participant does not cause any inconsistency in wave completeness checks. A recovering participant p starts sending messages after its RUT, so no waves between its RDT and RUT inclusive contain a message from p. Since p is added to the membership list at its RUT, it is considered for completeness checks for all succeeding waves. Thus, completeness decisions are consistent during the process of participant recovery. □

Theorem 5 *Alive processes have received all stable messages.*

Proof: A message m is declared stable when there is a message from all processes in ML following m in the context graph. Thus, all the processes that were alive at the time m was declared stable have received it. However, a process that was down at that time may not have received it. All the messages that are in the context graph of an alive process and that are missing in the context graph of the recovering process are retransmitted. Since message m is stable, it is retransmitted if it is not in the context graph of the recovering process. A recovering process starts participating in the conversation only after the retransmission of all the missing messages is complete, in particular after the receipt of message m. Thus, a failed process that rejoins the membership list has received all the messages that are declared stable. □

Before considering the next theorem, we observe that we can always consider the logical time at which a given membership check period started as the same in all alive processes. This follows from the property that consecutive membership check periods (or equivalently, sf-groups) can be combined into a single membership check period without changing the semantics. Thus, given a specific check period, we can successively merge earlier check periods as required in each process to obtain a single membership check period that starts at the same logical time everywhere. This iteration is guaranteed to stop since the initial membership check period must start at the same time in every process. Note that this merging is done only for the purposes of the proof, and is not actually reflected in the protocol itself.

In the following theorem we prove that if one process decides that a process p has failed, every alive process eventually decides the same.

Theorem 6 *Let L_p and L_r be the set of alive participants at the end of membership check period containing* $\langle q$ is down$\rangle$ *in views of participants p and r respectively. If $r \in L_p$ and $p \in L_r$ then $q \in L_p \Leftrightarrow q \in L_r$.*

Proof: Using the above observation, we logically combine sufficient previous membership check periods so that the periods in the two processes start at the same time.

Let $Q = L_r - L_p = \{q_1, ..., q_k\}$. That is, Q is the set of processes that are in membership list of process r but not p at the end of the membership check period.

Assume that $q \in Q$. Since q is in the membership list of r, the message $\langle q$ is down$\rangle$ is immediately followed by $\langle$Nack, q is down$\rangle$ from some participant (say s) in the view of participant r. Since q is not in the membership list of p, message $\langle q$ is down$\rangle$ is not followed by any $\langle$Nack, q is down$\rangle$ message in p's view. Since p receives every message from all participants in L_p, and has not received the $\langle$Nack, q is down$\rangle$ message from s, then s must belong to Q. Thus, we have the following:

> **[A] :** For every $q \in Q$, the $\langle q$ is down$\rangle$ message is followed by a $\langle$Nack, q is down$\rangle$ message from some participant $q_i \in Q$ in the view of r but not in the view of p.

Next, since $r \in L_p$, p has received every message from r. Since p has not received any of the NACKs, r must have sent $\langle$Ack, q is down$\rangle$ for every $\langle q$ is down$\rangle$ message where $q \in Q$. Thus, we have the following:

> **[B] :** r has sent $\langle$Ack, q is down$\rangle$ for every $\langle q$ is down$\rangle$ message for all $q \in Q$.

Now, since p has not received any of these NACKs and has received all of the $\langle q$ is down$\rangle$ messages, none of these NACKs are followed by a $\langle q$ is down$\rangle$ message. So, we conclude the following:

[C] : For any $q \in Q$, the message $\langle$Nack, q is down$\rangle$ is not followed by any $\langle$p is down$\rangle$ message where $p \in Q$.

In the following discussion, let s be some arbitrary process in ML. We say that "q is NACKed by s" if process r has received the message $\langle$Nack, q is down$\rangle$ from s, and this is the first NACK received immediately following the $\langle$q is down$\rangle$ message. Also, we say "q is freed before s" if q is NACKed before s; i.e., the first NACK message immediately following $\langle$q is down$\rangle$ is received before the first NACK message immediately following $\langle$s is down$\rangle$.

Let q be NACKed by s, where $s, q \in Q$. From [C], we see that $\langle$Nack, q is down$\rangle$ is not followed by $\langle$s is down$\rangle$ message. Since $\langle$Nack, q is down$\rangle$ is sent by s and [B] implies that r sent $\langle$Ack, s is down$\rangle$ in response to $\langle$s is down$\rangle$, r must have received the $\langle$s is down$\rangle$ message before receiving $\langle$Nack, q is down$\rangle$. Since r received the $\langle$s is down$\rangle$ message and later received $\langle$Nack, q is down$\rangle$ from s, s must have been freed before r received $\langle$Nack, q is down$\rangle$. This implies that s has been freed before q. So, we have the following:

[D] : If a participant q_i is NACKed by another process q_j, then q_j is freed before q_i for all $q_i, q_j \in Q$.

From [A], every participant in Q must be NACKed by some other participant in Q, and from [D] we see that every participant in Q has a participant in Q that was freed before it. Since Q is a finite set and "freed before" is a total relation, this is impossible for any finite non-empty Q. Thus, the set Q must be empty.

Therefore, $Q = L_r - L_p = \phi$. Thus $L_r = L_p$. So, $q \in L_r \Leftrightarrow q \in L_p$. $\square$

Theorem 7 *The removal of failed processes from the membership list follows the total order of sf-groups and all processes in one sf-group are removed simultaneously.*

Proof: Trivial. $\square$

In summary, we have shown the following. First, since a process starts accepting messages from a recovering process as soon as the recovering process is added to the membership list, by Theorem 3 we conclude that every alive process starts accepting messages from a rejoining process at the same time in the view. Theorem 6 confirms that the decision reached by the membership protocol about a failed (or suspected failed) participant is consistent at all alive processes. Thus, external consistency is ensured by Theorems 3 and 6. Similarly, the internal consistency is ensured by Theorems 2, 4 and 5. Finally, Theorem 7 guarantees that the order of the removal of failed processes is correct.

5 Concluding Remarks

This paper presents a membership protocol based on the partial ordering of messages provided by the Psync multicast mechanism. The protocol—along with a complete system that includes ordering, recovery, and detection protocols—is implemented on a network of Sun3 workstations running the x-kernel [16]. To demonstrate the system, we have also implemented a replicated directory object [15]. We are currently performing experiments to measure the overhead of our algorithm under different situations.

As mentioned in the Introduction, our membership protocol is a system-level protocol that is used to ensure the consistency of decisions made at that level. It does that by guaranteeing that stability and completeness decisions are made correctly and that such queries do not block even in the presence of processor failures and joins. Although not designed to provide information directly to higher levels, it can, however, be easily adapted to support such a facility by including membership change events in the portion of the context graph seen by the application. This can be done by simply transforming the message that prompts the current protocol to change the membership list into a message that is then propagated to the application. The receipt of this message would then be used by the higher levels to make decisions that require knowledge of the collection of functioning processors.

The protocol has some distinct advantages over others proposed in literature. One advantage is the way it handles simultaneous failures and recoveries. In other protocols that have been proposed [12, 8, 13, 10], the protocol is restarted if a subsequent failure or recovery is detected while the agreement algorithm is being executed. However, this is not in the case in our protocol—failures or recoveries that are detected while the protocol is in progress are taken into account incrementally by updating `SuspectUpList` or `SuspectDownList` appropriately.

A second advantage is the way our protocol relaxes the requirement that removal of a failed process from the membership list be totally ordered with respect to all other events. In particular, a process waits to update its membership list only until it has determined the last message sent by the failed process; it need not wait for other processes to update their membership lists. In contrast, other protocols force a process to wait until all alive processes have confirmed the failure.

A final advantage is that removal of failed processes from the membership list need not be done at the same time at all the processes. This results from the fact that sf-groups are created dynamically at each process, and these groups need not be the same at all processes. Thus, a process that does not have to merge two sf-groups will be able to remove the members of the first group before another process that does have to merge the two groups. This improves the efficiency of the application and simplifies the design of the protocol.

Acknowledgments

Discussions with F. Schneider and P. Verissimo helped to clarify various issues involving membership protocols. Also, the referees provided many useful comments on earlier drafts of this paper that helped improve its presentation.

References

[1] F. Cristian, "Probabilistic clock synchronization," in *Ninth International Symposium on DCS*, (Newport Beach, CA), pp. 288–296, Jun 1989.

[2] J. Y. Halpern, B. Simons, R. Strong, and D. Dolev, "Fault-tolerant clock synchronization," in *Third ACM Symposium on PODC*, (Vancouver, Canada), pp. 89–102, Aug 1984.

[3] H. Kopetz and W. Ochsenreiter, "Clock synchronizatin in distributed, real-time systems," *IEEE Transactions on Computers*, vol. C-36, pp. 933–940, Aug 1987.

[4] K. Birman and K. Marzullo, "The role of order in distributed programs," Tech. Rep. 89-1001, Department of Computer Science, Cornell University, 1989.

[5] H. Garcia-Molina and A. Spauster, "Message ordering in a multicast environment," in *Ninth International Conference on DCS*, (Newport Beach, CA), pp. 354–361, Jun 1989.

[6] P. Kearns and B. Koodalattupuram, "Immediate ordered service in distributed systems," in *Ninth International Conference on DCS*, (Newport Beach, CA), pp. 611–618, Jun 1989.

[7] L. Lamport, "Time, clocks, and the ordering of events in a distributed system," *Communications of the ACM*, vol. 21, pp. 558–565, July 1978.

[8] F. Cristian, "Agreeing on who is present and who is absent in a synchronous distributed system," in *Eighteenth FTCS*, (Tokyo), pp. 206–211, Jun 1988.

[9] H. Garcia-Molina, "Elections in a distributed computing system," *IEEE Transactions on Computers*, vol. C-31, pp. 49–59, Jan 1982.

[10] H. Kopetz, G. Grunsteidl, and J. Reisinger, "Fault-tolerant membership service in a synchronous distributed real-time system," in *International Working Conference on Dependable Computing for Critical Applications*, (Santa Barbara, California), pp. 167–174, Aug 1989.

[11] P. Verissimo and J. Marques, "Reliable broadcast for fault-tolerance on local computer networks," in *Ninth IEEE Symposium on Reliable Distributed Systems*, pp. 54–63, Oct. 1990.

[12] K. Birman and T. Joseph, "Reliable communication in the presence of failures," *ACM Transactions on Computer Systems*, vol. 5, pp. 47–76, Feb. 1987.

[13] J. Chang and N. Maxemchuk, "Reliable broadcast protocols," *ACM Transactions on Computer Systems*, vol. 2, pp. 251–273, Aug. 1984.

[14] L. L. Peterson, N. Buchholz, and R. D. Schlichting, "Preserving and using context information in interprocess communication," *ACM Transactions on Computer Systems*, vol. 7, pp. 217–246, Aug. 1989.

[15] S. Mishra, L. L. Peterson, and R. D. Schlichting, "Implementing fault-tolerant replicated objects using Psync," in *Eighth IEEE Symposium on Reliable Distributed Systems*, pp. 42–52, Oct. 1989.

[16] N. C. Hutchinson, L. L. Peterson, M. Abbott, and S. O'Malley, "RPC in the *x*-Kernel: Evaluating new design techniques," in *Proceedings of the Twelfth ACM Symposium on Operating System Principles*, pp. 91–101, Dec. 1989.

A Fault-Tolerant Architecture for the Intellectual Distributed Processing System

TOSHIBUMI SEKI, YASUKUNI OKATAKU, SHINSUKE TAMURA

Toshiba Corporation

Systems & Software Engineering Laboratory

70, Yanagi-cho, Saiwai-ku, Kawasaki, Kanagawa, 210 Japan

Abstract

The *Intellectual Distributed Processing System* (IDPS) is a system architecture to realize highly extensible and adaptive distributed systems. The reliability of the IDPS relies on replicated objects model, and this paper presents two new mechanisms for implementing that model on the IDPS. One is a fail-stop broadcast communication protocol, where replicated objects receive the same messages in the same order. The other is a commitment method, where each object receives only correct messages. By these mechanisms, an individual object does not need to be aware of the replication degree and the location of relevant objects. Moreover, each object can communicate with other individual objects without confirming message transmission. Therefore, the overhead for the fault-tolerant mechanisms can be maintained at a small level.

The IDPS fault-tolerant architecture makes it possible to realize highly reliable systems without reducing excellent extensibility, adaptability and processing efficiency.

Key Words: Fault-tolerant, distributed system, reliable broadcast protocol, replicated object, distributed operating system.

1 Introduction

The *Intellectual Distributed Processing System* (IDPS) [1, 2] is a declarative system architecture based on an object model. The IDPS consists of simply a set of objects, whose ability is declared independently from others. There is no centralized managing mechanism. Each object finds its own roles adaptively and dynamically through negotiations, so as to accomplish given jobs. Therefore, it becomes possible to construct highly efficient, adaptive, extensible, maintainable and reliable systems. This paper addresses a new fault-tolerant architecture for the IDPS.

As a method to realize highly reliable systems without spoiling the flexibility for distributed systems, the replicated objects model, in which replicated objects are invoked by broadcast messages, is well known. Because of broadcast communication, individual objects can behave independently from the existence and location of relevant objects' replicas. Replicated objects execute the same tasks in parallel and asynchronously. These properties make systems extensible, maintainable and so on. Also, because there is no necessity for synchronization among replicas, communication overhead among objects can be maintained at a small level, so as to be applicable to real time control systems. However, in order to guarantee the correct action of this model, replicated objects must receive the same messages in the same order, and must produce the same results. Although many methods have been proposed about replicated object models, in these methods the existence of a reliable broadcast communication is assumed in advance [3, 4, 5], or some time consuming synchronization mechanisms is required, such as two phase commitment protocol [6, 8], for assuring the correct behavior of replicated objects. Therefore, they are not applicable to systems where quick response is required.

This paper proposes new fail-stop broadcast communication protocol and first-*CN*-come method. The proposed broadcast communication protocol guarantees that normally operating replicated objects receive the same messages in the same order without exchanging acknowledgements. The first-*CN*-come method assures that all replicas of the same objects can reach the same correct decision. By these mechanisms, individual objects can work independently and asynchronously without knowing about the existence and location of replicated objects. Therefore, highly reliable systems can easily be realized, while retaining the system performance and flexibility high. For example, each object's replication degree can be adjusted in proportion to its required reliability.

2   IDPS Distributed System Model

The proposed fault-tolerant mechanisms are designed to work on the IDPS distributed computer system, which is connected by a local area network(LAN), as in Fig. 1. The communication unit in individual sites and the LAN can be replicated. Several kinds of objects are distributed on sites in the system. Each object is an encapsulated set of data and procedures. These objects can behave

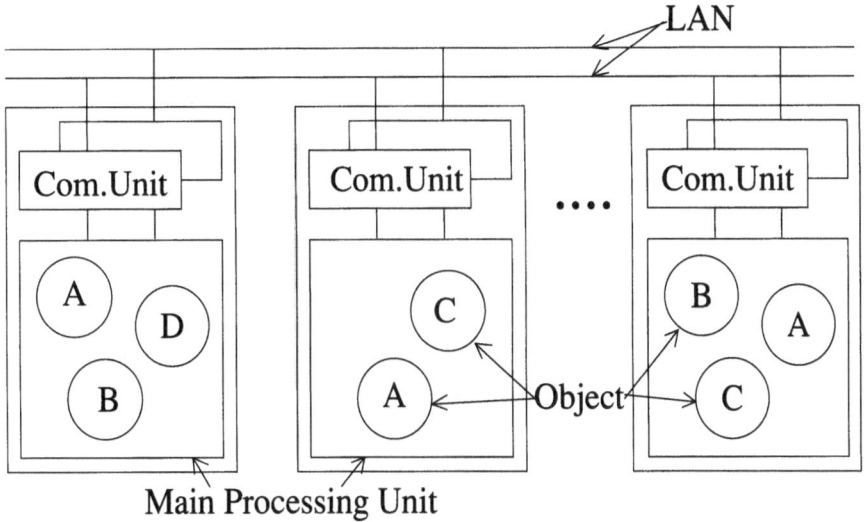

Figure 1: IDPS System Configuration

independently from their location and existence by exchanging information through broadcast communication. Therefore, it is possible to create objects at any site and to migrate anywhere, according to the system state.

An individual site is composed of two parts. One is a communication unit, in which the proposed fail-stop broadcast communication protocol is installed. The other is a main processing unit, in which application and system objects and the IDPS operating system kernel exist. These two units behave in parallel.

2.1 Replicated Objects

An active replication method [9, 3, 4], and **a passive replication method** (*primary-standby model*) [10, 5] exist for constructing highly reliable systems. In both methods, important objects are replicated on different computers. In the active replication method, each replica carries out the same operation in parallel. An example of an objects' behavior in the active replication method is shown in Fig. 2, where objects A, B, C and their replicas act in parallel. Individual $A_i(i=1,2,3)$ calls B1 and B2. Then, individual $B_i(i=1,2)$ also calls all $C_i(i=1,2,3)$. In the passive replication method, usually only one replica is active, and the other replica(s) is(are) in stand-by state. One of stand-by replicas

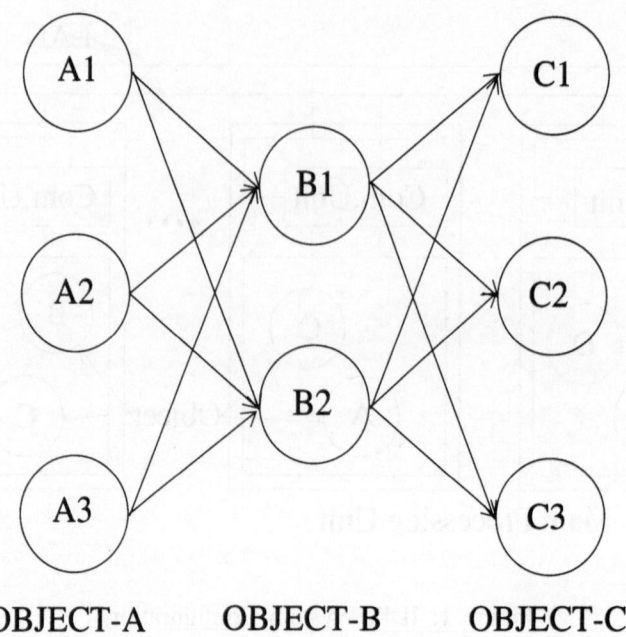

OBJECT-A   OBJECT-B   OBJECT-C

Figure 2: Active Replication Method

is activated, when a corresponding active replica fails. In both methods, the IDPS must preserve the following properties:

PR1) The reliable system must also maintain extensibility and adaptability. Namely, objects must be able to be replicated at any site and to any degree without affecting others.

PR2) Processing efficiency should not be reduced. Namely, computation and communication load, required for such as synchronization among replicas, must be minimized.

PR3) Even if any objects were to fail, the other objects must continue processing. This means that the behavior of replicas is governed by the replicas themselves.

In the passive replication system, hardware resources can be used effectively. However, a stand-by object has to receive the internal state information from its primary object at every checkpoint. Moreover, the time required for switching from stand-by to primary is not negligible in the system, which especially requires quick response. On the other hand, in the active replication system, objects can continue processing without any momentary stop, while at least one replica of each object is alive. IDPS application fields have to cover real time control systems, which require quick response. Therefore, in this paper, the IDPS fault-tolerant architecture for the active replication system is explained.

2.2 Active Replication Method Requirement

In order to satisfy the previous three properties, the IDPS fault-tolerant architecture adopts broadcast communication. Because replicated objects are invoked by broadcast communication, an individual object does not need to be aware of the existence and the location of replicated objects. Also, there is no need for replicated objects to synchronize with each other.

The IDPS fault-tolerant architecture realizes the following requirements distributedly, in order to assure that replicated objects produce the same correct results and maintain consistency among replicated objects, with small overhead.

RQ1) Replicated objects which are alive receive the same messages in the same order.

RQ2) Objects which are alive receive only correct messages.

In usual broadcast communication, an object which sends a message can not confirm that all relevant objects accept that message, without exchanging acknowledgments. Therefore, a large amount of communication overhead is necessary for satisfying the above requirements. In the IDPS fault-tolerant architecture, many synchronization mechanisms, such as exchanging acknowledgements, are removed. The first requirement(**RQ1**) is satisfied by the fail-stop broadcast communication mechanism, which is explained in Section 3. The second requirement(**RQ2**) is satisfied by the first-CN-come method, which is explained in Section 4.

These requirements should also be satisfied for the passive replication model, to guarantee correct behavior.

2.3 System Environment

The IDPS fault-tolerant architecture works on computer systems connected by a LAN, under the following system environment:

EN1) Each object behaves in a deterministic way. Namely, replicated objects produce the same results, if they receive the same messages in the same order.

EN2) A LAN conveys only one message at a time, and a communication unit on each site processes messages through the LAN in first-in first-out(FIFO) order. This item does not mean the exclusion of message collision on a LAN.

EN3) Communication environment is not reliable, so messages may be lost or doubly sent. However, network partitions do not occur.

EN4) Faulty objects may send erroneous messages to other objects.

EN5) Two kinds of faults may occur in a site. One is a temporary fault, such as message reception error, because of over-run error or buffer-overflow. The other is a permanent fault, such as site power down. Any uncontrollable state of a site, which disables any communication with other sites, also can be considered as a permanent fault. A site with a permanent fault can be regarded as fail-stopped, because it does not have any effect on other sites.

In the followings, the words *object's malfunction* and *site's malfunction* are used for the same meaning, because an object's malfunction is caused by a site's malfunction.

3   Fail-Stop Broadcast Communication

The fail-stop broadcast communication protocol is designed to accomplish requirement **RQ1**.

Although various reliable broadcast protocols have already been proposed, they can not have satisfied **RQ1** in a desirable way. In two phase commitment protocol [6, 8] and three-phase commitment protocol [7], each site or object has to recognize the number of sites in the system. Moreover, cumbersome extra operations are necessary. Chang [11] and Kaashoek [12] have proposed reliable broadcast communication by assuming the existence of a token site or a sequencer node. However, a token site or a sequencer node is a centralized manager, and it spoils the good properties of distributed systems. Mishra [13] and Smith [14] have proposed another protocol, in which each site holds communication logs and adds the log to its sending message. Although these protocols provide excellent distributed mechanisms, they entail the following difficulties. Each site must hold large communication logs and it requires a relatively long time for processing these logs. Moreover, message retransmission cannot be requested until the malfunctioned site sends the next message.

As mentioned above, there is no protocol yet which satisfies both the quick response requirement and the fully decentralized mechanisms.

3.1 Basic Protocol and Features

The proposed protocol is implemented on the LAN's MAC layer of each communication unit. The protocol is simple, that is, each communication unit is always counting the number of broadcast messages on the LAN, and maintains it as *Accumulated Message Number*(AMN). When an object broadcasts a message, a communication unit attaches the AMN value, which is stored in its memory, to the sending message. AMN corresponds to the message log used in the protocols proposed by Mishra and Smith [13, 14]. In order to simplify the following explanation, AMN is composed of the number of messages on the LAN in this section, but it is also possible to use accumulated message length or check-sum value as AMN. Because the LAN conveys the same message to all sites at a time, it is obvious that normally operating communication units' AMNs are identical. Therefore, each communication unit compares an AMN attached to the received message with the latest AMN stored in its memory when it receives a message. When two AMNs are not equal, it can be considered that either the sender or the receiver site is malfunctioned. However, the receiver can not determine which site is malfunctioned. Therefore, the receiver sends a *site fail* warning message, which means *you are wrong* to the message sender, and queues the received message at the same time. The malfunctioned site is determined in the following manner. When the communication unit receives the *site fail* warning messages from more than *a pre-defined number(PDN)* of sites within *a pre-defined period*, it recognizes that it has malfunctioned, and immediately stops processing or requests retransmission of the message. The constant number(*PDN*) is determined in accordance with the degree of the system reliability. Namely, by increasing the *PDN*, malfunctioning objects can be detected at any high accuracy level and can be removed from the system automatically, without propagating any adverse influence on other objects, which are working in normal conditions.

By this protocol, even the malfunctioned receiver can also be detected im-

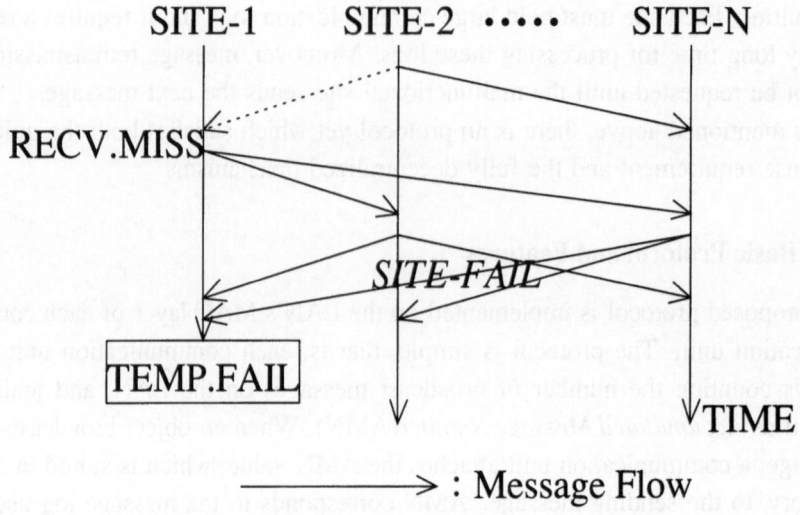

Figure 3: Message Reception Error Case 1

mediately, because the malfunctioned sites receiving the messages detect AMN difference and send a *site fail* warning message with the illegal AMN.

There is a case where a communication unit fails doubly, such that Site-A's sending message does not reach any other sites, and Site-A fails to receive a message from other sites. In such a case, AMNs in all sites coincide, and malfunctions can not be detected. However, the occurrence probability for such a situation can be decreased, by changing the AMN definition from the message number to the message accumulated check-sum value or something else.

Figure 3 and 4 show two examples. If a message reception error site broadcasts a message, it receives *site fail* warning messages from other sites and stops itself, as shown in Fig. 3. The reason is that the malfunctioning site's AMN is different from the normal site's AMN. In the same way, if a malfunctioned site receives a message, it sends a *site fail* warning message with illegal AMN. Then, it receives *site fail* warning messages from other sites, as shown in Fig. 4.

As mentioned above, the proposed protocol requires only an integer as the AMN, which is attached to a message. Therefore, extra overhead for fault-

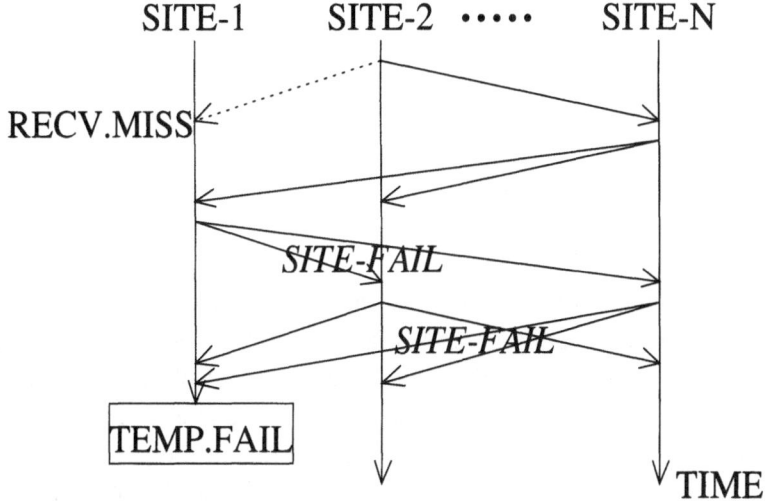

Figure 4: Message Reception Error Case 2

tolerant properties can be reduced drastically, compared with other methods [13, 14]. Also, because each communication unit maintains the AMN value independently, there are not any centralized managers nor any performance and reliability bottle-necks. Moreover, a failed site can be detected, not only by the message sending from the failed site, but also by messages from normal sites. Then, failure recovery operations become simple, because faults can be detected and the lost messages can be retransmitted immediately.

In order to preserve the reception order of messages among replicated objects, a communication unit sends the message to intra-site objects, after completion of sending out the message to the LAN.

When the removed site is repaired and re-established in the system as before, or when a new site is added, AMN value is copied from the other working site. That is, when the failed site is rebooted, the site broadcasts a starting message, which means *I'm ready*. Other sites, which receive the message, reply their current AMN value to the sender site. At the same time, the objects in the repaired site are dynamically copied from the normal sites, where their replicas are running.

The reliability of the proposed fail-stop broadcast protocol depends on *PDN*.

The probability that failed sites would continue operation or that normal sites would stop operation is less than:

$$
\begin{aligned}
&{}_N C_{PDN} * P^{PDN}(1 - P)^{N-PDN} +{}_N C_{PDN+1} * P^{PDN+1}(1 - P)^{N-PDN-1} \\
&+ \cdots +{}_N C_{N-1} * P^{N-1}(1 - P) +{}_N C_N * P^N \\
&< \quad P^{PDN} * ({}_N C_{PDN} +{}_N C_{PDN+1} * P + \cdots +{}_N C_N * P^{N-PDN}) \\
&< \quad P^{PDN} * (N - PDN) * N(N - 1) \cdots (N - PDN + 1)/PDN! \\
&< \quad N * (P * N)^{PDN}/PDN! \quad\quad\quad\quad\quad\quad\quad\quad\quad\quad\quad (1)
\end{aligned}
$$

where N is the number of sites, and P is the probability of each site's failure, which includes the probability of LAN's failure. *PDN* is defined, in advance in accordance with necessary reliability, by this expression. When high reliability is required, a large *PDN* should be selected. For example, consider the case where $P=0.001$, $N=10$, *PDN*=5 and *PDN*=3. The failure probabilities become less than 10^{-11} and 10^{-5}, respectively. The reliability of the communication unit can be locally improved by replicating it, in accordance with the required reliability.

3.2 Message Retransmission

When a communication unit detects a message reception error, it broadcasts a retransmission request for the message with the AMN list. The AMN list consists of a set of AMNs, which should have been attached to lost messages. It can be calculated easily by examining the AMNs discontinuity. A site, which receives the retransmission request, can determine whether or not it should retransmit the lost messages by comparing the requested AMN list with AMN history in its memory.

Therefore, each communication unit saves its sending messages with AMNs, until all sites receive the messages. However, in this protocol, only the storage for a few messages is sufficient. The reason is that, when a communication unit fails to receive messages, the *site fail* warning messages are sent to the LAN immediately. Namely, every communication unit sends at most one message from any site's failure detection time to the time which it receives the *site fail* messages from the other site, if no message contention exists. In an actual system, as message contentions may occur if Ethernet-like LANs are used, each communication unit must save more than two messages. However, the storage for five messages may be sufficient in practice.

Because one message may be sent before failure detection, there is a case where the malfunctioned site sends an extra-message before receiving the *site fail* messages from other sites. In such a case, there is a possibility that replicas receive messages in a different order. However, even in this case, it is sufficient that only the object, which is the destination of the extra-message, is removed from the system and is reloaded by copying the contents of the normal replica from different sites.

The proposed protocol also supports duplicated LAN. Therefore, individual objects can continue communicating with each other, without any momentary break, while at least one LAN acts normally.

3.3   Message Omission Failure

When a message omission failure happens, AMN of the sender site differs from AMNs of other sites. Therefore, an object which sends the omitted messages in the sender site is considered to be failed and stops operation. However, it is not failure. By the proposed protocol, only the replicas of the stopped object continue operation. However, it is not difficult to implement the message retransmission mechanism, if a sophisticated AMN, such as accumulated checksum, is adopted. That is, when a site receives AMNs which are less than the AMN of that site, and more than *PDN* sites show the same difference, the site can recognize the occurence of its sending message omission. The site can retransmit the omitted message in accordance with the difference of AMN, if the other site does not send any message before detecting the message omission. The site, which receives the retransmission messages, must send a message for stopping operation of the message retransmitting site, when it can not receive all messages which must be retransmitted or when it receives new messages before the completion of message retransmission. Otherwise, the message receiving order among sites will be inconsistent.

3.4   A Fail-Stop Broadcast Protocol

This section precisely describes the proposed fail-stop broadcast communication protocol.

Each communication unit logs in its memory the AMNs attached to the messages, which are sent or received by itself, as AMN history. An AMN history record consists of AMN value area and pointer area to a message buffer, as shown in Fig. 5. The latest AMN in the history indicates an AMN

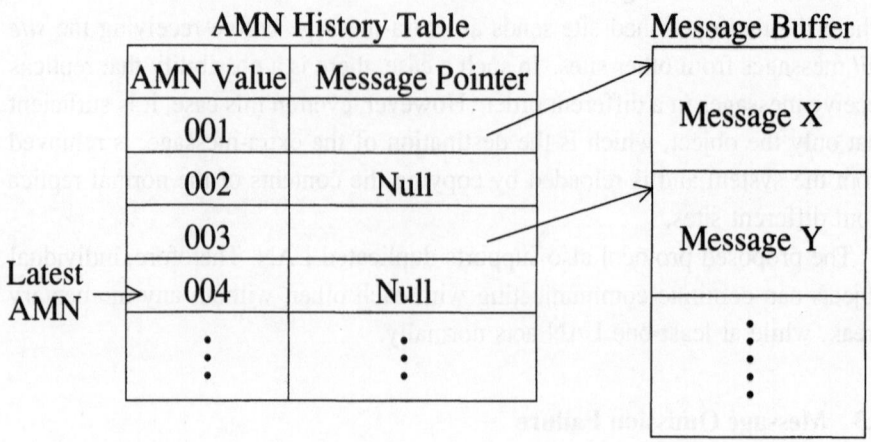

Figure 5: Relation Between AMN History and Message Buffer

to be attached to the message, which will be received or sent out next. The pointer area contains the buffer address for the message which was sent by the unit. The pointer area is used in case of message retransmission. The *site fail* message sender list(SFMSL) stores IDs of sites which sent the *site fail* messages to the unit. AMN history and SFMSL have a ring structure, and a communication unit writes new records over old records.

Figure 6 shows the state transition diagram for the communication unit. Each communication unit can send only *site fail*, *retransmission request*, and retransmission messages, in [INDEFINITE] and [TEMPORARY FAILURE] states. The fail-stop broadcast protocol works as follows:

1. A communication unit attaches the latest AMN stored in the history to its sending message. After completing the message transmission, it sets the message buffer address to the corresponding pointer area of the history. Then, the new record, whose AMN value is greater than the latest AMN value by just one and whose pointer area is NULL, is added to the history.

2. When a communication unit receives a message normally, it compares an AMN attached to the received message with the latest AMN in the history. If the values are equal, the received message is accepted, and the new record, whose AMN value is greater than the latest AMN value by just one and whose pointer area is NULL, is added to the history.

The communication unit also transmits the received message to the main processing unit, if the destination object exists in this site.

If the AMNs do not match, the communication unit queues the received message and sends a *site fail* message with the sender site ID of the received message as a failed processor ID. It also makes an AMN list which indicates the AMNs corresponding to the lost messages. Finally, it changes its state to [INDEFINITE].

3. When the communication processing unit receives a *site fail* message with the failed processor ID which indicates the unit, it changes its state to [INDEFINITE] and adds the *site fail* message sender's ID with the current time to SFMSL.

4. If the number of elements in SFMSL, which is received within a time duration $[C-T1,C]$, exceeds the *PDN*, the communication unit changes the state to [TEMPORARY FAILURE]. Here, C is the current time and T1 is a constant time. It also broadcasts a *retransmission request* message with the AMN list, in order to specify the lost messages. However, if the communication unit has sent extra-messages after failure, it kills the corresponding destination objects in the site.

5. All communication units, which receive the *retransmission request* message with AMN list, can find the messages corresponding to the AMN list by examining their AMN history. Therefore, they retransmit these messages. Then, they change their states to [NORMAL], and process the queued messages.

6. When the communication unit in [TEMPORARY FAILURE] state receives all the retransmitted messages corresponding to the elements of AMN list, it processes the retransmitted messages and the queued messages. Then, it initiates the AMN history and records the latest AMN value in the history. Finally, its state returns to [NORMAL]. However, if the unit can not receive all the retransmission messages within a definite time, it's state becomes [PERMANENT FAILURE].

Figure 7 shows the message format.

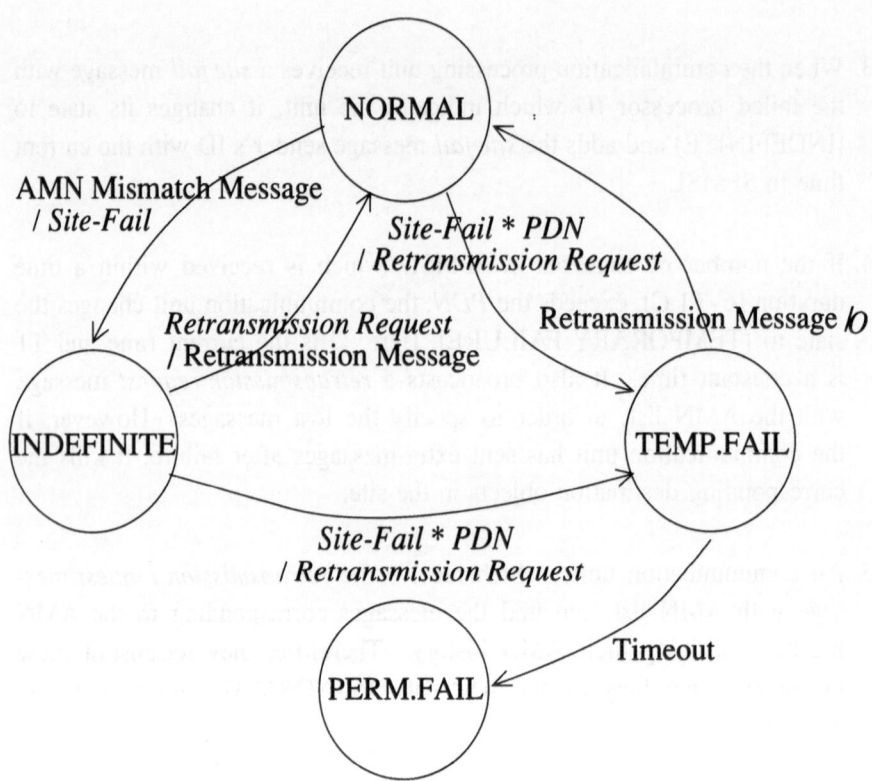

Figure 6: State Transition Diagram for Communication Unit

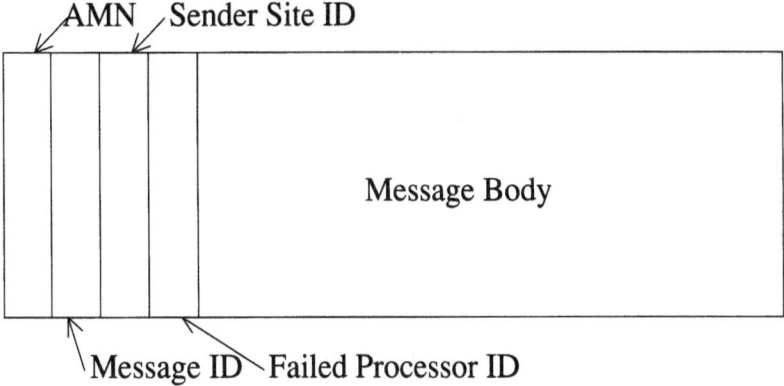

Figure 7: Message Format

4   First-*CN*-Come Method

The first-*CN*-come method is designed to satisfy the requirement **RQ2**. *CN* stands for confirmation number, which is determined in accordance with required reliability of objects. In order to avoid erroneous message reception, the IDPS compares messages sent from replicated objects.

4.1   First-*CN*-come method Mechanism

In conventional comparison mechanisms, voting methods are used. Therefore, each object must know the replication degrees of its relevant objects and wait for the messages from all the replicated objects, so as to determine correct messages.

In the IDPS, each object independently declares its own *CN*, in accordance with the required reliability. When an object receives *CN* messages with the same contents, it accepts the contents. Messages which are received after receiving identical *CN* messages are discarded. Therefore, an individual object does not need to know the replication degree of its relevant objects. Each object only needs to know its own confirmation number(*CN*) and it can behave asynchronously with other replicas. The duration required for message acceptance can also be reduced more than that for the majority based decision method.

Of course, if *CN* is greater than half of the replication degree of relevant objects, the correct message is uniquely determined. Even if the declared *CN* is less than half of the relevant objects' replication degree, an individual object can accept the same message by the first-*CN*-come method. The reason is that fail-stop broadcast communication protocol guarantees that every replica receives the same messages in the same order. The probability that each replica receives the erroneous message or can not receive the correct message also can be calculated by Eq.(1). But in this case, *N* is the replication degree of each object, and *P* is the probability of each object's failure, and *PDN* correspond to *CN*.

If an object's replication degree is smaller than *CN*, which is independently declared by the message destination object, the message can not be accepted within a definite time. At that time, a communication unit finds the message sender's replication degree, and generates new replicas by copying that object, if an application system optionally requires to continue working.

4.2   Message Identifier

In order to realize the first-*CN*-come method, each object must have a naming mechanism for extracting identical messages, which are sent from replicated objects. That is, all replicas of an object in the same state should issue the same message identifier(ID) to their broadcast messages. Moreover, they should issue these IDs independently from others. By these IDs, each object can discriminate received messages for comparison.

The message ID issuing mechanism should work in the following conditions.

CO1) There are several threads in an object, and they act concurrently.

CO2) Although replicated objects activate these threads in the same order in accordance with message reception order, they can not complete these threads in the same order because of the difference in time slice timing.

If only one thread in an object can be active, a message ID, which consists of object name and its sending message count, is sufficient. However, it is not sufficient under the previous conditions. Therefore, IDPS adopts the following message ID.

Message ID = { object name,
                            object activated count,
                            thread's message sending count } ;

The object name identifies the name of the message sending object. The object activated count indicates the number of object's invocations after the system starts up, and the thread's message sending count indicates the number of messages which are sent by that thread. As the replicated objects receive messages and are invoked in the same order by the fail-stop broadcast communication protocol, each replica can specify the thread, which sends the message, by the object activated count. The thread's message sending count specifies the timing of messages sent by the thread.

Therefore, each object can identify messages in the same context by the proposed message ID. Moreover, each replica can issue a message ID by itself, without any synchronization among replicas. Therefore, the object autonomy can be maintained.

5 Implementation and Experience

The proposed fault-tolerant mechanism has been implemented on the IDPS-OS with a fully distributed architecture. It works on 32-bit microprocessors for main processing units and 16-bit microprocessors for communication units connected by duplicated 10 Mbit local area networks. The communication delay for 256 bytes message exchange among objects on different sites is 3.9(ms), using the fail-stop broadcast communication architecture. The overhead, required for attaching and comparing AMNs, is less than a few micro-seconds. It can be considered small enough. Although the communication delay(3.9ms) is larger than V-kernel(2.54ms) [15] and Amoeba(1.4ms) [16], this is mainly due to the hardware performance and the hardware configuration. For example, in this implementation the communication unit must switch memory banks, when it reads a sending message from the main processing unit's memory and writes a received message to the main processing unit's memory. The switching requires about 0.74(ms). Also, a simple replacement of communication unit, through changing the CPU from a 16-bit microprocessor to a 32-bit microprocessor, will improve the communication's speed by a few times.

As an application of the fault-tolerant IDPS architecture, a train traffic control system, which traces the location of trains and controls their directions, so as to insure safe and punctual operation, was developed. It was connected to the railway system simulator in order to measure its performance.

The train traffic control system consists of four different kinds of objects. Those are ADMAP object, TRAIN object, STATION object and SECTION ob-

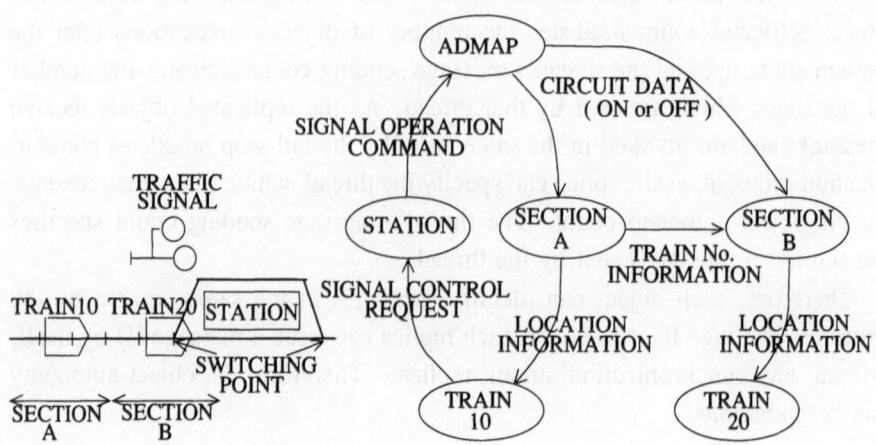

Figure 8: IDPS Train Traffic Control System

ject. Figure 8 shows the relations among these objects. SECTION corresponds to a straight portion of a railway and traces train locations by exchanging messages among its neighboring SECTIONs. It also notifies the location information to the relevant TRAINs. TRAIN is dynamically generated in duplicate on the two least loaded sites, when the train's starting time comes. TRAIN contains its diagram and sends signal control request message to the next STATION, when it reaches the specified location. STATION produces the control signals for the apparatuses in the station in response to the request from TRAINs and sends these signals to ADMAP. Finally, ADMAP sends control signals to the physical apparatuses. ADMAP also detects any change in railway state and sends this information to relevant SECTIONs. Individual TRAIN, STATION and SECTION are fully duplicated. Therefore, even if a site fails and/or a LAN fails, the system continues working normally.

System designers can arbitrarily decide upon each object's replication degree, in proportion to its required reliability. An object's replication degree can also be dynamically increased, by copying the corresponding object to a lower loaded site during operation, so as to improve the system reliability.

The total system performance has been evaluated under the following situation. The railway length is 124(km). There are 60 STATIONs and 624 SECTIONs. 166 TRAINs exist at the same time. All objects are duplicated, and installed on IDPS distributed system, which consists of four sites. In this case, the IDPS fault-tolerant system can control this system under 50% load ratio, and obtain a sufficiently quick response time. Before the proposed fault-tolerant architecure was introduced, message reception errors happened a few times per hour, because of over-run error. However, this system can continue working normally without message reception error by using the proposed fault-tolerant architecture. Moreover, the features for fault-tolerance and recoverability by shutting down one of the sites and then restarting it without affecting the running application are confirmed.

6 Conclusion

The IDPS fault-tolerant architecture has been described. The proposed architecture is composed of two mechanisms. One is to guarantee that replicated objects receive the same messages in the same order. The other is that individual objects receive only correct messages. These mechanisms are realized as the fail-stop broadcast communication protocol and the first-CN-come method. By these mechanisms, replicated objects do not need to be aware of relevant objects replication. Also, because no synchronization among replicated objects is necessary, the overhead for fault-tolerant mechanisms becomes very small.

Both the object model and broadcast communication adopted in the IDPS ensures the complete location and existence transparencies which lead to the excellent extensibility and adaptability of potentials of distributed systems. And by introducing the two mechanisms for fault-tolerant architecture to the IDPS, it becomes possible to realize highly reliable systems without reducing excellent extensibility, adaptability and processing efficiency.

References

[1] S. Tamura, Y. Okataku, T. Endo, and Y. Matsumoto, "IDPS: Intellectual Distributed Processing Systems," in *Proc. Pacific Computer Communications Symposium*, Seoul, Korea, pp. 129–133, 1985.

[2] S. Tamura, Y. Okataku, and T. Seki, "Distributed computer systems in FMS -Intellectual Distributed Processing System-," in *Proc. Int. Conf. Manufacturing Systems and Environment*, pp. 497–502, 1990.

[3] D. Powell, G. Bonn, D. Seaton, P. Verissimo, and F. Waeselynck, "The Delta-4 approach to dependability in open distributed computing systems,"

in *Proc. 18th Int. Symp. Fault Tolerant Computing*, pp. 246–251, June 1988.

[4] R.F.Cmelik, N.H.Gehani, and W.D.Roome, "Fault tolerant concurrent C: A tool for writing fault tolerant distributed programs," in *Proc. 18th Int. Symp. Fault Tolerant Computing*, pp. 56–61, June 1988.

[5] N. A. Speirs and P. A. Barrett, "Using passive replications in Delta-4 to provide dependable distributed computing," in *Proc. 19th Int. Symp. Fault Tolerant Computing*, pp. 184–190, 1989.

[6] M. Takizawa, "Cluster control protocol for highly reliable broadcast communication," in *Proc. IFIP WG10.3 Working Conf. Distributed Processing*, pp. 431–445, Octorber 1987.

[7] S.-W. Luan and V. D. Gligor, "A fault-tolerant protocol for atomic broadcast," in *Proc. 7th Symp. Reliable Distributed Systems*, pp. 112–126, October 1988.

[8] K. P. Birman and T. A. Joseph, "Reliable communication in the presence of failures," *ACM Trans. Comp. Syst.*, vol. 5, no. 1, pp. 47–76, February 1987.

[9] E. C. Cooper, "Replicated distributed programs," *Proc. 10th ACM symp. on Operating system Principles, Operating Systems Review*, vol. 19, no. 5, pp. 63–78, 1985.

[10] J.F.Bartlett, "A nonstop kernel," *Proc. the 8th Symp. on Operating Systems Principles, Operating System Review*, vol. 15, no. 5, pp. 22–29, 1981.

[11] J.-M. Chang and N.F.Maxemchuk, "Reliable broadcast protocols," *ACM Trans. Comp. Syst.*, vol. 2, no. 3, pp. 251–273, August 1984.

[12] M. F. Kaashoek, A. S. Tanenbaum, S. F. Hummel, and H. E. Bal, "An efficient reliable broadcast protocol," *ACM Operating System Review*, vol. 23, no. 4, pp. 5–19, October 1989.

[13] S. Mishra, L. L. Peterson, and R. D. Schlichting, "Implementing fault-tolerant replicated objects using Psync," Tech. Rep. TR 89-3, Department of Computer Science, The University of Arizona, 1989.

[14] P. M. Melliar-Smith and L. E. Moser, "Fault-tolerant distributed systems based on broadcast communication," in *Proc. 9th Int. Conf. Distributed Computing System*, pp. 129–134, June 1989.

[15] D. R. Cheriton, "The V distributed system," *Comm. ACM*, vol. 31, no. 3, pp. 314–333, March 1988.

[16] S. J. Mullender, G. van Rossum, A. S. Tanenbaum, R. van Renesse, and H. van Staveren, "Amoeba: A distributed operating system for the 1990s," *IEEE COMPUTER*, pp. 44–53, May 1990.

Validation

Validation

A New Approach of Formal Proof: Probabilistic Validation

G. FLORIN [+], C. FRAIZE[+*], S. NATKIN [+]

+ *CEDRIC - Centre d'Etudes et de recherche*

en Informatique CNAM

292 rue Saint-Martin 75141 Paris Cedex 03, France

* *GEC-ALSTHOM Tour Neptune La Defense Paris*

Abstract

This paper presents a new concept of validation of distributed or safety-critical systems. The main problem of existing methods is related to the exponential growth of the analysis complexity with the model size. Our method relies on a state transition model which includes a description of operation duration and frequency of events (stochastic Petri nets). The aim of our work is to develop a new approach based on a partial exploration of the reachability set. At the end of the partial exploration we can demonstrate, for a given period of operation, that assertions about the behaviour of the system are verified with an acceptable probability level.

Key Words: Distributed systems, formal proof, probability, safety critical systems, simulation, stochastic Petri nets, temporal logic, validation.

1 Introduction

The validation of complex systems such as distributed systems or safety-critical systems is a fundamental goal of dependable system design. It relies on a formal model of the system behaviour and the demonstration of a set of assertions on this formal model.

Safety-critical systems

Safety analysis relies generally on an inductive or deductive model of failure consequences. FMEA (Failure Mode and Effect Analysis), PHA (Preliminary

Hazard Analysis) and fault tree models are typical examples of these methods. They have two major drawbacks:

They generally use classical logic which is not able to describe dynamic phenomena. Hence consequences of operations sequencing and synchronization are difficult to express.

They rely on an enumeration of "bad" events, which can be combinatorially complex. Even when this work is tractable omission and errors are possible.

Distributed systems

Complex distributed computer system analysis uses sophisticated models which explicitly describe dynamic system behavior.

In this case, validation methods can be separated into three classes :

The first one is an axiomatic approach where axioms are the specifications of a system. Axioms are specified using temporal logic, hence many synchronization properties can be taken into account. Validation relies on the application of formula rewriting. Correctness proofs of several simple communication protocols have been achieved using this method [1]. Some authors have proposed to apply temporal logic to safety analysis [2]. But the complexity of formula rewriting limits this approach to the analysis of simple systems.

The second class defines models in term of state transition systems [3]. This class contains Petri nets and synchronized automata. Validation is based on an exploration of the reachability graph. But these methods encounter difficulties mainly because the state space grows exponentially with the model complexity.

The last class is based on simulation of the specifications [4]. The result of a simulation cannot be considered as a proof of a system property. Like testing, it can help to discover misbehaviours and to increase confidence in the system. As the exhaustive analysis is not the goal, this technique is easier to apply than the two first ones. In counterpart the relationship between the result of a simulation and the reliability of the system in operation is not established.

Principles of probabilistic validation

As it has been stated exhaustive analysis is limited by the exponential growth of the model complexity. Simulation does not give an indicator of the validation level.

The aim of our work is to develop a new approach based on a partial analysis of a system model. This approach can be compared, from a complexity point of view to the simulation approach. But the model and the analysis method must demonstrate, for a given period of operation, that assertions about the behavior of the system are verified with an acceptable probability level.

Our analysis relies on a state transition model which includes a description of operation duration and frequency of events. As the use of a system is often not deterministically known and as the model must include randoms events such as failures, the model is necessarily stochastic.

Probabilistic validation tries to use the fact that "with a very low probability in a given period" is sometimes simpler to prove than "never".

From a user point of view a system is sufficiently dependable when undesirable events are never observed during an operational period. For example distributed protocols are proven to be correct for a given failure level [5]. A higher level failure is supposed to occur during the operational life of the system with a very low probability. Hence an operational point of view of validation should try to prove that "bad events" will occur with a sufficiently low probability during a given period. This point of view relies on an explicit expression of time and operation frequencies. It is already accepted that operational faults are tolerated with a given probability level. We think that a similar approach must be taken in consideration for design faults.

Several works are related to our approach. Some authors try to relate simulation and proof [4]. Our approach is also related to approximate solution of stochastic models and in particular stochastic Petri nets [6], [7]. Principles and methods able to evaluate complex user oriented goals, called performability measures, of a system have been developed by Meyer [8]. Maxemchuk presents a computation method for protocol probabilistic validation [9].

But to our knowledge a general framework of probabilistic validation has not been presented yet. This is the subject of the next section of this paper. In the third section several analysis methods are proposed. Then a short example illustrates these principles. In conclusion problems encountered and future developments are mentioned.

2 Probabilistic validation

2.1 Time dependent validation

Most of the validation techniques, even temporal logic, do not take into account the "real time". In temporal logic or state transition models the time is expressed as ordering (concurrency and synchronization) conditions between events. Operations duration and frequency are not explicitly described.

It has been formally proved [10] that several deterministic distributed protocols do not tolerate failures without an explicit use of operations duration. Practically timers and clocks are basic tools of dependable systems design. Hence models of such systems must include a real time representation.

Several authors have added to state transitions systems an explicit timing behaviour [6]. Their approaches try to validate assertions by a complete examination of the timed state space. This allows an exact validation of systems under timing constraints but the complexity of computations is often much greater than in non timed models.

Moreover numerous operation durations are not known deterministically. Hence, the timed behaviour of the transition systems must be stochastically defined.

Stochastic Petri nets are, to our point of view, a good tool for this kind of modeling [11] [12].

Petri nets allow to describe at a high level fashion, representation of concurrency, synchronization and parallelism. The property to be verified can be expressed as an assertion on the markings reached (state assertion) or on the transitions firing sequences (trajectory assertion). The property can be expressed using temporal logic operators.

The stochastic timing introduces naturally the random duration of phenomena involved in the model behavior.

2.2 Stochastic Petri nets

The concepts and the notation used in this paper are detailed in [13],[12]. We just recall in this section some importants aspects.

Petri nets

In this paper the underlying Petri nets are the classical "valued" Petri nets [14]. The underlying Petri net is denoted $R(P, T, V)$ where :

- P is the set of places (with cardinal $|P|$)

- T is the set of transitions (with cardinal $|T|$)

- V is the set of arcs between places and transitions.

Let M_0 be the initial marking of the Petri net. The behaviour of the net is completely defined by the incidence matrices C^- and C:

- condition : a transition is firable from a marking M_i if $M_i \geq C^-(.,j)$
- action : the firing of t_j from M_i leads to a new marking M_k such that $M_k = M_i + C(.,j)$.

Let $s = (t_1, t_2, \cdots, t_{j_n})$ a sequence of transitions fired from a marking X_0 and $X_1, X_2, \cdots, X_n$ the sequence of successive reached markings in s. The characteristic vector of a sequence s is an integer vector $\bar{s}$. The jth component of $\bar{s}$ is equal to the number of transition t_j occurences in the sequence s. Hence for any reached marking X_n from X_0 the fundamental equation of the net is: $X_n = X_0 + C.\bar{s}$.

A p-invariant is a non-negative integer vector f such that its transposed vector f^t satisfies the linear system of equations: $f^t.C = 0$. So in this case $f^t.X_n = f^t.X_0$.

Timed Petri nets

According to a set of hypothesis for the timed behaviour it is possible to built the trajectory space of the timed net i.e. the set Ω of the trajectories ω defined by the infinite sequence of couples $(X_n(\omega), \tau_n(\omega))$ where $X_n(\omega)$ is the reached marking number n from $X_0(\omega) = M_0$ and $\tau_n(\omega)$ is the date of arrival in the marking from $\tau_0(\omega) = 0$.

General definition of stochastic Petri nets

A stochastic Petri net (SPN) is a timed Petri net with a random timed behaviour. A formal definition of SPN must be such that all assumptions at a time t on the markings and the firing sequences are measurable with regard to a random space. The classical way is to define a probability measure of the trajectory space Ω. A study of the general definition of SPN can be found in [13] [12].

Markov stochastic Petri nets

We consider the class of stochastic Petri nets such that the marking at time t is an homogeneous Markov process with continuous time. This is the case if the probability of firing a transition t_j between t and $t + dt$ in any marking M_i is equal to $a_{j,M_i}.dt$. The real positive number a_{j,M_i} is called the firing rate of the transition t_j in the marking M_i.

2.3 Probabilistic validation of a property : from "always" true to "probably" true

For a system modelled using stochastic Petri nets, the goal is to prove that a given property Q is satisfied. The property can be defined by an assertion on the state space of the model and more generally on the trajectory space.

An assertion on the state space is defined as a boolean function g on the markings of the reachability graph. For example $g(M) = (M(P1) \geq 1)$ may represent that at least one equipment is always available.

An assertion on the trajectory space may express that some boolean functions are true on visited states.

For example, let us consider the following proposition : a system is working correctly if periodically a given resource becomes available. This property is true when in a stochastic Petri net model a given place P1 is periodically marked that is to say $g(M) = (M(P1) \geq 1)$ is true infinitely often.

Taking a classical stochastic process notation, an assertion on the trajectory space may be considered as a predicate on trajectories ω, marking sequence number n and assertions on the markings $g(X_n(\omega))$.

For example the preceding trajectory assertion infinitely often $g(M)$ can be written as:

$$\forall \omega\, ,\ \forall i\, ,\ \exists n\, ,\ n \geq i\ :\ g(X_n(\omega))$$

To simplify notations and to build a bridge between probabilistic and usual validation methods we will use a set of temporal logic operators to express these assertions.

Temporal logic allows to describe logical properties of event sequences in state transition systems.

A variety of systems of temporal logic have been proposed [15]. These systems offer different modalities and notations. However, they are usually classified in two groups : linear time logics and branching time logics.

Linear time temporal logic allows expression of assertions on a given sequence of events of a transition system. These assertions are built up from ordinary logical operators formula ($\neg\ \lor\ \land\ \longrightarrow\ \longleftrightarrow$) and the linear time temporal operators, $\Box$ (always) $\Diamond$ (eventually).

The formula $\Box\ Q$ (always Q) means that the property Q is true now and will remain true forever.

The formula $\Diamond\ Q$ (eventually Q) means that Q is true now or will become true.

These two operators correspond respectively to the universal and existential quantifier applied to state sequence numbers of a given sequence.

$M(P1) \geq 1$ is infinitely-often true on a given state sequence may be written as :

$$\Box \Diamond (M(P1) \geq 1)$$

Validation of properties on nondeterministic state transition systems relies on branching time temporal logic. It introduces operators which allow to manipulate quantifiers over the set of possibles trajectories. There are two basic operators A and E which respectively correspond to the universal and existential quantifier over the trajectory set. Hence a branching probability formula is a quantifier operator followed by a linear temporal logic formula.

Exact validation must prove that the property is true for all trajectories ω ($X_n(P1, \omega)$ denotes the mark of the place $P1$ in the X_n marking of the trajectory)

$$\forall \omega , \forall i , \exists n , n \geq i \; : \; X_n(P1, \omega) \geq 1$$

This expression can be stated in term of branching temporal logic as follows:

$$A \Box \Diamond (M(P1) \geq 1)$$

We can now state one of the fundamental differences between exact and probabilistic validation. From an operational point of view it is sufficient that the property remains true over the set of non null measure trajectories.

$$Prob\{\omega / \forall i, \exists n, n \geq i : X_n(P1, \omega) \geq 1\} = 1$$

In probability theory the property is said to be almost sure. A property can be false and also almost sure. A system may travel through sequences of states such that a given property is not verified. For an exact validation method the property is false. But if all the trajectories built from this sequence have a null measure then the user can be confident in the system.

Probabilistic validation selects trajectories according to their probability measure. Then an assertion classifies trajectories in two classes. Linear time logic is sufficient to express such property. The example property may be written as: $Prob \{\Box \Diamond M(P1) \geq 1\} = 1$

At last from a practical point of view it may be sufficient that the property remains true with a probability greater than a given value: $Prob\ \{\Box \Diamond M(P1) \geq 1\} = 1 - \epsilon$

In this case we say that the property is epsilon sure.

Another principle of probabilistic validation is to try to prove ϵ sure properties, as it is generally simpler to prove that something is probably sure rather than it is always true.

2.4 Solving the model

The following remarks are used to limit the model analysis.

A system which is submitted to a formal validation procedure is supposed to be correctly designed. An incorrect behaviour is a very rare event. If such an event may be observed it is generally the consequence of a complex operations sequence, which is out of the standard behaviour of the system.

Safety-critical systems tend to have reliability requirements ranging from 10^{-5} to 10^{-12} over a given time period. For example, NASA has a requirement of 10^{-9} chance of failure over a 10 hours flight.

Moreover it is well known that a very few number of system behaviours covers most of the operational situations.

For example, the requirement for interoperability between implementations of standard protocols is usually satisfied by demonstrating that these implementations are sufficiently correct to transfer files with a reasonable reliability level. The tests try to cover only situations encountered in a file transfer procedure, and not all pathological network and user behaviour.

Basic principles of the method take into account these aspects:

- From a structural analysis of the model we make the best possible classification between correct and incorrect event sequences.

- Then we examine "rare" sequences leading possibly to incorrect situations.

- Another efficient strategy is to examine a few "good" situations, which have a very high probability.

In that spirit, our idea is to consider first the structural behavior of the Petri net. From the classical theory of Petri nets, much information on the model behavior can be derived. For example, place invariants limit a priori the reachability set, and transition invariants lead to knowledge about information on firing sequences. The complexity of this kind of structural analysis is much less than the reachability analysis.

At the end of such analysis: either we can conclude that the property Q is true or false, or it is not possible to characterize completely the set of possible trajectories that might lead to "bad" situations.

In the second case, which will be the most frequent, it is necessary to build the reachability graph of the net. The following principles are used for the exploration. First, the results of the net analysis are used. Sequences of events leading to a "good" situation are visited as rarely as possible. Situations having a "good" probability to lead to a "bad" event are visited first.

If ϵ is the required precision in results, at the end of the computation, we must be able to conclude that the probability for Q to be true is greater than $(1 - \epsilon)$.

At a given step of the analysis, the behavior of the system may be partitioned in three classes. The set of visited sequences such that Q is true has a probability $\Pi 1$, the set of visited sequences such that Q is false has a probability $\Pi 2$, and the set of non visited sequences has the probability $(1 - \Pi 1 - \Pi 2)$. The exploration is performed until: $either\, \Pi 1 > (1 - \epsilon)$ the property is ϵ sure, $or\, \Pi 2 > \epsilon$ the property is ϵ false.

Analysis of efficient techniques able to take the best advantage of the partial exploration is the goal of our current research which is in progress.

Three classes of methods are considered. The two first ones are easy to apply when the space of trajectories is finite. An example of stochastic Petri net modeling for embedded non repairable systems is given in [7] [16].

Such a system evolves in the following way: a state that has been reached and left will never be reached again. As a consequence of this definition, the reachability graph has no cycle. This implies that there are no sequences leading from a marking to the same marking.

For such systems, we have proved that the reachability graph can be generated by levels. The evaluation of the graph and the parameters computation can be done during the generation step without storing the whole graph and performing a lot of useless computations. Some criteria are used in order to stop the generation as soon as the set of markings that are not yet generated have a very low probability to be reached.

This is the basic idea of the first method for partial reachability analysis with probability evaluation. This method visits the reachability graph in a width first search traversal.

The second method is associated with a depth first search traversal of the graph. In this case we try to drive the graph generation in order to reach as quick as possible "bad" situations (steepest descent). A branch and bound algorithm can be efficiently used in that case [17].

To treat cyclic reachability graphs some modifications of these basic principles have to be studied.

The third method is able to solve cyclic systems. It combines the two first ones with monte-carlo simulation. But since the probability of "bad" sequences is relatively small, it has been observed that simulation takes a large amount of computational time. A technique which may be applied to reduce drastically the simulation duration is to use importance sampling variance reduction [18]. The goal of importance sampling is to increase frequency of "interesting" events.

In probabilistic validation the notion of bad event is not as clear as in reliability analysis. We will show in the next section how the structural analysis can be used to define the events which frequency must be increased.

2.5   Probabilistic validation of acyclic stochastic Petri nets

Let us consider a system modelled using an acyclic stochastic Petri net and Q the property to be verified. "Not Q" represents an "undesirable" event. In the following the property Q is a marking property. The method can be slightly modified to analyze trajectory property.

The method relies on the following principles.

- Let F be the set of integer vectors $\in N^{|P|}$ such as $(\neg Q)$ is verified. The first step is to consider the structural behavior of the Petri net. So we compute Petri net invariants, i.e. a set of vectors f such as $f^T.C \leq 0, f \geq 0$. A reachable marking verifies the following property: $f^T . M \leq f^T . M_0$, if M_0 is the initial marking.

- Let L be the set of integer vectors $\in N^{|P|}$ verifying the invariant property.

Either $F \cap L = \phi$ so we conclude that there is no undesirable marking and the property Q has the value true for all sequences of the net.

Or $F \cap L \neq \phi$ we cannot conclude if undesirable events exist in the system. But if such a marking exists, it is such that:

$$M \in L \cap F , M = M_0 + C \bar{s}$$

where $\bar{s}$ is the characteristic vector of the sequence s leading from the initial marking to the undesirable event.

- To characterize the set of possible trajectories that might lead to a bad situation, we solve the system: $M = M_0 + C . x$

- If a sequence s exists leading to an undesirable marking, then its characteristic vector $\bar{s} \in X$.

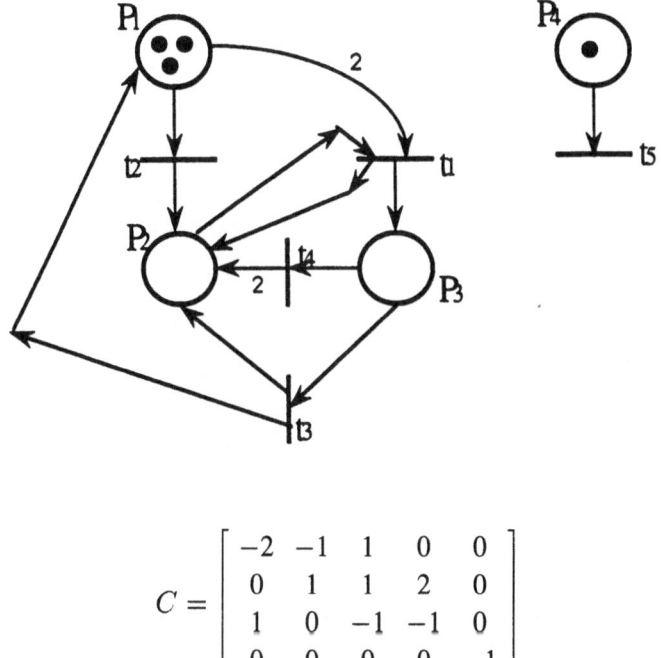

$$C = \begin{bmatrix} -2 & -1 & 1 & 0 & 0 \\ 0 & 1 & 1 & 2 & 0 \\ 1 & 0 & -1 & -1 & 0 \\ 0 & 0 & 0 & 0 & -1 \end{bmatrix}$$

Figure 1: **A simple non acyclic Petri net and its incidence matrix**

- We can then use one of the three methods for partial reachability analysis with probability evaluation.

3 Example

3.1 Presentation

In this section we illustrate the concepts presented in this paper on a very simple example. The model represents the behaviour of a computer system composed of three computers used in active redundancy. The failures are detected either by a self checking algorithm performed on each computer or by a difference in the results computed by each operational computer. This model is described precisely in [7].

Figure 1 represents the Petri net model and its incidence matrix C and figure 2 the complete reachability graph.

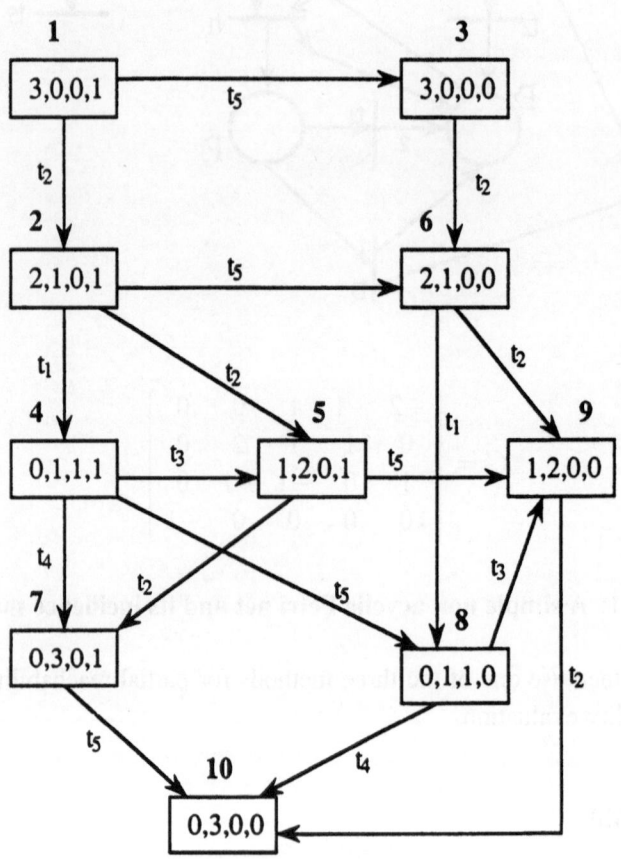

Figure 2: **Complete reachability graph**

The assertion to be validated is:

$$g(M) = \neg((M(P2) = 3) \wedge (M(P4) = 1))$$

Let us consider the following transitions rates. Transitions t1 and t2 represent a failure of a computer. Transitions t3 and t4 represent the diagnosis procedure, t5 the end of the mission. Transitions rates (generally denoted λ_{ti} are: $\lambda_{t1} = \lambda_{t2} = 5.10^{-5}$, $\lambda_{t3} = \lambda_{t4} = 36,000$, $\lambda_{t5} = 0.2$. The precision required on the assertion validation is 2 10^{-7}.

3.2 Structural analysis

Let F be the set of undesirable markings. F is a set of integer vectors such as (using the transposition operator):

$$F = \{M/M = (a, 3, b, 1)^t , \; a , b \in N\}$$

Now, let us consider the structural behavior of the Petri net. The computation of invariants, that means integer vectors f such as $f^t . C = 0$ give the results: $f^t = (1, 1, 2, 0)$, that means that $(M(P1) + M(P2) + (2 * M(P3)))$ is an invariant of the net and in this example is equal to 3.

So, the set of integer vectors L of reachable markings of the net is such as:

$$L = \{M/M = (M(P1), M(P2), M(P3), M(P4))^t ,$$

$$M(P1) + M(P2) + 2M(P3) = 3\}$$

Here, the intersection of the two sets of vectors is not empty, that is: $F \cap L \neq \phi$.

As a matter of fact, the vector $(0, 3, 0, 1)^t$ is in the set F of undesirable events and is also in the set L of reachable markings of the net. So we cannot conclude only on the Petri net structural analysis.

But, if such an undesirable reachable marking exists, it is such that: $M \in F \cap L$.

To characterize the set of possible trajectories that might lead to bad situations, we solve the system: $M = M_0 + C . x$

Let X be the set of integer solutions of this system. If a sequence s exists leading from the initial marking M_0 to M, then its characteristic vector $\bar{s} \in X$.

Solving the equation $M = M_0 + C.x$ leads to six solutions:
$x_1 = (0, 3, 0, 0, 0)$, $x_2 = (1, 1, 0, 1, 0)$,
$x_3 = (1, 2, 1, 0, 0)$, $x_4 = (2, 0, 1, 1, 0)$,
$x_5 = (2, 1, 2, 0, 0)$, $x_6 = (3, 0, 3, 0, 0)$.

We can then use one of the methods presented for partial reachability analysis with probability evaluation.

3.3   Width first search traversal

In this method, markings must be generated by level [7]. As computation of the parameters can be done during the generation, we will stop the generation as soon as a sufficient precision about results is found.

From the initial marking numbered 1 (3,0,0,1) in figure 3, two transitions can be fired: t_2 and t_5.

t_2 leads to the marking number 2 (2,1,0,1) in which the mission is in progress.

t_5 leads to the marking number 3 (3,0,0,0). We know using characteristic vectors that t_5 cannot lead to an undesirable event. So we stop the generation of that sequence and compute the probability of this marking.

At this step the precision ϵ obtained is equal to 2.49 10^{-4}. That means that probability of this marking is equal to $1 - 2.49$ 10^{-4}. This precision is not sufficient. Hence we continue the generation from marking 2.

From this marking three transitions can be fired t_1, t_2 and t_5. The same approach can be used for the marking reached by transition t_5 and we can compute its probability.

At this step, precision obtained is equal to 1.24 10^{-7}. That means that others states not yet visited have a probability less than 1.24 10^{-7}. In particular we can conclude that we have not encountered "bad" markings but if such markings exist they have a probability smaller than 1.24 10^{-7}.

This result has been found generating 6 markings instead of the whole reachability graph.

Results on this small example are not sufficiently significant but this method has been applied to industrial models [16] for which graphs about more than one million of states have been evaluated by generating only about 2,500 markings.

3.4 Depth first search traversal

This method relies on a different idea. We try to reach "bad" situations as quickly as possible.

Characteristic vectors found by the analysis of the structural behaviour of the Petri net will help us in the steepest descent.

In the example given, the six characteristic vectors found do not contain transition t_5. So we know that sequences leading to undesirable markings do not contain that transition. All other transitions are included in the characteristic vectors.

From marking 1 (3,0,0,1) two transitions can be fired t_2 and t_5 but we are only interested in firing transition t_2 and we do not generate markings obtained by firing transition t_5. We can compute the probability of this trajectory (firing t_2 instead of t_5) which is equal to $\lambda_{t2}/\lambda_{t2}+\lambda_{t5}$. We go on this steepest descent until reaching an undesirable marking, computing at each step the probability of the trajectory.

In the example we find three sequences leading from the initial marking to the marking numbered 7 in the reachability graph which is equal to (0,3,0,1). In this marking the mission is in progress and the three computers have failed.

These sequences are : $s_1 = (t_2, t_1, t_4)$, $s_2 = (t_2, t_2, t_2)$, $s_3 = (t_2, t_1, t_3, t_4)$, which have the following probabilities : for $s_1, 3.12 \ 10^{-8}$, for $s_2, 1.56 \ 10^{-11}$, for $s_3 : 7.80 \ 10^{-12}$.

As we know that there are not others undesirable trajectories, we can give the probability of reaching the undesirable marking which is equal to the sum of the probability of each trajectory leading to it. This probability is equal to $3.12 \ 10^{-8}$.

This result is compatible with the one obtained by the first method, as $3.12 \ 10^{-8}$ is smaller than $1.24 \ 10^{-7}$. This last value is the probability of markings not visited in the first method.

3.5 Simulation using importance sampling variance reduction

This method is a classical stochastic simulation such that the frequency of rare events is increased. The method used to build estimators is presented in [18] for the markovian case. We are working to adapt it to a more general class of stochastic Petri net. But the main difference between importance sampling simulation applied to reliability analysis and probabilistic validation is related to the knowledge of rare interesting events. In reliability analysis these events are user defined generally as equipment failures. In probabilistic validation the sequences of undesirable events are not known. It is even possible that such sequences do not exist. Hence the frequencies of event are altered dynamically during the simulation taking into account the static analysis. We artificially increase the probability measure of trajectories the characteristic vector of which tends to the one computed by the static analysis.

We consider now the preceding example. As the stochastic process associated with this example is not regenerative, hence a sequence of simulation started from the initial marking must be performed n times to reach a given confidence level. According to the value of the firing rates a standard simulation leads to firing t_5 much more frequently than t_2 in the initial marking. But from the structural analysis we show that all sequences leading to a failed situation include a firing of t_2. So the rate of t_2 must be increased, for example up to the same value as the rate of transition t_5. As soon as t_5 is fired we can stop the current simulation: bad situations cannot be reached. If t_2 is fired during a simulation then marking $M2$ is reached. The same approach is applied.

4   Conclusion

Stochastic Petri nets in design and analysis of safety-critical systems and distributed systems have been introduced and shown to be useful. But complete behaviour of such systems can be very large and hence untractable. Our approach of probabilistic validation develops methods mainly associated with partial searching in very large state space. In this paper a general framework of probabilistic validation and our first results on this work in progress have been presented.

Nevertheless numerous problems remain to be solved. They concern both theoretical aspects (formal logical expression of probabilistic proof, relationship between temporal logic assertions on a Petri net and the invariant method ...) and algorithmic aspects.

References

[1] B. Halpern and S. Owicki, "Modular verification of computer communication protocols," *IEEE Transactions on Communications*, vol. com-31, no. 1, 1983. 1983.

[2] J. Gorski, "Design for safety using temporal logic," in *Safecomp*, (France), 1986.

[3] A. Cavalli and F. Horn, "Proof of specification properties by using finite state machines and temporal logic," in *IFIP*, 1987.

[4] A. Cavalli and E. Paul, "Exhaustive analysis and simulation for distributed systems, both sides of the same coin," *Distributed Computing*, vol. N2, 1988.

[5] F. Christian, "Understanding fault-tolerant distributed systems," Research Report RJ 6980(66517), IBM Research Division, 1990.

[6] G. Juanole and J. Roux, "On the pertinence of the extended time petri net model for analyzing communication activities," in *PNPM*, 1989.

[7] K. Barkaoui, G. Florin, C. Fraize, B. Lemaire, and S. Natkin, "Reliability analysis of non repairable systems using stochastic petri nets," in *FTCS*, June 1988.

[8] J. Meyer and W. Sanders, "Performability evaluation of distributed systems using stochastic activity networks," in *International workshop on Petri nets and performance models*, August 1987.

[9] N. F. Maxemchuck and K. Sabnani, "Probabilistic verification of communication protocol," *Distributed Computing*, vol. N3, 1989.

[10] M. Fisher, N. Lynch, and M. Paterson, "Impossibility of distributed consensus with one faulty process," *JACM*, vol. 32, 1985.

[11] G. Florin and S. Natkin, "RDPS: a software package for the evaluation and the validation of dependable computer systems," in *Safecomp*, (France), 1986.

[12] G. Florin, C. Fraize, and S. Natkin, "Petri nets: properties applications and tools," *Microelectronic and Reliability*, 1990. also Cedric Research Report.

[13] G. Florin and S. Natkin, *Les reseaux de Petri stochastiques; theorie, techniques de calcul, applications*. PhD thesis, Universite de Paris VI, Paris, June 1985.

[14] T. Murata, "Petri nets: properties, analysis and applications," in *Proceedings of the IEEE*, 1989.

[15] E. Audureau, P. Enjalbert, and L. Farinas Del Cerro, *Logique temporelle, Semantique et validation de programmes paralleles*. Masson, 1990.

[16] C. Fraize, "Les reseaux de petri stochastiques a graphes de marquages sans circuit : theorie et application a l'analyse de la fiabilite et des performances des systemes informatiques," *memoire d'ingenieur, CNAM*, 1988.

[17] M. Bouissou, "Automatic generation and quantification of event sequences leading to repairable system failure," tech. rep., EDF — Bulletin de la Direction des Etudes et Recherches, 1987.

[18] A. Conway and A. Goyal, "Monte carlo simulation of computer system availability reliability models," in *Proc. 17th IEEE International symposium on fault tolerant computing*, 1987.

Testing Programs to Detect Malicious Faults*

RICHARD HAMLET

Computer Science Department

Portland State University

Portland, OR 97207, USA

Abstract

Program testing has traditionally been of two kinds: for fault finding (debugging), and for establishing operational reliability (confidence). We investigate the question of using traditional methods to determine the dependability of a program, under two assumptions: (1) the only sources of failure are inadvertent mistakes in design, coding, etc., and the program developers cooperate in trying to eliminate such faults. (2) the source of failure is sabotage — malicious code is inserted in the program and cleverly concealed. Paradoxically, it appears to be easier to detect sabotage than subtle unintentional mistakes, in the off-line situation where the sabotage takes place during development, and must be detected prior to program release. Furthermore, the very situations that can make traditional testing a nightmare, for example, real-time constraints, actually may help a tester trying to detect sabotage.

1 Introduction

Malicious software security violations raise disturbing questions for a researcher in program testing. Testing research has addressed the problem of finding program bugs that arise from innocent, well intentioned actions, and of measuring the quality of a system where accidental flaws may be hidden. Software is often buggy, but it's problems have been taken to be a natural consequence of problem complexity, of the limits of human ability, of human communication breakdown, etc. The resulting errors are "random" rather than "systematic" in the sense of scientific error.

Program viruses, worms, Trojan horses, etc., represent a very different situation, which has not been considered from the viewpoint of testing. Suppose that

---

*Work supported by NSF grant CCR-8822869.

a tester had the problem of finding not "random" bugs that could be anywhere and produce any sort of failure, but rather "systematic" bugs, purposefully inserted with malicious intent. Would different methods be appropriate to test for purposeful bugs?

In conventional program testing, there is a controversy about disclosing the test arsenal to the developer. One side argues that by devising and publicizing tests as software is developed, mistakes are avoided, requirements sharpened, questions asked that positively influence all phases of development [1]. On the other side, if all the tests have been "used" in this way, they are unlikely to then expose any problems, and it is difficult to attribute any significance to their success. The tester has delivered up his cartridges to the other side, and should not be surprised if they fizzle in action [2]. In security discussions this controversy is mirrored in the question of whether security information should be openly available (to aid those who seek to protect themselves), or hidden (to avoid informing those who might exploit it). The analogy is often used that police should (or should not) reveal tricks used by burglars.

When trying to detect malicious faults, there is no similar dispute. The "developer" as saboteur is certainly not to be helped; no one involved needs protection information. In the police analogy, one might believe that police should publicize methods used by burglars. However, police should not disclose the methods they use to identify and apprehend burglars — it could help no one but criminals. The analogy even suggests techniques for the tester and honest developer: they might use the software equivalent of a "sting" operation to trap saboteurs.

In Section 2 we summarize the methods used in conventional software testing, categorizing them according to whether they try to find bugs, or seek to measure quality. These methods are then examined in Section 3 for applicability to program sabotage, with the somewhat surprising results that it is easier to detect tampering than to catch subtle unintended mistakes, and that the appropriate methods are not those usually associated with software reliability. Section 4 suggests steps to enhance the security of systems by testing.

2   Testing Methods: Debugging vs. Quality Assessment

In reviewing conventional testing methods, the distinction between _failure_ and _fault_ frequently arises, and we will try to speak precisely here, using the IEEE definitions of these terms. A _failure_ is an observable program behavior judged to be wrong. This implies knowledge of what the program is supposed to do, that is, its specification. A *fault* is the part of a program to which a failure is traced, the cause of that failure. Although "fault" is the more natural idea (which intuitively corresponds to "bug" or "defect"), it is imprecise. Failures do not have a unique cause, and different people may well fix the same failure in radically different ways, making "the" fault ill-defined. In past published research, the word "error" is often used for both "fault" and "failure," with the result that it is now taboo in the testing literature. However, it has been used [3] to describe the important idea of a wrong internal program state, which may not (perhaps ever) be manifest as a failure. (The more detailed specification needed to identify such states is seldom available; nevertheless, the intuitive idea is important.) In this paper program faults are thought of as inserted on purpose, so the idea is well defined.

Tests of a program "succeed" in two opposite ways: (*debugging*) the tested program fails, and this "successful" test has therefore exposed a bug; (*quality assessment*) the tested program executes properly, and this "successful" test increases confidence in the program's quality. Characteristically, debugging methods involve far fewer tests than quality-assessment methods. This section presents a review of conventional testing methods as a prelude to using those methods to detect sabotage.

2.1 Testing for Debugging

It has long been known that a proper mindset is essential to practicing the art of program debugging, and that this mindset is pessimistic: the successful debugger must be enthusiastically seeking failures and mistakes [4]. It is therefore not surprising that most of the testing methods in use today are really debugging methods, designed to expose faults. Testing for debugging is based on the twin ideas of *coverage* and *surprise*. Portions of the program are isolated (in different ways for different methods), and it is arranged to consider these portions under unusual circumstances. The intuition behind such methods is that bugs could be lurking anywhere (hence the coverage, preferably systematic), and that unusual cases are the most likely to expose these bugs.

There may be as many different debugging methods as there are people doing the testing, but the following are representative:

Walkthroughs and inspections [5]. The program text is examined by hand by a group of experts, each playing a role designed to discover faults. The great strengths of this method are that people can bring their intuitive expertise to bear, and that it can be applied to incomplete programs or designs early in development. The rather surprising cost effectiveness of inspections is the result of early application, but also indicates that many bugs are shallow, and experienced professionals spot them easily.

Specification-based, blackbox testing [6]. Tests are chosen by breaking down the specifications into cases, and selecting a small number of points to invoke each case. The tests can thus be devised before any program is written, and the method has the virtue that it can detect "missing logic" faults. However, its flaw as a debugging method is that a detected failure may be difficult to trace to a corresponding fault.

Control-flow coverage testing. In its simplest variant, this program-based technique is called *statement testing*. To satisfy the method, tests must be selected that cause each statement of a program to be executed. Programmers like it because it is systematic, and instrumentation to measure statement coverage is very easy to implement efficiently by preprocessor techniques. Statement testing has been invented over and over throughout the history of programming.

A slightly different control-flow coverage method is probably the most widely used today. In *branch coverage*, test data must force each conditional statement to take both of its possible branches.

A more sophisticated method (or really, collection of methods) is *dataflow coverage* [7]. The required data must invoke all paths on which certain interesting relationships hold between values of program variables. For example, covering DEF-USE paths for a variable means executing paths on which the variable is first defined, then referenced. The intuition is that such a path represents one of perhaps many different ways in which the variable is employed.

There are many other variants of path testing, restrictions of the requirement to cover all possible execution paths. For programs containing loops or recursive procedure calls, the number of potential paths is infinite, and the variants each attempt to isolate an interesting but finite subset. All path-testing variants are relatively easy to implement using worst-case static flow analysis, and most are easy to use systematically.

Infeasible paths represent the only theoretical flaw in control-flow testing.

It is an unsolvable problem to decide in general if a control point can in fact be reached, so it can happen that the tester trying to cover a program may be attempting the impossible. In practice, this difficulty is not experienced as severe, however.

Expression-coverage (mutation) testing. The analog of path testing for data coverage is not so satisfactory, because it is harder to understand, and implementation is difficult and inefficient. An expression is covered by test data that precludes it taking alternate forms [8]. The classic example of data that does not cover is the value 2 for variable X in $X + X$ – with this data the expression could as well be $X * X$ or $X * *2$, among others. Expression coverage explores the range of values variables may assume when a statement is encountered. The name *mutation testing* arose in a slightly different system [9] with somewhat different motivation.

It is observed that expression coverage in combination with even simple forms of control-flow coverage is hard to fool. It seems to expose bugs that are not really in the expressions covered. This is probably because it directly addresses the weakness of all forms of path testing, namely, that a path may be taken with such trivial data that the "coverage" detects nothing. By in addition requiring expression coverage, the data reaching each expression must be varied.

Expression-coverage testing is little used, because its implementations work by brute-force trial of all the alternate expressions. Perhaps a greater impediment to its adoption is the difficulty of finding and understanding the necessary test data. Like control-flow coverage, expression coverage has a theoretical limitation: there is no way to tell in general if an alternate expression is equivalent to the one being testing. In practice, this equivalent-expression problem causes the tester considerable difficulties.

A restricted form of expression coverage called *weak mutation testing* [10] is efficient. In this method very narrow faults (for example, wrong relational operator in a conditional expression) are detected with certainty. The drawback is that the method requires a specification at the level of each expression, and inserted internal probe points to check it. (In the terminology of [3], the method looks for errors rather than failures.)

The methods described above are traditionally used at the unit-test level, with individual routines or modules. Those below (and also blackbox testing) are used at the integration- or system-test level.

Fault-tree analysis. Conventional safety-engineering fault analysis has an analog for software [11]. By tracing imagined failures from the functional
   level, tests may be devised that are likely to excite those failures. The
   tests seem to perform well in combination with other methods, finding
   bugs that have resisted less directed testing.

Stress testing. For software, "stress" means forcing the most difficult system
   behavior, on the plausible assumption that when the going gets tough, the
   software will fail. The idea applies best to real-time software, where difficult behavior means complicated concurrent processing and heavy system
   loads that conflict with stringent timing constraints.

"Random" (haphazard) testing. True random (or statistical) testing is a quality-assessment technique, to be discussed in Section 2.2. But by far the
   most common debugging technique is its unscientific cousin, haphazard
   testing, in which test points are chosen unsystematically. That haphazard
   testing works at all is a tribute to good luck and the deplorable quality of
   some software.

Debugging methods seem to work in practice, perhaps because they systematically direct human attention to parts of the program, in a specific test
situation that makes it easy to think through possible problems.

2.2   Testing for Quality Assessment

The methods described in Section 2.1, when they fail (that is, when they find no
failures!) should not be expected to carry much weight as quality-assessment
methods [12]. They are not probabilistic, as a reliability method must be, and
viewed as sampling methods, their tests are few, and chosen to satisfy criteria
that compromise sample independence. From the complexity of programs and
their executions as the objects to be sampled, we would expect the number of
failure-free samples needed to support confidence to be large. Unfortunately,
the questions of sample distribution and sample independence have proved
difficult to address.

Conventional reliability theory [13] assumes a program "fault rate" to which
the long-term average ratio of failed to total executions tends, when test inputs
are drawn from a distribution characteristic of the program's use. By selecting
N independent tests from this distribution, the absence of failure guarantees

that the fault rate is below p, with confidence $1 - e$, if

$$(1 - p)^N \leq e.$$

For example, 90% confidence that a program will fail no more than one time in 10^6 executions requires

$$N = \frac{\log(.1)}{\log(1 - 10^6)} \approx 2.3 \text{ million tests.}$$

Bad as this is, it has been argued [14, 15] that reliability cannot capture the software property which Parnas calls "trustworthiness." We can rely on software only when it is safe to deviate from "normal" usage, and thus no test distribution can be appropriate. Furthermore, it is impossible to guarantee independence of the test samples, because their position within the input domain in no way corresponds to behavior within the program. It may very well happen that two quite different inputs are not distinguished by a program, and hence no additional confidence can result from testing both.

These deficiencies of the conventional theory appear in its results: it is counterintuitive that the number of tests needed to guarantee a low fault rate does not depend on program size, nor on size of the input domain. Although the input space is the one that tests must sample, that space is the wrong one for determining trustworthiness. Instead, we should be sampling the state space of the executing program. Any failure is associated with a point in this state space, a control point at which some program variable first takes a wrong value, eventually leading to a failed execution. If we could sample this space, a uniform distribution of tests would be appropriate, and i t is possible to estimate the number that would be required. Suppose we wish to guarantee that a program of size M lines and domain size D has no more than f faults/line. The number of tests required N is proportional to MD/f, with a proportionality constant of roughly 10^{-4} for 99% confidence and large D/f [15]. For example, a medium-sized program of 50,000 lines on a million-point domain, requires at least

$$N = \frac{10^{-4} \times 50000 \times 10^6}{.001} = 5 \text{ billion (USA) tests}$$

to guarantee less than 1 fault per thousand lines. Although fault density 0.001 is good by industry standards, it is not acceptable in critical applications. Larger domains, larger programs, and acceptable fault densities require testing with so many points that it simply cannot be carried out in practice. (Thus it is a moot point that we do not know how to perform appropriate sampling of the state space!)

To summarize, the techniques that have been invented for debugging involve relatively few tests, and seem to be useful in practice. (Perhaps only because we define their usefulness as limited to the time when they find bugs!) On the other hand, conventional reliability testing requires a vast number of tests, whose significance is in doubt. We do not know how to sample for trustworthiness, and estimates of the number of samples required indicate that the problem is infeasible for large programs. The difference between the debug and quality-assessment situations seems to be that in the former, we are testing within a small space connected with programmer mistakes, and hence a few clever tests are effective; in the latter, the failures are literally needles in a haystack, so only random testing is effective, and then only when far more tests are used.

3 Testing for Software Sabotage

If testing for faults that arise accidentally and despite the best efforts of the program developers is difficult, it might seem that detecting malicious faults — sabotage — would be impossible. The "developer" is trying to hide the "fault," which might be expected to invalidate the usual debugging tests, and make the needle even harder to find. But sabotage is purposeful, and this gives malicious faults a structure that aids detection.

Some ground rules are essential to this study. We wish to consider a narrow, technical problem, and thus must exclude many realities of sabotage. For example, it is simple and effective to attack any system, but particularly computer systems, from the inside. Instead of using technical sophistication, buy or steal a password; or, physically break into a room containing an unprotected machine. (There is a story about a DEC PDP-6 timesharing system placed in a "fishbowl" room in Maynard to show off its flashing lights. The room was attended by a receptionist, and customers could walk in to use the system. The console terminal logged in as [1,1] with all privileges was in this room. A customer did walk in and print the password file for the entire PDP-6/10 organization on the convenient line printer, tore off the listing, and walked out with this "sample output.") We must ignore such practical cases, to investigate what testing can accomplish.

We assume that the program tested is the one in which the bug resides, and that the bug is observable at the source level. That is, studying the program text could in principle reveal the sabotage. One way to guarantee this is to assume that programs must be written in a given language, that the only compiler for that language is secure, and that its object codes cannot be modified. This kind of security was used on the Burroughs B5000 systems to control powerful machine operations whose misuse could compromise virtual memory boundaries. (As an aside, one somewhat bizarre failure of the compiler assumption was a bug in the Burroughs COBOL compiler that created system-crashing object files from empty source programs. The bug was exposed (more than five years after the system was put into commercial use!) when a long-time COBOL programmer tried to learn ALGOL and used his habitual compile command by mistake. The ALGOL was parsed as a COBOL comment.) Of course, the bulk of program sabotage occurs outside these assumptions. In particular, systems that attempt even minimal security try to allow users to execute arbitrarily faulty programs, without harming anything outside their own protection space. Hence the would-be saboteur of such a system concentrates on the operating system or special security-maintenance programs, usually without the ability to modify and install new sources. In that case the roles are reversed: the saboteur is more like a tester, looking for a hole in the running system to exploit. Such a hole, if found, is a program failure of the "random" kind, representing an inadvertent mistake rather than the malicious action subsequently taken by the saboteur.

An immediate consequence of these assumptions is that if a program under test communicates with another program (or executes one), that second program is subject to the same assumptions as the first. We will assume for simplicity that there are no such second programs, their substance being incorporated in the original. It is also necessary to assume that a special test environment is available, so that the tests themselves do not become part of the problem. It does no good, for example, to identify a file-destruction program if in the test the real files are destroyed. Finally, as in all testing theory, we must assume the existence of an *oracle*, a means of judging the program execution. Detecting sabotage often means observing a peculiar peripheral effect, and we must have solid information about which side effects are spurious. It is evident that extreme care is needed to notice strange behavior, which in sabotage may be disguised or minimized. In practice, program testers are not very good at noticing even obvious program problems; but again we must assume that everything goes perfectly in order to make our point.

In short, we consider only the "off-line" case in which source code is maliciously devised, and the detection devices are limited to program testing.

To make the discussion concrete, consider a particular kind of sabotage: the bug is a Trojan horse. That is, the program under test is to be vetted for the possibility that it can, on command or in response to some unusual circumstance (the "trigger"), take illicit action. One of the modes of the Internet worm was based on such a bug in the UNIX "sendmail" program. We ask the question: can a Trojan horse be detected by testing? The answer partly depends on the definition of "illicit action." We must assume that the sabotage is worthwhile to its perpetrator, and hence its effects are observable. Then the testing problem is precisely that of forcing the illicit action to appear. (It is our strongest assumption, difficult to realize in practice, that the tester will necessarily notice any illicit action that arises during test.) Again for concreteness, suppose that the illicit action is a single parametrized operating-system call statement. Such a statement could be used to delete a file, to change protection information, etc. Once this action was taken, other programs, perhaps innocent by themselves, might be used to accomplish a more complicated illicit task.

It is apparent that quality-assessment methods are no better at exposing sabotage than they are at detecting unintended faults. The saboteur can easily hide the Trojan-horse trigger in the input domain, in the sense that the fraction of operational inputs that excite it is near zero. Then although there is always a chance that sampling will hit on the illicit behavior, the saboteur can make that chance arbitrarily small. Of course, testing for trustworthiness by adequately sampling the state space would expose *any* fault with high probability; but, the number of tests required is impractical, and again the saboteur can control his chances by making the internal state more complex.

Debugging methods seem to have more promise. Evidently, using a coverage technique can expose a Trojan horse precisely because if its trigger is not activated, some code is *not* covered and is thus exposed to scrutiny. Indeed, cleverness may work against the saboteur, because obscure code that apparently never is used will be studied diligently once the test pinpoints it. How well the saboteur hides his work from test-coverage exposure depends on his knowledge of what test methods may be applied to catch him. We use Pascal examples to illustrate. Suppose that the illicit operating-system service the saboteur is trying to invoke is represented by the call

```
GreeksOut(17955472)
```

Imagine that nothing about this call is syntactically significant, so that it can be detected only by pernicious effects during test executions. Similarly, when we place helpful comments in the code examples, these are for the current reader, and not available to the security staff. The straightforward code:

```
if NighTime {trigger}
   then GreeksOut(17955472){violation}
```

where `NighTime` represents the trigger condition, will be detected by statement-coverage testing. If the test happens to make `NighTime` true, the call will be made, and we assumed that the violation will be caught. If not, then it remains unexecuted, and will either be noticed, or some test will be added that executes and detects it.

However, suppose the saboteur does know about statement testing. He might try:

```
{Placed in a loop}
Spears:=0;
Shields:=0;
if NighTime1   {first part of trigger}
   then Spears:=17950000;
if NighTime2   {second part of trigger}
   then Shields:=5472;
GreeksOut(Spears+Shields)   {violation only if
                                both triggers}
```

The call is made with harmless arguments unless a two-part trigger has occurred. All of these statements can be executed without the violation occurring; indeed, the two triggers will be made difficult to activate together. However, dataflow testing will detect the unexecuted path through both conditionals to the violation. The "data context" coverage of Laski [16] is designed for precisely this situation. Joseph uses essentially this example in his thesis [[17], Section 5.2.5]. He makes the point that only full path testing (requiring an arbitrary number of paths) could be sure to catch such subterfuges. On the contrary, Laski's simple method will do the job. The reason is that the saboteur must correlate the triggers with the violation, and data-context testing looks for that correlation.

As Joseph notes, however, the saboteur might very well abandon control-flow tricks altogether, using something like the following:

```
Achilles:=Formula;      {trigger on Formula=17955472}
GreeksOut(Achilles)
```

Formula represents an expression used to produce legitimate calls, which can be forced to take the trigger value by an odd combination of its terms. There is a single path through this program so no variant of path testing is likely to expose the fault. But expression-coverage testing will do so. In order to adjust Formula so that it can attain the trigger value, it will have to be made more complex than need be for the legitimate cases, and if testing is confined to those cases, a simpler expression will be flagged as an expression-coverage failure. Again, the need for the sabotage to "make sense" lets the testing method succeed, where in general it might not.

The reader can no doubt now suggest a means for a saboteur to circumvent expression-coverage testing, then a further coverage technique that will catch him out yet again, and so on. The general unsolvability of many testing problems shows that this is precisely the locksmith/lockpick problem. Each defensive method can be circumvented, and each circumvention can in turn be detected, only in turn to fall to a new assault, and so on. In the lock game the locksmith is at a disadvantage. His handiwork is available for study, but he must wait until the lock is picked (and guess how it was done) to try again. In the off-line variant of the security problem we are considering, the saboteur is at the disadvantage, since his code is available to the tester, and he cannot adjust it in response to tests. However, it is all-important that the saboteur not know what testing methods will be applied, or it puts the shoe on the other foot. (Just as it does in the second round of the lock game if the locksmith knows how the last lock was picked, and improves it.) But there is no reason to expose test methods to saboteurs. Mutation testing, for example, has several variants, each difficult to perform without computer assistance. The saboteur who has no access to the testing system is liable to miscalculate and be caught.

The difference between detecting sabotage and detecting any possible failure is precisely that sabotage has a purpose, and that purpose is helpful in looking for the damaging code. It must be admitted that the Trojan horse is one of the easiest to detect, because it's purpose involves a coordinated trigger and action, to which methods like dataflow- and expression coverage apply. At the other extreme would be sabotage that mimics unintentional errors: it intends only to make the software fail, but not in a directed way. The situation is similar to a testing technique called "fault seeding" in which known faults are planted to help in counting unknown faults. Fault seeding is seldom used because it has proved difficult to find "real" faults to seed, and now this difficulty works against the saboteur.

Insofar as a saboteur succeeds in mimicing "random" failure, his work will be harder to find using debugging methods. The tester will have to turn to sampling, but again there is a twist that works against the saboteur. If the failure is fairly common, conventional reliability testing may well discover it; if not, the saboteur might be hoist by his own petard: the failure might be so isolated in the input space that it does not arise in practice! Again, sabotage is easier to locate because of its purpose.

Integration-level testing methods are also difficult for the would-be saboteur to counter in advance. Fault trees help the tester to work top-down from unwanted behavior to tests that are likely to excite that behavior. The saboteur can attempt to second-guess the tester by applying his own fault tree, but to no avail. If the fault tree identifies a unique series of conditions necessary to a failure, and produces a test meeting those conditions, no malicious fault can hide from that test.

Stress testing is normally a frail reed in exposing failures of a real-time system, because the process interactions are so complex. But once again peculiar circumstances work against a saboteur. If he fails to grasp the complexity, his carefully hidden faults may become glaringly evident, and a stress test is just the mechanism to excite them. Joseph [[17], Section 5.2.3] noted this phenomenon in trying to cause coincident failures in a multi-version voting system.

4 Discussion and Recommendations

The testing methods usually used for debugging show promise for detecting program sabotage. The simplest coverage techniques (e.g., statement testing) are effective against crude attacks. (Statement testing would have found the problem with "sendmail" involved in the Internet worm.) A testing method like expression coverage, which is difficult to simulate by hand and has many incomparable variants, if kept hidden, is likely to be effective across the board. (In a sense, hiding the testing method poses the potential saboteur the same problem usually faced by security programmers: how much effort is it worth putting into protection that may never be needed? For the saboteur, how much effort is it worth putting into dodging tests that may never be used?)

A program to be used on a secure system should be tested as follows:

1) Development testing should begin with debugging (coverage) methods at the unit level, particularly those like dataflow- and expression coverage that are attuned to the kind of code a saboteur must write. At the system level, fault-tree analysis is effective. All the tests thus determined should be documented and saved.

2) At the conclusion of debugging (when no more failures are found), the largest affordable collection of random tests should be used at the unit level. While less effective than debug methods at catching sabotage like Trojan horses, random tests are most effective against attempts to simulate "real" faults, and are the only form of test that a saboteur cannot circumvent.

3) When the software is installed and when it is updated, the tests of 1) should be augmented and rerun, but with the mindset of seeking security violations. A fault-tree analysis can concentrate on failures that might have been introduced by the new installation/update.

4) The methods used in 1) and 3) should be kept secret and changed from time to time.

The existence of hidden test sets for each program reduces the effort required to retest for security, because the tests can be documented to describe the many peculiar events they cause that are not related to security, and to alert the tester to possible anomalies that might indicate a violation. Although this reduces the time required for retesting, it also raises yet another security issue, because access to the hidden test set and its documentation would be invaluable to a saboteur.

The user community of a relatively secure system has a responsibility to act as a field-test group for software, continuing the process of seeking security violations. This means that they should be on the lookout for strange happenings, and should have a method for reporting them accurately and easily. How many Internet users knew about "sendmail" but did nothing? Just as a potential saboteur spends time probing and studying existing software for "holes" that he can exploit, so the legitimate users can be on the lookout for those holes to forestall their later use.

There are many things that the designers and coders of secure programs can do to enhance the efficacy of testing methods in detecting malicious bugs. For one, programmers can avoid obscure code, because each legitimate puzzle in a program diverts the test analysis from other potential problems, and provides hooks for a saboteur to exploit. "Obscure" is defined relative to the test methods that might be used. For control coverage, apparently infeasible paths are obscure, and anomalous use of variables such as multiple-DEF paths on which there are no intervening REFs. For expression coverage, it is obscure to use a complex expression unnecessarily, or to use expressions with many equivalent forms. In the debugging phase testing tools will identify these situations, and the tester should attempt to eliminate them. It may be that what has been discovered is sabotage, or simply honest mistakes, but obscurity should be eliminated in any case.

Legitimate programmers can lay testing traps in code for a saboteur. For example, the documentation for dataflow-coverage tests can note long DEF-REF paths that should not be disturbed, in places where sabotage seems likely. (Fault-trees can be used to identify these places.) Programs that control other programs, notably operating systems and security monitors, are in a sense performing a continuous field test of software, and their logs can be analyzed just like any test. For example, when a system call is rejected for illegal parameter values, the log would evidently aid the tester trying to find a Trojan-horse bug as described in Section 3. The traps in system and programs, to be set off by tests for sabotage, are "sting" operations. Just as "design for testability" is essential for code that seeks to minimize unintentional failures, so thoughtful preparation for the tests that may detect sabotage is also essential. Of course, to be effective, this design must also be hidden.

For the future, three ideas have promise in testing, and even more promise in the detection of malicious faults. They are:

Self-checking code. Leveson [18] has reported favorably on internal-state checks as an alternative to N-version programming. The programmer writes redundant code to check results (particularly in the form of consistency checks on internal data structures) following computations. The presence of such checks makes the task of the saboteur more difficult, because not only must malicious faults be inserted, but the checking code altered to ignore them. In an ideal scenario, the checking code would be so simple and clear that it could be inspected for tampering, and then allowed to serve as watchdog on other, more complicated calculations. Theoretical work by Blum and others [19] has demonstrated simple situations which are even more startling. Using probabilistic checks following a computation, they show that even though the computation is in error some of the time, it can be cleverly used to produce a result that is arbitrarily close to correct. The important point is that the checking/correcting code is simple and easily examined for faults. In an ideal case, if this code remains intact, a saboteur would find that although he succeeded in modifying a program's calculation, the final result did not change!

Fault-based testing. A promising theory [20] combines testing and proving techniques, to show that a program *cannot* contain a certain class of faults, on the evidence that a single test succeeds. The faults precluded are so far very simple ones, and the proofs are too difficult to imagine an early practical use of the theory. As usual, the application to detecting sabotage is more promising than the general case. Instead of having to prove that the test precludes *any* fault, it should be easier to show that it precludes the more structured, interconnected faults that represent sabotage. An extension of this idea [21] measures the ability of a program's locations to hide faults; this "sensitivity" to testing is also a measure of a program's line-by-line resistance to sabotage. Sensitivity analysis has the considerable virtue that it eliminates the need for an oracle.

Data-state-coverage testing. In [15] it is suggested that the quality of tests should be measured by their statistical penetration into internal data states. That is, a test is good if it's coverage of internal data structures is uniform and extensive. Although the idea has forbidding theoretical difficulties, in practice it is easy to measure this coverage [22]. The application to detecting sabotage is a variant on the idea presented in Section 3. The outcome of a data-state-coverage test is described by statistical measures of coverage at each structure and control point. These measures are impossible to predict and control by clever coding, so changes in them might be used to detect sabotage.

References

[1] D. Gelperin and W. Hetzel, "The growth of software testing," *CACM*, vol. 31, pp. 687–695, June 1988.

[2] R. Hamlet, "Editor's introduction, special section on software testing," *CACM*, vol. 31, pp. 662–667, June 1988.

[3] A. Avizienis and J. Kelly, "Fault tolerance by design diversity: concepts and experiments," *Computer*, vol. 17, pp. 67–80, August 1984.

[4] G. Myers, *The Art of Software Testing*. Wiley, 1979.

[5] G. Weinberg and D. Freeman, "Reviews, walkthroughs, and inspections," *IEEE Trans. Software Engineering*, vol. SE-10, pp. 68–72, January 1984.

[6] J. Goodenough and S. Gerhart, "Toward a theory of test data selection," *IEEE Trans. Software Engineering*, vol. SE-2, pp. 156–173, 1976.

[7] S. Rapps and E. Weyuker, "Selecting software test data using data flow information," *IEEE Trans. Software Engineering*, vol. SE-11, pp. 367–375, April 1985.

[8] R. Hamlet, "Testing programs with the aid of a compiler," *IEEE Trans. Software Engineering*, vol. SE-3, pp. 279–290, July 1977.

[9] R. DeMillo, R. Lipton, and F. Sayward, "Hints on test data selection: help for the practicing programmer," *Computer*, vol. 11, pp. 34–43, April 1978.

[10] W. Howden, "Weak mutation testing and completeness of test sets," *IEEE Trans. Software Engineering*, vol. SE-8, pp. 371–379, July 1982.

[11] N. Leveson, "Software safety: why, what and how," *Computing Surveys*, vol. 18, pp. 125–163, 1986.

[12] R. Hamlet and R. Taylor, "Partition testing does not inspire confidence," *IEEE Trans. Software Engineering*, vol. SE-16, pp. 1402–1411, December 1990.

[13] J. Duran and S. Ntafos, "An evaluation of random testing," *IEEE Trans. Software Engineering*, vol. SE-10, pp. 438–444, July 1984.

[14] D. Parnas, A. van Schouwen, and S. Kwan, "Evaluation of safety-critical software," *CACM*, vol. 33, pp. 636–648, June 1990.

[15] R. Hamlet, "Probable correctness theory," *Info. Proc. Letters*, vol. 25, pp. 17–25, April 1987.

[16] J. Laski, "Data flow testing in STAD," *Journal of Systems and Software*, vol. 12, pp. 3–14, 1990.

[17] M. K. Joseph, "Architectural issues in fault-tolerant, secure computing systems," Tech. Rep. CSD-880047, UCLA.

[18] N. Leveson, S. Cha, J. Knight, and S. T., "The use of self checks and voting in software error detection: an empirical study," *IEEE Trans. Software Engineering*, vol. SE-16, pp. 432–443, April 1990.

[19] M. Blum, "Designing programs to check their work," Tech. Rep. TR88-009, International Computer Science Institute, Berkeley, November 1988.

[20] L. Morell, "A theory of fault-based testing," *IEEE Trans. Software Engineering*, vol. SE-16, pp. 844–857, August 1990.

[21] J. M. Voas and L. J. Morell, "Applying sensitivity analysis estimates to a minimum failure probability for software testing," in *Proc. 8th Pacific Northwest Software Quality Conference*, pp. 362–371, October 1990.

[22] A. Babbitt and S. Powell, "Building prototype testing tools," in *Proc. 8th Pacific Northwest Software Quality Conference*, pp. 264–280, October 1990.

Signatures

On-Line Signature Learning and Checking

HENRIQUE MADEIRA, JOÃO G. SILVA

University of Coimbra

URB Boavista-Lotel-1

3000 Coimbra, Portugal

Abstract

This paper presents a new approach to concurrent error detection in multiple processor systems using on-line signature analysis. In this new technique, called On-line Signature Learning and Checking (OSLC), the block identification and the reference signature generation are performed at run time. Many hardware control signals are included in the signatures, which improves the error detection coverage, and the alterations and/or extensions in the compilers, assemblers and loaders are avoided. In OSLC the signatures are stored in the local memory of a watchdog processor, the Checker, which is based on a new principle that reduces the storage requirements of control flow information to less than 2% of the signature overhead. Furthermore, the Checker is very simple and can check several processors concurrently. A demonstration system of this technique has been designed and built. Results of fault injection experiments have shown that 99.4% of instruction type faults can be detected by OSLC with a very short latency (26 μSec). The coverage for general faults is 94.5% and the average latency is 464 sec..

1 Introduction

Several studies indicate transient and intermittent faults as the main cause for computer failures. Even in optimum environments it is estimated that more than 90% of the failures in computers are caused by transient faults [1], [2]. The detection of these faults must be carried out with the system in its operational environment and performed concurrently with the application task. A number of well-known techniques are available for *concurrent error detection.*

Techniques based on structural duplication and voting are quite effective for transient faults but they require a very large hardware overhead [3], [2].

Error detecting codes have relatively low overhead, but they are effective only for regular structures like memories.

Software error-detection mechanisms such as duplication of the processes and time-limits can also be used. However, they lead to complex synchronization problems and represent a very high performance overhead [4].

In the last years there have been many proposals for the use of a watchdog processor (or circuit) to perform concurrent error detection [5]. A watchdog processor is a small device meant to detect errors by monitoring the behavior of the system. The watchdog processor is previously provided with information about the system to be checked. At runtime it uses that information to detect deviations from the correct behavior of the system.

Several aspects of the system behavior can be monitored in order to have error detection: memory access behavior [6], control flow [7, 8, 9, 10, 11], control signals [12], or reasonableness of results [13]. Many behavior based error detection techniques use concurrent signature analysis as a way to compress the large amount of information associated with the description of the several system behavior aspects.

In this paper a new signature monitoring technique called On-line Signature Learning and Checking (OSLC) is presented. In this approach the block identification and the reference signature generation are performed at run time during the normal final test of the application programs. The main advantages of this technique are the detection of a broad range of errors and the use of standard compilers, assemblers and loaders. A simple and effective watchdog processor, which does not require storage for information about the control flow of the monitored programs, is also presented.

Our presentation begins with a summary of the existing signature monitoring approaches In section 2 the general configuration of a system using OSLC is presented. The OSLC is introduced in section 3. The feasibility of this technique is discussed in section 4. In section 5 the new concept of signature verification is described. Sections 6 and 7 discuss the memory overhead and coverage of OSLC. The hardware complexity in the implementation of OSLC is discussed in section 8 and demonstration system is presented in section 9.

1.1 Signature Monitoring: Background

Several studies [14], [15] compared various behavior features (abstractions) and found that program control flow gives the best error detection coverage. Techniques based on signature analysis seem to be the most general and effective approaches for control flow error detection. The basic idea is to partition the application program in blocks at assembly time. Each block has only one valid and unchangeable sequence of instructions. It is then possible to compute a deterministic signature as a function (e.g. a cyclic redundancy code algorithm) of the instructions within the block. At runtime the signature of each block is calculated by special hardware and errors can be detected by comparison between the runtime signature and the precomputed signature. Several proposals have improved this basic idea.

Namjoo [6] proposed a scheme called Generalized Path Signature Analysis (GPSA) using justifying signatures to decrease the number of signatures, thus reducing the memory overhead and the performance losses.

A technique called Signature Instruction Streams (SIS) [9] proposed by Shuette and Shen uses Branch Address Hashing to reduce the number of signatures. Both GPSA and SIS increase the length of the blocks, which leads to a larger latency time. Sosnowski [11] proposed a technique using justifying instructions, i.e. selected processor instructions that act as justifying signatures.

Wilken and Shen [10] proposed a technique called Continuous Signature Monitoring, combining traditional signatures and horizontal signatures. In another paper [16] they proposed a technique combining signature monitoring with encryption, which extends the sources of possible errors detected to computer viruses.

All the signature monitoring techniques referred to above require the signatures to be stored within the code of the application program, which causes performance overhead because the main processor wastes time avoiding the execution of the signatures. This overhead can be removed by storing the signatures in the local memory of the watchdog processor. However, storing the signatures in the local memory of a watchdog processor requires storage for information about the control flow of the monitored program. This information is the program control flow graph. Each node in the graph represents an instruction block and each arc represents a possible transition between blocks. Control flow errors can be detected by the watchdog processor checking if the sequences of instructions executed by the processor correspond to a valid path in the program graph.

All existing approaches in which the signatures are stored in the local memory of the watchdog processor [7], [8], [17] require the explicit or implicit storage of program control graph information, which leads to large memory overhead. In the approach presented in this paper the signatures are stored in

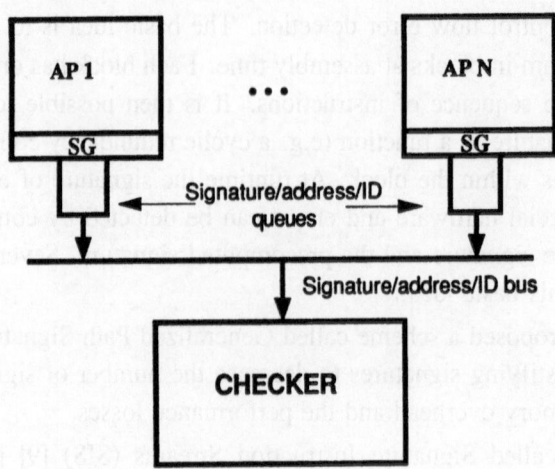

Figure 1: General Configuration for OSLC.

the local memory of the watchdog processor but, as will be seen, the storage of
the control flow graph is avoided and the memory overhead is greatly reduced.

Namjoo [7] proposed a watchdog processor called Cerberus-16 which is
tightly synchronized with the main processor and for that reason can only
check one processor. This watchdog processor executes an image program that
has the same control flow structure as the main processor program.

An asynchronous approach was proposed by Eifert and Shen in [8]. In this
approach the signatures of the programs running on each application processor
are stored as a linked list data structure (the signature graph). The watchdog
processor checks the signatures received from each processor according to the
current position in the respective signature graph.

Shen and Tomas [17] proposed a special purpose architecture called Roving
Monitoring Processor (RMP) for the technique described in [8].

2   Configuration of a System using OSLC

The hardware configuration of a system using OSLC is presented in Figure
1. Near the application processor there is a small added circuit called the
Signature Generator (SG). This circuit detects the beginning and the end of
each program block executed by the application processor (AP) and sends the
computed signature to the Checker.

The Checker checks the correctness of the signatures according to a new principle: the signature verification. In fact, the Checker is not a processor in the sense that it does not have an instruction set and, consequently, it does not require to be programmed. Furthermore, it will not be necessary to develop assemblers or other support software for this watchdog processor. The Checker (detailed in section 9) is just a set of sequential circuits, always performing the same operations.

The Checker is asynchronous in what regards the application programs and it can easily be extended to a system containing several application processors (AP1 to APN).

3 Signature Learning

An obstacle to signature monitoring dissemination is the requirement for special compilers/ assembler and loaders to generate a signature encoded program or a control flow graph. Furthermore, relatively complex watchdog processors such as Cerberus-16 and RMP, which have a specific instruction set, require a package of support software. In spite of the complexity of the special software required by all previous approaches, this point has often been neglected in the literature.

In OSLC the block identification and reference signature generation are carried out during the normal execution of the application program, in a phase called learning phase. This phase is included in the final test of an application program. In this test the program will be submitted to a large number of test vectors in order to obtain a test (and a signature learning) as comprehensive as possible. The essence of signature learning is as follows:

- Every time the Signature Generator detects the execution of a control flow altering instruction (from here on referred to as branch instruction) it assumes that this branch instruction is the last instruction of the current block;

- The instruction executed after a branch instruction is the first instruction of another block, which ends at the next branch instruction;

- The SG is able to determine whether if a block is being executed for the first time or not. If it is so, the signature computed at the end of that block is assumed as the reference signature of that block. This signature, which is considered learned, is then stored somewhere.

Several places for storing the signatures during the learning phase can be considered: an external computer, the system where the signature learning is being performed or in the watchdog processor itself. In the experimental system designed the signatures are stored in an external personal computer.

The detection of the first run of a block is carried out with a column tag bit memory. Before the learning phase the tag bit memory is zeroed. Then, every time a block is executed with the tag bits of the first and/or the last instruction at "0" it means that the block is being executed for the first time. Then, the SG changes the tag bits of the block to "1" (by a read-modify-write alike cycle) which means that the signature of that block has already been learned.

After the learning phase the tag bit memory can be read by the monitored processor for purposes of supervision of the learning process.

The duration of the learning phase depends on the length of the final test of the application program (or module). During this phase the program is exposed to a number of test conditions as large as possible. It is reasonable to expect that most of the paths of the application program should be executed at least once during that test. This is the phase of spontaneous learning.

After that phase the portions of the program not yet executed can be identified through the reading of the tag bit memory. Then, with the help of a software tool, the execution of those program portions can be forced and the signature learning will be completed. Once the learning has been completed the signatures can be stored in a file and no further learning will be required.

4   Feasibility Analysis of the OSLC

After all the signatures have been learned the achievement of OSLC does not seem questionable. Then, the feasibility of OSLC depends only on the feasibility of the learning phase. The main question is: is it possible/easy (necessary) to obtain 100% of signature learning?

To validate this point a tag bit memory and a package of software was built for an IBM PC-AT [18]. The objective was to reproduce in the PC-AT the scenario of the learning phase. Several programs were analyzed and data about the results of the signature learning w. gathered.

The circumstances of the experience were as follows: each selected program was submitted to a number of tests meant to cover all their features, as stated in their functional specification. During the tests the tag bit memory tags all instructions that have been executed at least once. The results of the

Program	Source	O.S.	Compiler	Source code	Object code	Code exec-
				(bytes)		uted
MCALC	Borland Inc.	MSDOS 3.2	Turbo C	60K	17634	91%
mkdir	Univ. Amst.	Minix 1.1	ACK	1319	554	79%
cat	Univ. Amst.	Minix 1.1	ACK	1398	488	85%
cp	Univ. Amst.	Minix 1.1	ACK	2888	1093	74%
ls	Univ. Amst.	Minix 1.1	ACK	12783	5634	83%

Univ. Amst. is University of Amsterdam, Holland and

ACK is Amsterdam Compiler Kit (minix)

Table 1: Percentage of Executed Code with Functional Test Suits.

experiments are summarized in table 1:

Although the test suits used were not as complete as the ones used to validate the programs at the end of their development, it is noteworthy that the code coverage obtained was quite high. Still, a total coverage was needed. A software tool was then used to identify the missed source code (by reading the tag bit memory), leading to 100% code coverage test suit.

Our experiments have shown that it is quite easy (and fast) to reach 100% signature learning for small program modules. This suggests that the signature learning should be performed by small modules instead of the whole program, which is in accordance with structured programming concepts.

A error detection technique such as OSLC has its field of application in systems for critical applications such as industrial process control, railway control systems, and avionics. The development of software for this kind of applications is quite different from non-critical software. Two important features (concerning the feasibility of the signature learning) should be stressed in the reliable software development:

i) The software does not change very often, i.e. once the development of an application program is finished it remains unchanged for a long time. Compilations and run of new programs are very uncommon in fault tolerant computers, specially in industrial control and mission oriented computers. This suggests that signature learning is not a frequent operation and it can be isolated (and hidden) in the final test of the application programs.

ii) The test of critical applications software is far more complete and exhaustive than the test of traditional software. For example, test techniques such as Branch Coverage Test and Control Flow Coverage Test [19], [20] will easily fulfill the requirements of the signature learning. In this context the signature learning does not mean an extra procedure overhead, as it is performed taking advantage of the mandatory final test required by all (reliable) software.

Another aspect being investigated consists of the following: the learning of the signatures corresponding to seldom executed program paths may not be necessary, which releases the requirement for an exhaustive signature learning. Errors occurring during the execution of a program block whose signature was not learned cannot be detected at the end of this block. However, in most of the cases the erratic behavior will proceed and the error will be detected further on when a signature learned block is reached. This suggests that the error latency is more affected by a non-exhaustive signature learning than the error coverage. Fault injection experiments are planed to assess the effect of an incomplete signature learning on the error coverage and latency.

5   No Need for Control Flow Information

This section describes a concept called signature verification and shows how the need for control flow information storage can be avoided.

Signature verification is based on the following principles (see also Figures 2 and 3):

a) The monitored program is logically divided in small sections in such a way that there is only a small number N of signatures (i.e. blocks) for each program section. The number N should be small compared to the total number of possible signatures for a given signature size. For example, choosing 16 bit size signatures some possible choices for N are 64, 128, or 256 signatures per section;

b) Each program section includes as many contiguous program instructions as required to have N blocks/signatures, i.e. the program sections are not fixed size;

c) For each monitored program section there is a correspondent segment in the Checker memory where all signatures of that program section are stored;

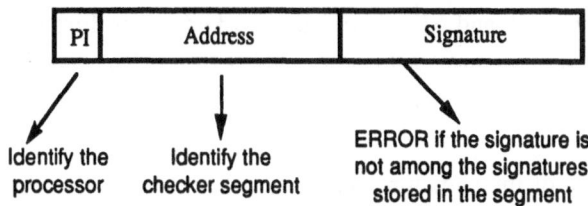

Figure 2: Information Format Received by the Checker.

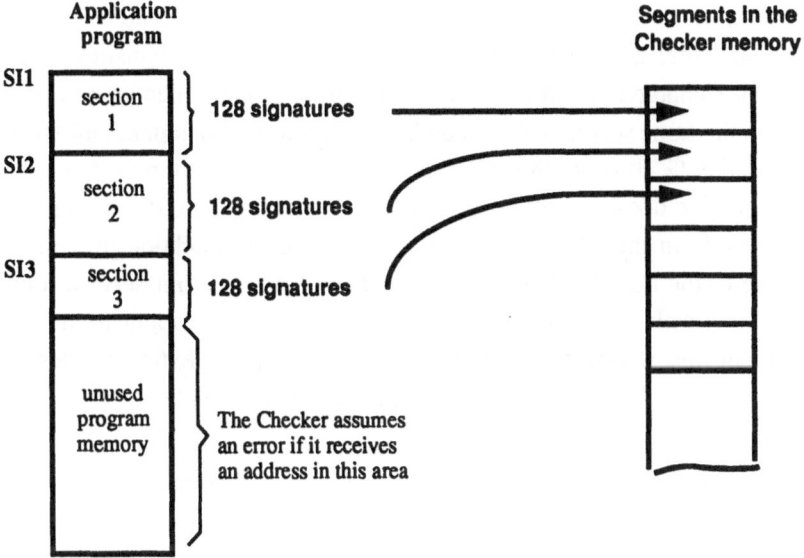

Figure 3: The Checker Principle.

d) Every time the Signature Generator sends a signature to the Checker it also sends the address of the last instruction of corresponding block, and so the watchdog processor can identify the segment that signature belongs to (see fig. 3);

e) A signature is considered correct by the Checker if it is present among the signatures stored in its segment. As the number N is small this verification can be done quite fast using a binary search. Within each segment the signatures must be stored in a sorted way because of the binary search.

The Segment Identifiers (SI1, SI2, in fig. 4) are in fact the address of the first signature in each program section. It is very easy to identify the Checker segment a given signature belongs to. For example: all signatures having addresses between SI1 and SI2 should be checked in the first segment of the Checker memory, signatures with addresses between SI2 and SI3 should be checked in the second segment, and so on. All the segment identifiers are stored in the Checker.

The basic idea of the signature verification lies the fact that the probability of a wrong signature being equal to any of the other signatures in the same section is very low, if a random distribution for the signatures is assumed. The theoretical error detection probability of the Checker is approximately given by $1 - S/2^w$, where S is the number of signatures per segment and w is the width of the signatures in bits. For instance, for sections with 64 signatures each and for 16 bit signatures, the probability of a wrong signature becoming one of the others in the same section is less than 0.0009, representing a potential error coverage of 99.9 %. It is worth saying that in OSLC the occurrence of an interrupt service routine does not represent a particular circumstance demanding special processing from the Checker, as happens with other watchdog processors. In fact, due to the non-deterministic occurrence of an interrupt service routine it cannot be directly represented in the program graph of the application program. Because of this, traditional watchdog processors requires extra schemes to deal with interrupts which increase the complexity of the watchdog processor.

In the presented approach the interruptions are completely transparent to the Checker, where the signatures of an interrupt service routine are verified exactly in the same way as the program signatures. The interruptions are handled only by the Signature Generator: when an interrupt is accepted the current value of the Parallel Linear Feedback Shift Register (PLFSR) is pushed in a signature stack and the PLFSR is initialized. At the end of the interrupt service routine the value in the signature stack is poped into the PLFSR and the computation of the signature of the interrupted block is resumed.

Taking advantage of the small average size of the blocks there is an alternative strategy for interrupt handling. It consists of intercepting, at the external interrupt management circuit level, the interruption requests in order to delay the interrupt acceptance until the current block reaches its end. In this way the interruption requests would only be recognized during the execution of the last instruction of each block and the signature stack would not be necessary. It is worth saying that this strategy also supports interrupts in a multiprocessing environment, where returns from interrupts do not return the processor back to the interrupted routine. Traditional approaches require very complex multiple signature stacks and additional context information to support multiprocessing environments. However, delaying the interrupts acceptance might not be reasonable for some hard-real-time applications.

6. Reduced memory overhead

In the techniques in which the signatures are stored in the local memory of the watchdog processor the memory overhead is formed by two contributions: the signature overhead and the control flow overhead. In all proposed techniques the overhead required by the control flow information storage is the dominating part.

In Cerberus-16 [7] the image program has an instruction for each program block which requires storage for: 1) the opcode of the Cerberus-16 instruction, 2) the size of the block, 3) the address of the next block (in case of control flow alteration), and 4) the signature of the block. The storage requirements for 1), 2), and 3) are greater than for the signatures.

In [8] the signature graph is stored as a linked list data structure and, for example, for a block ending in a two way branch it is necessary to store the signature of the block and three more words which are the type of branch and the pointers for the next elements in the linked list. Furthermore, storage is also required for the signature checking program.

In RMP [17] the signatures are embedded in the watchdog program in order to reduce the signature checking time. This technique, called compiled graph tracing, increases the storage requirements even more.

It is important to stress that as the signatures are not stored within the application code the above mentioned techniques can not use signature reduction techniques such as the ones based on justifying signatures. In this way, all existing techniques storing the signatures in the local memory of the watchdog have very high global memory overhead.

The Checker storage requirements are strictly for the signatures and for the segment identifiers. Table 2 shows the storage overhead required by the segment identifiers for a Checker with a signature memory of 64K words and for several values of signatures per segment. A 20 bit address bus is assumed for the application processor. The right column indicates the segment identifiers overhead as a percentage of the signature overhead.

The segment identifiers overhead is perfectly neglected if 256, 128, or 64 signatures per segment are used. Thus the memory overhead of OSLC is in fact reduced to the signature overhead, as control flow information storage is not required. In this way OSLC has the lowest memory overhead among the techniques storing the signatures in the local memory of the watchdog

Signatures/segment	Segment identifiers	
256	256 × 20 bits	0.49 %
128	512 × 20 bits	0.98 %
64	1024 × 20 bits	1.9 %
32	2048 × 20 bits	3.9 %

Table 2: Storage Overhead Required by the Segment Identifiers.

processor. Several studies [9], [21], suggested typical block sizes from 4 to 10 words which represent a signature memory overhead of 10 to 25 percent.

6 Increased Error Detection Coverage

In OSLC the reference signatures are computed at run time, which allows the inclusion of hardware signals in the signatures increasing the error detection potential. All hardware signals that give unvarying signatures can be included.

The inclusion of these internal control signals is not feasible in previous proposals because the reference signatures are generated at compile or assembly time when hardware internal control information is not available. The modifications/ alterations required of the compilers in order to accommodate the inclusion of internal hardware control signals in the signatures are very complex. These modifications would have to include a detailed model of the hardware, which is not trivial. Furthermore, the compiler would be dependent of the different hard-ware implementations of the same instruction set.

In OSLC the signatures include the following information: 1) instruction fetch, which gives control flow and instruction code information, 2) hardware control signals, which give significant information about the processor/system internal behavior, and 3) address bus values during instruction fetch, which gives additional control flow information.

The inclusion of address information in the signatures should be carefully considered for each application, as it causes the signatures to depend on the particular location of the application program in the memory. For many fault-tolerant systems, such as process control computers, in which the software is very static, this signature dependency of the program location is not important.

The inclusion of all this information in the signatures improves the error detection coverage in various ways:

i) **Improved coverage through the detection of more types of errors.** In previous techniques the behavior abstraction is formed by control flow and instruction code information. In OSLC this abstraction is augmented with additional control flow information and with internal hardware control sequences.

ii) **Improved coverage by the inclusion of information collected during the execution of data access instructions.** Traditional techniques do not put any information into the PLFSR during the execution of data access instruction because data sequences do not produce constant signatures. Obviously, the data information can not be included in the signatures but the various internal control signals produced during the execution of those instructions can be included. Furthermore, "1"s can be forced into the PLFSR (instead of the data bus contents) during the reads and writes in the data segment, giving additional information about the size of the operand accessed.

iii) **Improved coverage by the elimination of intermediate signature correlation.** The intermediate signature correlation (due to blocks with few instructions in which the PLFSR is initialized with the same constant value) decreases the coverage [22]. In OSLC this problem is removed. The instruction fetch, the address, and internal hardware control information are put in the PLFSR in such a way that for each instruction code word the PLFSR is filled with two more words: an address word and a hardware control signals word. This increases the virtual size of the blocks which reduces the correlation.

Given a good intermediate signature uncorrelation in OSLC it is fair to assume a random distribution in the signatures. Then, the error detection coverage of OSLC is limited by the theoretical error detection probability of the Checker. This probability is $P = 1 - S/2^w$, where S is the number of signatures per checker segment and W is the width of the signatures in bits. For example, for 16 bit wide signatures and with 64 signatures per Checker segment the error detection probability is 99.9%.

7 Hardware Complexity in the Implementation of OSLC

The hardware complexity in the implementation of OSLC should be analyzed individually for each part of the system: the Signature Generator (SG) and the Checker.

The hardware complexity of the SG depends on the signature monitoring technique itself and on the features of the application processor(s). As OSLC uses the basic signature monitoring technique the SG is reduced to its simplest form. All that is required is to detect the end of each block, to compute the signatures, and send them to the Checker. This can be implemented by a simple instruction decoder, a PLFSR, and some logic. Furthermore, the interrupt requests can be handled by the method proposed at the end of section 5, which avoids the need for a signature stack. In all other approaches (using embedded signatures or not) the generation of the signatures is more complex because these proposals use special techniques such as justifying signatures or branch address hashing, which increases the SG complexity.

The features of the application processor have a great impact on the complexity of the SG. Particularly, the complexity of the SG depends on the information about the internal state of the application processor available at the IC package pins. For example, if the application processor signals to the outside the start of each instruction fetch the SG will be simpler because circuits for opcode fetch detection are not necessary. Other processor features such as the prefetch queue or the instruction cache have also great impact on the SG complexity. This is also true for other signature monitoring techniques. As the recent microprocessors have internal caches and deep prefetch queues, actual implementation of the SG tend to be included into the processor chip, where all the information required for the generation of the signatures is available.

The Checker, which does not depend on the application processor, is much less complex than other proposed watchdog processors. For example, the RMP [17] uses 90 SSI/MSI packages (excluding memory) while the implementation of the Checker described in section 9 only uses 14 SSI/MSI packages (excluding memory and external interface). However, the signature bus required by OSLC is wider than in other techniques, because the address of each block has to be sent with each signature. Globally, OSLC have lower hardware overhead than previous similar techniques.

8   The Experimental System Designed

A system aimed at experimental evaluation of OSLC has been designed and built. It is formed by the application system and the Checker. The application system is a small computer with a Signature Generator (SG) circuit. It includes a CPU, memory, input/output, address decode, interrupt control circuits, and auxiliary circuitry such as buffers and a clock generator.

The SG includes a PLFSR, an opcode detector, a signature stack, control logic, and a tag bit memory. The inclusion in the computed signatures of information from the data bus, the hardware control signals, and/or the address bus can be selectively enabled/disabled by special PLFSR input circuitry. The SG uses only 17 MSI/SSI packages.

The Z80 was chosen as the application processor for the experimental system. The main reasons for choosing a small CPU like Z80 are:

i) The Z80 does not have internal error detection mechanisms. In this way, it will not be required to decouple the effect of the internal error detection mechanisms (as was the case described in [9]). These internal mechanisms would interfere with the evaluated technique, leading to wrong (or non-complete) results.

ii) The Z80 has some features that simplifies the SG. Two of the most important are that it signals to the outside the beginning of an instruction fetch and it does not use prefetch. In this way, the Z80 seems to be a good choice for OSLC evaluation because it keeps the experimental system small and simple while being complex enough to give significant results. This last aspect is corroborated by recent fault injection experiments where a processor of about the same complexity of the Z80 (the 6809) was used [14]. The Z80 has also been used by Schmid et al. [15] as target processor in fault injection experiments.

The Checker

The Checker is a two stage pipeline machine (see Figure 4). The Address-Segment Converter is formed by a small memory (for the segment identifiers), a successive approximation register, and some control logic. For each signature address it finds a correspondent segment identifier. This stage behaves like an associative memory in which the matching function "equal-to" was replaced by an "inside-the-interval-of" function.

The second stage (Signature Verification) receives the segment identifier that defines the Checker segment where the correspondent signature must be checked: if the signature matches any of the signatures in its segment then the signature is correct. In the second stage each memory segment behaves like an associative memory, and the segment identifier acts as a key indicating which segment should be checked.

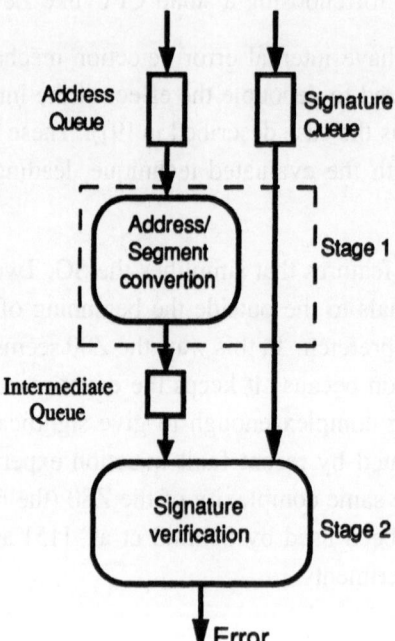

Figure 4: Checker Organization.

These types of associative memories are implemented, in both stages, by a binary search over a conventional memory. The incoming data is compared with the information on the memory data bus and the result of that comparison controls a successive approximation register whose output drives the memory address bus for the next cycle. When a match is attained, in the first stage, the result on the data bus is the segment identifier. In the second stage a match means that the signature is correct. If the mismatch occurs at the first stage it means that the address received does not fit any Checker segment, which indicates a control flow error (with violation of the code segment bounds). If the mismatch occurs at the second stage it means that a wrong signature was received, notifying an error in the monitored processor.

The signature and the address identifier memories can be accessed in two ways: by the Checker during normal signature checking, and by an external computer for initial loading of the signatures and address identifiers. In fact, at the learning phase the signatures+addresses are collected by an external computer. Once the learning is completed the signatures are sorted out and loaded into the Checker. The implementation of these dual-ported memories is easy as there is no concurrent access.

Figure 5 show the implementation of each Checker stage. In the present design the size of the signature memory is 32 K words. The number of signatures per segment is 128 which results in 256 segment identifiers 16 bit wide each. The number of signatures per segment is a power of two because the signature check is based on a binary search.

The signature verification is performed at an average rate of 5 memory cycles (from 1 memory cycle minimum to 6 cycles maximum), and the address-signature conversion at an average rate of 6 memory cycles. The average signature rate of the Checker is the rate of the slower stage, because both stages work in paralell when the pipeline is full. However, as the storage requirements for segment identifiers are very small, a fast memory can be used in the first stage and then the overall signature verification rate is limited by the signature verification stage. Using ordinary 100 nanoseconds access memories at this stage an average signature checking rate is of 2,000,000 signatures per second.

With this rate it is easy to accommodate the monitoring of several application processors in a fashion similar to the "roving concept" defined in [8] and [17]. For instance, for a medium/high speed microprocessor with an average instruction execution time of 1.25 μSec. and for typical blocks between 3 and

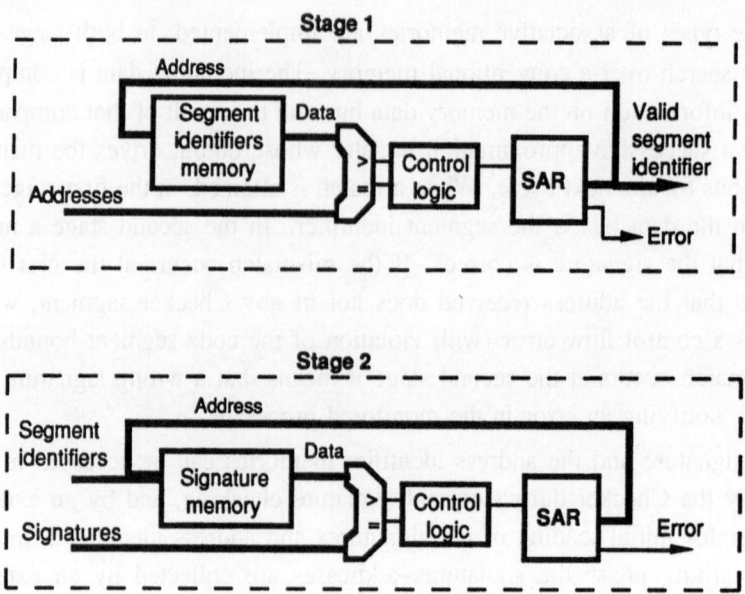

Figure 5: Stages Organization.

4 instructions, signature production rates from 200,000 to 270,000 signatures per second can be expected. The experimental Checker can check up to seven application processors like this.

In this first phase only one application processor was considered, as the ability of the Checker to monitor several processors does not seem to be questionable. As the Checker is very modular it will be easy to extend the present design to monitor more than one processor. To accomplish that, several input queues and a queue scheduling strategy should be will have to be added to the present design.

The stages are asynchronous, each stage being triggered by its input data following the data-flow model. While the Checker is not a real data-flow machine, it can use some of the data-flow model advantages, namely straight-forward parallelizing. In fact, it would be possible to replace each stage by a pack of identical stages working in parallel which would result in a quasi-linear increase in speed with the number of stages in parallel.

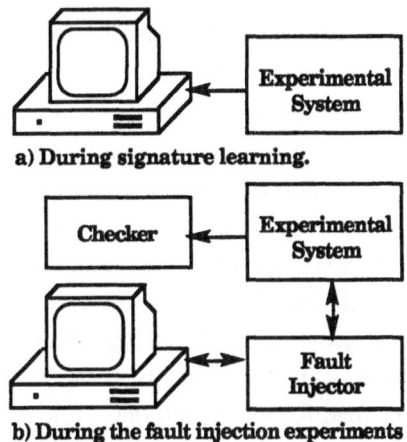

a) During signature learning.

b) During the fault injection experiments

Figure 6: Experimental System Configuration.

9 Experimental Evaluation

A detailed evaluation of OSLC is described in [23]. In this section the experimental setup used for OSLC evaluation is briefly described and the most important results are summarized.

The configuration showed in Figure 6a) was used in the signature learning phase. All new signatures are collected by a personal computer. To do this the SG adds a bit of New Signature to the signature + address everytime a block is executed for the first time.

The program executed by the target system is constituted by five small benchmarks [24] meant to be as representative as possible of the various types of program structures executed by computers. Examples of benchmarks used are a pseudo-random number generator, a quick sort, and a Sieve prime number generator. All these programs were written in C and are executed one after each other in an endless While loop.

The benchmarck programs were previously executed in order to have a complete signature learning. In the present case the signature learning is quite easy because all benchmarcks work on an invariant set of data. In this way, all the signatures are generated in only one benchmark run. Once the signature learning is finished the signatures are organized in program sections using the address information. After this the signatures belonging to each program section are sorted out and stored in the correspondent Checker segment.

In the checking phase the configuration in Figure 6b) is used. The faults are injected in the application system bus and the results (error detected or non-detected and the latency) are collected by the personal computer.

A fault injector has been designed and built [24]. It consists of a fault injector module which is connected to a personal computer by a 24 bit input/output parallel port. The personal computer also has a board with six programmable counters where the error detection latency of up to six error detection methods can be measured simultaneously (see experiments described in [24] and [23]).

The faults injected are divided in instruction cycle faults and data cycle faults. The former are injected during the reading of the instruction bytes and the latter are injected while a data operand is being read or written. As the duration of the faults can span for more than one cycle many faults are mixed faults, i.e. they affect both instruction cycles and data cycles. In terms of OSLC coverage analysis mixed faults should be regarded as instruction type faults, because both instruction type and mixed type faults can be directly observed by the signatures.

The faults were injected according to the following criteria:

i) The locations in the benchmarks code where the faults are injected are chosen at random;

ii) The fault duration is a random value between 1 and 4 memory cycles of the target system;

iii) The bus line affected is chosen at random;

iv) Only one bus line is affected at each fault.

A total of 5963 faults were injected for general OSLC evaluation. The coverage for instruction type faults is quite high (see Table 3) and the average latency is very low.

The low latency obtained for instruction type faults means that the great majority of the faults were detected at the end of the block that was being executed at the time the fault was injected.

The coverage obtained for instruction type faults injected in the data bus is lower than the coverage obtained for faults injected on the other bus lines. It was found that this effect was caused by some faults injected in the data bus, which causes the processor to execute a Halt instruction, which stops the processor and so the end of the current program block will never be reached (see [23] for details). This situation can be detected directly in the Halt pin of

Lines Faulted	Cover.	Lat. (μSec)	Std. Dev. (μSec^2)
Address	99.9 %	25.9	22.8
Control	100 %	19.9	19.1
Data	94.9 %	29.9	22.9
Total	98.6 %	25.6	22.3

Table 3: Results for Instruction Type Faults.

Lines Faulted	Cover.	Lat. (μSec)	Std. Dev. (μSec^2)
Address	94.0 %	511	6209
Control	99.6 %	19.8	18.7
Data	91.3 %	503	7068
Total	94.5 %	464	6915

Table 4: Results for General Faults.

the Z80 or by a watchdog timer. If the errors detected by the activation of the Halt pin were considered the total coverage for instruction type faults would be 99.4%.

Table 4 shows the results considering all faults, i.e. faults that affect randomly data cycles and instruction cycles. The total coverage is 94.5%, which is a very high coverage considering that pure data cycle faults are difficult to detect by signature monitoring techniques. The coverage for faults injected in the control bus is still very high (99.6%) because the control bus information are included in the signatures in all bus cycles.

The average latency is much larger than for instruction type faults and the standard deviation is very large too. This was caused by a small number of faults that were detected with a very high latency (hundreds of milliseconds). It was found (see [23]) that these faults are pure data faults that affect the stack content without disturbing the control flow. These errors may stay latent for a long time and become effective when the mistaken stack area is used after a RET instruction. Chart 1 shows the latency distribution.

Table 5 shows the coverage variation with the fault duration. As expected, the faults with duration of one and two cycles are more difficult to detect and the average latency and standard deviation are larger.

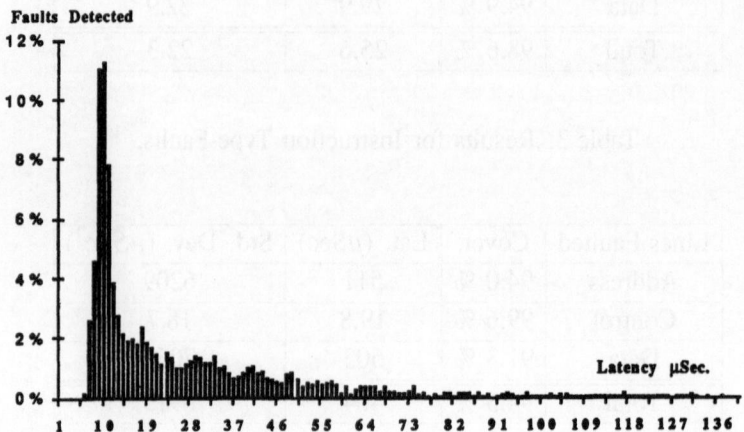

Chart 1: Latency distribution for general faults.

Duration	Cover.	Lat. (μSec)	Std. Dev. (μSec^2)
1 cycle	86.3 %	1524	12557
2 cycles	94.8 %	409.6	6757
3 cycles	98.5 %	24.1	21.3
4 cycles	98.6 %	24.9	24.3

Table 5: Coverage Variation with the Fault Duration.

The coverage results obtained by OSLC show that the inclusion of hardware control signals and address bus information into the signatures is of utmost importance. Fault injection results obtained by Schuette and Shen for the technique Signature Instruction Streams (SIS) [9] (in which the signatures include only the data bus information during instruction and operand fetches) suggests that the coverage of SIS is substantially lower than the coverage of OSLC. This is particularly true for general type faults where the coverage obtained with SIS was only 82% (93.1% for OSLC in similar circumstances; see [23] for details) and for faults of small duration (i.e. one memory cycle faults) where the coverage of SIS was only 66% (86.3% for OSLC).

Summary

In this paper we have presented a new approach to concurrent error detection in multiple processor systems called On-line Signature Learning and Checking (OSLC). In this new signature monitoring technique the block identification and the reference signature generation are performed at run time taking advantage of the normal test of the application programs. OSLC does not require alterations and/or extensions in the compilers, assemblers and loaders. Since the reference signatures are generated at run time it is feasible to include in the signatured information many hardware control sequences which increases the error detection coverage. A new principle for building a watchdog processor,the Checker, is described. The Checker is very simple and does not require the storage of extensive control flow information about the application program.

A demonstration system of this technique has been built. Results of fault injection experiments have shown that 99.4% of instruction type faults can be detected by OSLC with a very short latency (26 μSec). The coverage for general faults is 94.5% and the latency is 464 μSec. These results represent a significant improvement relatively to previous results and suggest that the inclusion in the signatures of hardware control information and address information is of utmost importance.

Future work will include a detailed study of the non-detected faults in order to identify other relevant information to include in the signatures.

References

[1] D. Siewiorek and L. Lai, "Testing of digital systems," in *Proc. IEEE*, pp. 1321–1333, October 1981.

[2] D. P. Siewiorek and R. S. Swarz, *The Theory and Practice of Reliable Design*. Digital Equipment Corporation, Bedford, Massachusetts: Digital Press.

[3] P. K. Lala, *Fault tolerant and fault testable hardware design*. New York: Prentice Hall International, 1985.

[4] A. Daam, "The effectiveness of software error detection mechanisms in real time operating systems," in *Fault Tolerant Computer Symposium*, pp. 171–176, July 1986.

[5] A. Mahmood, "Concurrent error detection using watchdog processors - a survey," *IEEE Transactions on Computers*, vol. 37, February 1988.

[6] Namjoo, "Techniques for concurrent testing of vlsi processor," in *Proc. of the International Testing Conference*, pp. 461–468, Novemeber 1982.

[7] M. Namjoo, "Cerberus-16: An architecture for a general purpose watch-dog processor," in *Digest of Papers, 13th Int. Symp. on Fault-Tolerant Computing*, pp. 216–219, June 1983.

[8] J. B. Eifert and J. P. Shen, "Asynchronous signature instructions streams," in *Proc. of 14th Int. Conf. on Fault-Tolerant Computing*, pp. 20–22, June 1984.

[9] M. A. Schuette and J. P. Shen, "Processor control flow monitoring using signatured instruction streams," *IEEE Transactions on Computers*, vol. 36, March 1987.

[10] Wilken and Shen, "Continuous signature monitoring: Efficient concurrent-detection of processor control errors," in *IEEE 18th Int. Test Conf.*, 1988.

[11] J. Sosnowski, "Detection of control flow errors using signature and check-ing instructions," in *18th Int. Test Conf.*, pp. 81–88, 1988.

[12] S. F. Daniels, "A concurrent test technique for standard microprocessors," in *Proc. of Compcon Spring 83*, pp. 389–394, 1983.

[13] A. Mahmood, A. Ersoz, and E. J. MaCluskey, "Concurrent system level error detection using a watchdog processor," in *Proc. Int. Test Conf.*, pp. 145–152, November 1985.

[14] U. Gunneflo, J. Karlsson, and T. J., "Evaluation of error detection schemes using fault injection by heavy-ion radiation," in *Fault Tolerant Computer Symp.*, pp. 340–347, June 1989.

[15] M. Schmid, R. Trapp, A. E. Davidoff, and G. M. Masson, "Upset exposure by means of abstraction verification," in *Proc. of the Fault Tolerant Computing Symp., FTCS-12*, pp. 237–244, June 1982.

[16] K. Wilken and J. Shen, "Concurrent error detection using signature monitoring and encryption," in *1st Int. Working Conference on Dependable Computers in Critical Application*, August 1989.

[17] J. P. Shen and S. P. Tomas, "A roving monitoring processor for detection of control flow errors in multiple processor systems," *Microprocessing and Microprogramming*, vol. 20, pp. 249–269, 1987.

[18] J. G. Silva and H. Madeira, "A tool for the determination of code coverage of test suits," tech. rep., Univ. of Coimbra, October 1989.

[19] R. Hamlet, "Testing programs to detect malicious faults," in *2nd Int. Working Conf. on Dependable Computing for Critical Applications*, pp. 162–169, February 1981.

[20] S. Rapps and W. E., "Selecting software test data using data flow information," *IEEE Transactions on Software Engineering*, vol. SE-11, pp. 367–374, April 1985.

[21] A. Mahmood and E. J. MaCluskey, "Watchdog processor: Error coverage and overhead," in *Proc. of 15th FTCS*, pp. 214–219, 1985.

[22] K. Wilken and J. P. Shen, "Embedded signature monitoring: Analysis and technique," in *Proc. of Int. Test Conf.*, pp. 324–333, 1987.

[23] H. Madeira and J. G. Silva, "On-line signature learning and checking: Experimental evaluation," in *IEEE CompEuro*, May 1991.

[24] H. Madeira, G. Quadros, and J. G. Silva, "Experimental evaluation of a set of simple error detection mechanisms," *Microprocessing and Microprogramming*, vol. 30, pp. 513–520, August 1990.

A New Efficient Signature Technique for Process Monitoring in Critical Systems *

SHAMBHU J. UPADHYAYA, BINA RAMAMURTHY

Department of Electrical and Computer Engineering

State University of New York at Buffalo

Buffalo, NY 14260, USA

Abstract

A new simple, inexpensive and time/space efficient signature technique for process monitoring is presented. In this technique, signatures are accumulated to form an m-out-of-n code and the corresponding locations in the memory are tagged. During the run-time, the generated signatures at the tagged locations are monitored using a simple hardware to determine whether they form m-out-of-n codes. This approach offers flexible error latency and high coverage. Results of an experiment conducted to verify the latency are presented. The main advantage of our approach is the requirement of no reference signatures unlike existing techniques leading to both memory and run-time execution time savings. The flexible latency and the low cost hardware will make this technique suitable for critical applications where quick detection of errors is deemed important.

Key Words: Branch-free interval, error coverage, error latency, signature analysis, watchdog processor.

1 Introduction

Concurrent checking of processor operation by signature technique is an effective mechanism for the reliable operation of computer systems. Signature analysis by on-line monitoring of some key program attributes has been found useful in the detection of transient errors in computer systems. Signature analysis using watchdog processors [1] is gaining ground as a means of error

---

*Research supported in part by National Science Foundation Grant CCR-89-09243.

detection in harsh environments like radiation where transient errors are generally prevalent. Error detection forms the basis for initiating recovery in fault tolerant systems.

Low cost watchdog processors have been proposed for the detection of two types of errors known as *sequencing errors* and *bit errors* [2] in computer programs. Sequencing errors are caused by illegal control flow of programs whereas bit errors are normally due to the mutilation of the instruction opcodes. It is generally assumed that such errors are caused by transient malfunctions. Data errors on the other hand, are detected and corrected by error detecting/correcting codes, thereby not requiring any sophisticated hardware such as a watchdog processor.

In order to detect sequencing errors using watchdog processors, the *branch-exit* and *branch-entry* points are first identified and a program graph is prepared where a node represents a collection of *branch-free* intervals and the edges represent the control flow. Certain *messages/signatures* are assigned at the beginning and/or end of the nodes. During normal execution, these messages/signatures are checked for consistency. Any inconsistency indicates an error. In order to detect bit errors, the signatures are generated dynamically during the execution of the program and then compared with the corresponding reference signatures that are generated during the compilation phase. Justifying signatures are added at branch merge points so as to reduce the space overhead when multiple paths exist between two points in a program [3].

The existing methods for signature monitoring can be classified into two categories on the basis of how the signatures are incorporated and used by the program. In the first category, a signature graph is prepared during the compilation phase and is used as a reference at run-time. This approach has two main drawbacks: (1) memory overhead to store the signature, and (2) slower execution due to the additional fetching of the signatures. Techniques of [4] and [5] are examples of such systems. In the second category, signatures are embedded into the instruction stream [6], [7], [8]. This approach requires the use of a mechanism such as tag bits to distinguish the signatures from regular instructions. System speed slows down due to the execution of NOPs (No Operations) during the processing of signatures.

The space and time overheads of incorporating signature analysis in on-line error detection are generally determined by the number of signatures and the frequency of comparison. Storage overhead can be reduced by generating an *all-1* or *all-0* signature [2] at the signature comparison points using standard justification methods [3]. This leads to storing only one type of reference signature. However, the signature comparison points are often fixed or determined by the size of the nodes. Thus, the available schemes are not flexible and the latency is fixed. Critical applications incorporating fault tolerance require flexible latency and reduced run-time speed degradation to support the recovery process. We present a scheme that does not require a reference signature at all for signature analysis. This has the advantage that the latency can be minimized by increasing the frequency of comparison. There will be no increase in the memory overhead nor a degradation in the run-time performance because no reference signatures need to be stored nor fetched. The proposed scheme is flexible in that it can offer a controllable latency. The hardware required is only the addition of a single bit tag memory and a simple hardware unit for run-time signature generation and evaluation. The details of the scheme are given in Section 2. A technique based on continuous signature monitoring by Wilken and Shen [9], [10] also offers very low latency and has many similarities with our scheme. In Section 3, we compare these two schemes and show that our scheme is much simpler and less expensive than the continuous signature monitoring scheme of [9]. Some experimental results and a discussion appear in Section 4.

2   The New Scheme

One of the most common functions used for signature generation[1] is the modulo-2 sum on the bits of the instruction opcodes. A signature is derived for a node (block of instructions) during the compilation phase and stored as the first instruction of the node [6]. A signature is generated during the program execution and is compared with the precomputed signature before exiting the block. This technique requires the storing of one signature per node. Thus, in a program with large number of nodes, the storage requirement becomes very large. Two tag bits are also required to distinguish the precomputed signature from regular CPU instructions and to recognize the last instruction of a node [6]. The tag bits can be eliminated by hashing [9] or by using software delimiters and a monitor [7]. This may however affect the run-time performance or compromise on the hardware overhead.

We now present a new technique which requires one tag bit and no reference

---

[1]Other functions such as a checksum are also suggested [11].

signature at all for error detection in sequential codes.[2] This approach will eliminate the storage problem that existed in the previous schemes. Given a sequential code, a signature is derived by applying a signature generating function on the opcode successively until the signature forms an m-out-of-n code for a specified m and n.[3] The location in the memory that corresponds to an m-out-of-n code is tagged as a checkpoint for comparison. During the normal execution, the generated signature at a tagged location is checked to determine whether it forms an m-out-of-n code. If it fails to form an m-out-of-n code at the designated checkpoint, an error is indicated.

Sequencing errors can be handled easily by modifying the sequential code checking method as follows: The instruction at the destination location of a branch is made to form an m-out-of-n code thereby forcing a signature checkpoint at the branch destination. This can be done by applying the signature generating function successively on the branch instruction opcode and the instruction at the destination location and *adjusting* the accumulated signature to form an m-out-of-n code. Such an adjustment will however require one additional byte per branch. Certain modifications are also required in the compilation and execution phases in order to incorporate the proposed signature analysis.

2.1 Compilation Phase

The process that takes place during the compilation of the code to incorporate signatures is explained by an algorithm. An initial signature $INIT$ such as an *all-0* or *all-1* is assigned to the intermediate signature $INTSIG$ at the start of the program. The following steps are then executed for each instruction of the program in order to generate the signatures.

Algorithm 1

1. $SIG \longleftarrow F(INST, INTSIG)$; /* Let $INST$ be the opcode of the instruction to be processed next, F, a signature generating function */

2. if SIG forms an m-out-of-n code then

---

[2]A set of instructions is called a sequential code if there is only one entry and one exit point for the entire block of instructions.

[3]An m-out-of-n code is one which has m 1's in an n-bit stream. m-out-of-n codes have been used for microprogram control checking; but the overheads here are quite high due to decoding and re-encoding [2]. Our scheme does not have any such overheads.

$TAG(INST) \longleftarrow 1$; /* set a tag to indicate a checkpoint */
$INTSIG \longleftarrow INIT$;

3. else

$INTSIG \longleftarrow SIG$;
$TAG(INST) \longleftarrow 0$;

endif;

4. if $INST$ a BRANCH then /* if the instruction is a branch */

(a) $SIG \longleftarrow F(INTSIG, B\_DEST)$; /* B_DEST is the opcode of the instruction at the branch destination address */

(b) $B\_EXIT\_BYTE \longleftarrow$
$F(SIG, ADJUSTER)$; /* embed the resulting signature as B_EXIT_BYTE after the branch instruction */

(c) $TAG(B\_DEST) \longleftarrow 1$; /* set a tag to indicate a signature checkpoint */

endif;

At Step 2 of the algorithm, the intermediate signature is monitored by a software counter to verify whether it forms an m-out-of-n code. At Step 4, if the instruction is a branch, then a checkpoint is forced at the branch destination location by adjusting the intermediate signature to form an m-out-of-n code. This is done by using a predetermined byte called $ADJUSTER$ and storing the resulting signature as B_EXIT_BYTE in the program code. The $ADJUSTER$ byte is essentially an m-out-of-n code. The entire algorithm can be incorporated into the compilation phase and therefore, no special hardware is required for its execution.

2.2 Execution Phase

During the execution phase, the signature at a tagged location is verified for being an m-out-of-n code. Besides, the following operation will take place when a branch instruction is encountered. The signature generating function is applied to the branch instruction and the signature accumulation is continued with the instruction at the branch destination location if the condition for branch is met and control transfers to the branch destination. Otherwise, signaturing continues with the next sequential instruction. If the branch has taken place, then the byte B_EXIT_BYTE that is fetched concurrently is applied to the intermediate signature at the branch destination location. If the resulting signature forms an m-out-of-n code, the branch will be treated as correct. The complete checking process during the execution is carried out by a hardware unit whose function is described by the following algorithm.

When the program is started, an initial signature such as an *all-0* or *all-1* is used as an intermediate signature. Algorithm 2 is executed by the hardware unit for each instruction encountered during the program execution.

Algorithm 2

1. $SIG \longleftarrow F(INST, INTSIG)$; /* generate signature */

2. if $TAG(INST) = 1$ then

   (a) if SIG does not form an m-out-of-n code then

   ERROR $\longleftarrow$ TRUE:EXIT

   else ERROR $\longleftarrow$ FALSE;
   endif;

   (b) $INTSIG \longleftarrow INIT$;

3. else /* it is not a checkpoint yet */

   (a) $INTSIG \longleftarrow SIG$;

   endif;

4. if $INST$ a BRANCH and CONDITION TRUE then /* if the instruction is a branch and branch condition is met */

   (a) wait until the instruction at B_DEST is fetched;

   (b) $SIG \longleftarrow F(INTSIG, B\_DEST)$;

   (c) $INTSIG \longleftarrow SIG$;

   (d) $SIG \longleftarrow$
       $F(INTSIG, B\_EXIT\_BYTE)$;

   (e) if SIG does not form an m-out-of-n code then

ERROR ⟵ TRUE:EXIT

else ERROR ⟵ FALSE;
endif;

(f) $INTSIG \longleftarrow INIT$;

endif;

Remarks

(1) When sequencing error and bit error checking schemes are combined as above, there may be an inconsistency in the signature analysis along the path that will be taken if the branch conditions are not met. This path when meets the branch merge point at $DEST\_LABEL$ shown in Figure 1(a) should be made to have an m-out-of-n code in the signature. Otherwise, the paths continuing beyond $DEST\_LABEL$ will have inconsistent initial signatures. This problem can be easily solved by having an extra byte prior to the branch merge point to adjust the signature to an m-out-of-n code. The same effect can also be obtained by having an unconditional branch instruction prior to the instruction $DEST\_LABEL$ as shown in Figure 1(b), which should branch to the very next location. Algorithms 1 and 2 are still applicable without any modifications.

(2) It may be noted that multiple branch merging at a single location as shown in Figure 1(a) is handled by our scheme very elegantly. There is no need to have any special signaturing nor there is a need to execute several NOPs at the branch merge points as in [6].

(3) In Algorithm 1, there are several options for storing the extra byte (B_EXIT_BYTE) used for the signature adjustment. This byte can be augmented to the branch instruction in which case, it should be fetched concurrently with the branch instruction. If the branch condition is not met, this byte can be discarded. As an alternative, B_EXIT_BYTE can be stored elsewhere in the memory, but its fetching then has to be well coordinated whenever a branch instruction is encountered during the execution of the program.

(4) In our technique, while considering a sequential code, once a tagged location is reached, the signature is reset to the initial signature $INIT$. Instead, the signature accumulation can be continued without a reset until it forms an m-out-of-n code, at which point the location is tagged again. This variation for the accumulation of signature will have a significant advantage in terms of error coverage as will be described in the next section.

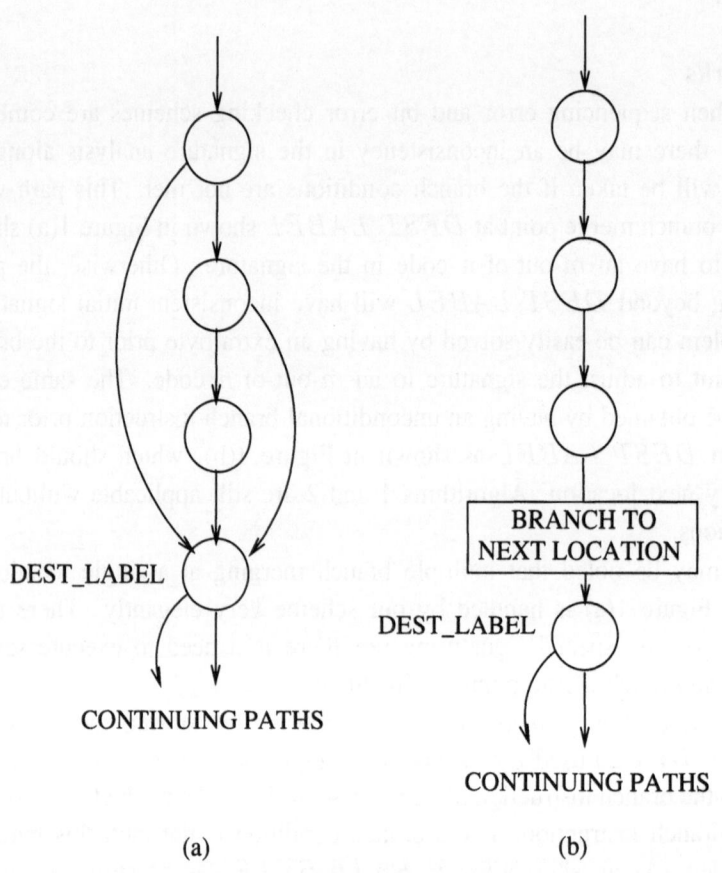

Figure 1: Cases of branch merging.

Hardware Overheads

The hardware consists of a signature generator, an m-out-of-n checker and a branch detector as shown in Figure 2. If the number of bits considered for the signature generation is n, the signature generator will consist of n D-type latches and n exclusive-or gates. An m-out-of-n checker can be realized using a simple counter or adder and a comparator. Branches can be detected by monitoring the jumps in the address sequence or by a special circuitry. Since the fetch and execute of instructions take several clock cycles, the signature checking hardware can monitor each instruction concurrently with the main processor.

3 Performance Analysis

Algorithm 2 suggests that the sequencing errors can be modeled as bit errors for the purpose of performance analysis.[4] This is so because, the addition of the byte B_EXIT_BYTE following the branch instruction to force a checkpoint at the branch destination location makes the entire code look like a continuous sequence of instructions. Therefore we consider only the bit errors in our analysis.

We define error latency as the average number of instructions between two consecutive tagged locations. If an error occurs during the interval between the tagged locations, it will be detected at the next tagged location. However, it is important that the error coverage be large enough so that an error is detected at a tagged location with very high probability. Therefore, it is essential to consider both latency and coverage in order to determine the performance of the signature technique. Next we derive expressions for the latency and coverage of our scheme.

3.1 Error Latency

Consider a segment of a program (sequence of instructions) where n bits of the opcode are used to derive the signatures. Let the signature generating function be a modulo-2 sum. Signatures are derived by making the exclusive-or of the n bits of consecutive opcodes in the program. The probability that an n-bit

---

[4]Such an equivalence has also been considered in [12].

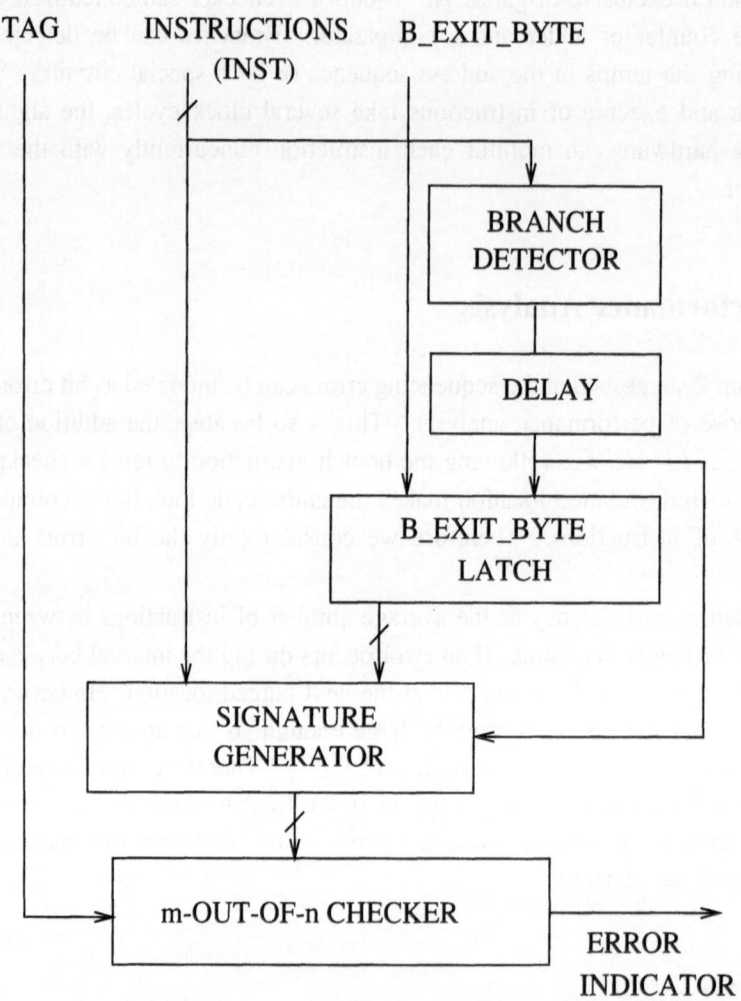

Figure 2: Hardware organization.

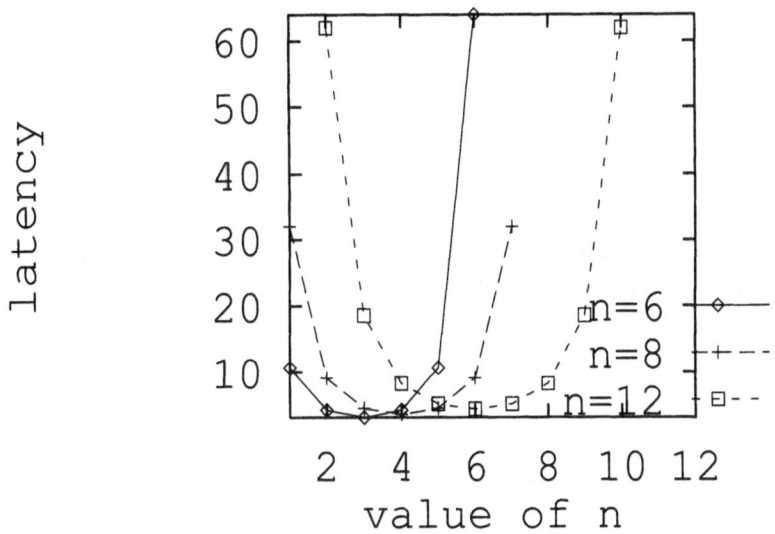

Figure 3: Variation of latency with n and m.

signature at any given point is an m-out-of-n code is

$$\frac{n!}{m!(n-m)!2^n} \tag{1}$$

assuming that the bits 1 and 0 are evenly distributed. Since there will be a checkpoint corresponding to an m-out-of-n coded signature, on an average, error checking will be done at an interval of $\frac{2^n m!(n-m)!}{n!}$ instructions. One can see that a flexible error latency can be obtained by this scheme since both n and m are controllable. For instance, if n is chosen as 8 and m is say 3, the average latency is 5.3 instructions. A number of curves are plotted in Figure 3 showing the dependency of error latency on n and m. If $m = 0$ or $m = n$, we have the case of *all-0* and *all-1* signatures in which case, the latency is 2^n. Obviously, the error latency is minimum at $m = n/2$.

3.2 Error Coverage

Our scheme offers a good error coverage. All single errors between the checkpoints are detected by the m-out-of-n coded signature. The multiple error coverage is dependent upon the relative values of m and n. Multiple errors that go undetected are the ones that change the signatures in such a way that an erroneous signature correlates with another valid m-out-of-n code. This event can occur in $(n!/(m!(n-m)!)) - 1$ ways. There are $2^n - 1$ ways that a single code of n bits can change under error condition. Out of these $(n!/(m!(n-m)!)) - 1$ cannot be detected. Therefore, the coverage factor is

$$1 - \frac{n!/m!(n-m)! - 1}{2^n - 1} \tag{2}$$

One can verify that the coverage is maximum when $m = n$ or $m = 0$.

A low coverage is obtained if expression (2) alone is considered. For instance, the coverage is only 73 % for $n = 8, m = 4$. This is not the case when latency is also considered. If the signatures are not reset at a checkpoint as noted in Remark (4), our technique gives a much better coverage. This is because an error that is not detected at a checkpoint may eventually be detected at the next checkpoint with a high probability. Let the unconditional coverage at a checkpoint given by (2) be represented by p. Now, given that an error goes undetected at the 1st, 2nd, and $(l-1)^{th}$ checkpoint in a sequential code, the probability of this error being detected at the lth checkpoint is:

$$p_l = p \cdot \sum_{j=1}^{l} (1-p)^{j-1} \tag{3}$$

It can be verified that for the case of $n = 8, m = 4$, if an error goes undetected at a checkpoint ($p_1 = 0.73$), it will be detected at the next checkpoint with a probability of $p_2 = 0.93$.

Latency and coverage can be improved by checking the signature to form a non-m-out-of-n code at the nontagged locations in the program code. This will require the checking at the end of each instruction. Such a modification can offer better latency and coverage while utilizing the m-out-of-n checker hardware more efficiently.

3.3 Comparison

The comparison of this technique with the existing signature analysis techniques is based on the number of reference signatures per node, error latency, coverage and the overhead factors. A radical difference of the proposed scheme is the simplicity of the approach, simplicity of the hardware and the reduced overhead and performance degradation.

First of all, our scheme does not need any reference signatures at all. As far as the error latency is considered, every branch is guarded by a signature checkpoint. Sequential code is guarded by the checkpoints at the tagged locations. Whereas the latency of the regular signature based techniques is equal to the number of instructions in the sequential block [11], our scheme offers a much flexible latency. Latency of the order of 3 or 4 instructions can be achieved in the new scheme as evidenced in Figure 3. Although a low latency is also achievable in Saxena and McCluskey's method [11] by checking at the end of every instruction, the overhead of their scheme is quite high. For instance, one extended checksum signature is required for every sequential block in [11]. Further, not all branches are exclusively guarded in [11] and so, there is a potential for multiple sequence errors in their approach.

The multiple error coverage of the proposed scheme is somewhat less than the coverage of the existing schemes [3], [8] and [11] although, single errors are completely covered. Multiple error coverage of our scheme can be enhanced by decreasing the value of m, however, a proper value for m should be arrived at by considering the latency issue also.

One of the drawbacks of the proposed scheme is the requirement of one extra byte per branch. But this memory overhead is much less than the overheads of the existing schemes because, no reference signatures need to be stored in our scheme unlike the others. These extra bytes can be fetched concurrently with the branch instruction. Further, since there are no NOPs to be executed anywhere in the program, there is no degradation in the normal run-time performance by employing our scheme.

Since our technique bears a resemblance to Wilken and Shen's continuous signature monitoring scheme (CSM) [9], we make a special comparison with this scheme. Since the CSM scheme uses a two dimensional signature, it has a better latency and coverage. It provides a latency of one instruction cycle because it uses a horizontal signature in addition to the conventional vertical signature. If only a single bit is used for horizontal signature in CSM scheme, both the CSM and the proposed schemes can be implemented without the use of the tag bit memory. The parity bit of the memory can be used for the purpose of storing the horizontal signature in CSM and to indicate the checkpoint in the proposed scheme. If multiple bits are used for the horizontal signature in CSM, the entire memory has to be padded with as many bits as in the horizontal signature. This means a large memory overhead in CSM whereas no such problem exists in the proposed scheme.

Another conspicuous problem in CSM is the information overload on the parity bit. The parity bit is loaded with three types of information: horizontal signature, cumulative horizontal signature for the interval and the indicator bit for the vertical signature. To extract the useful information from the parity bit, hashing and rehashing is required which results in increased latency and poor run-time performance. No such problem exists in our technique.

The number of vertical signature required in CSM and the number of extra bytes in the proposed scheme are nearly equal. The overall memory overhead of the proposed technique is expected to be smaller than the CSM technique because no reference signature is used in our technique. The latency and coverage of our technique is well balanced and the simple hardware make it reasonably good to be practically usable.

4 Experimental Study

In order to obtain simple expressions for error latency and coverage, we assumed that the occurrence of 1's and 0's is equally likely in a given opcode. However, in practice the average latency will depend upon the frequency of the occurrence of the various instructions in a program. We have performed a computer analysis on a reasonably sized program to demonstrate this point. Three example programs written in various languages were considered. The first program written in Modula-2 is a class project simulating a virtual memory management system. The second is a database decomposition program written in Pascal, whereas the third one is a RISC processor simulator written in C language. All the programs were converted to VAX-11 assembly code at the time of compilation. All the blocks of sequential code (nodes) are identified and the formation of the m-out-of-n code is traced for various values of m. The modulo-2 sum function is used for signature generation.

The tables (1, 2 and 3) give the data for only a set of sample nodes of reasonable size. In these tables, the first column gives the size of the nodes in terms of the number of instructions. The tables list the frequency of the m-out-of-n codes for various values of m. The last row in the tables give the sum of the frequency of the various codes indicating the average frequency of occurrence. From the tables, it can be seen that controllable latency can be achieved by selecting a proper value for m. For instance, in Table 1, if m is chosen as 3, the average latency is much smaller than that corresponding to m equal to 2. We also note that due to the nonuniform usage of the various

node size	frequency of m-out-of-n code								
	0	1	2	3	4	5	6	7	8
17	2	4	5	6	5	0	0	0	0
12	0	0	0	3	3	1	1	0	0
10	0	1	1	1	2	1	0	0	0
20	2	1	5	13	6	2	0	0	0
23	2	1	6	15	6	2	0	0	0
12	0	1	1	1	3	2	0	0	0
12	0	1	1	3	2	1	0	0	0
8	0	0	3	5	1	1	1	0	0
9	0	0	4	6	2	1	1	0	0
sum	6	9	26	53	30	11	3	0	0

Table 1: Error latency in sequential code – Modula-2 program.

node size	frequency of m-out-of-n code								
	0	1	2	3	4	5	6	7	8
11	0	1	1	1	4	3	0	0	0
13	0	0	4	6	3	2	0	0	0
9	0	0	0	1	1	1	0	0	0
8	0	0	1	1	1	1	0	0	0
10	0	0	1	1	2	2	0	0	0
7	0	0	2	2	3	1	1	0	0
6	0	0	2	2	2	2	0	0	0
16	0	0	1	2	2	10	0	0	0
5	0	1	0	0	1	1	0	0	0
sum	0	2	12	16	19	23	1	0	0

Table 2: Error latency in sequential code – Pascal program.

node size	frequency of m-out-of-n code								
	0	1	2	3	4	5	6	7	8
11	0	1	1	1	4	3	0	0	0
13	0	0	4	6	3	2	0	0	0
14	0	0	1	3	6	3	0	1	0
8	0	0	1	1	1	1	0	0	0
10	0	0	1	1	2	2	0	0	0
16	0	0	1	2	2	10	0	0	0
10	0	0	0	1	2	0	0	0	0
sum	0	1	9	15	20	21	0	1	0

Table 3: Error latency in sequential code – C program.

instructions in practical programs (biased usage), the lowest latency does not occur when $m = n/2$. In all the examples considered, the lowest average latency occurs in the neighborhood of the mid-value.

It may be noted that the problems existing in other known techniques for handling precompiled libraries, interrupts and system calls continue to exist in the proposed signaturing scheme also. In summary, our technique has several good features: simplicity in the hardware, low and flexible error latency, minimum run-time performance degradation and low memory overhead. Low memory overhead is achieved by not requiring to store any reference signatures unlike all existing signature techniques. This is one of the unique features of our technique. Flexible latency and low run-time performance degradation will make it attractive as an error detection tool in critical systems employing fault tolerance.

References

[1] A. Mahmood and E. McCluskey, "Concurrent error detection using watchdog processors - A survey," *IEEE Transactions on Computers*, vol. 37, pp. 160–174, February 1988.

[2] T. Sridhar and S. Thatte, "Concurrent checking of program flow in VLSI processors," *Proceedings of the International Test Conference*, pp. 191–199, 1982.

[3] J. Sosnowski, "Detection of control flow errors using signature and checking instructions," *Proc. 18th International test Conference*, pp. 81–88, 1988.

[4] J. Eifert and J. Shen, "Processor monitoring using asynchronous signatured instruction streams," *Proc. 14th Fault Tolerant Computing Symposium*, pp. 394–399, 1984.

[5] M. Namjoo, "Cerberus-16: An architecture for general purpose watchdog processor," *Proceedings of the 13th Fault Tolerant Computing Symposium*, pp. 17–20, June 1983.

[6] M. Namjoo, ""Techniques for concurrent testing of VLSI processor operation," *Proc. International Test Conference*, pp. 461–468, November 1982.

[7] M. Schuette and J. Shen, "Processor control flow monitoring using signatured instruction streams," *IEEE Transactions on Computers*, vol. C-36, pp. 264–276, March 1987.

[8] K. Wilken and J. Shen, "Embedded signature monitoring: Analysis and technique," *Proc. 17th International Test Conference*, pp. 324–333, 1987.

[9] K. Wilken and J. Shen, "Continuous signature monitoring: Efficient concurrent detection of processor control errors," *Proc. 18th International Test Conference*, pp. 914–925, 1988.

[10] K. Wilken and J. Shen, "Continuous signature monitoring: Low cost concurrent detection of processor control errors," *IEEE Transactions on Computer-Aided Design of Integrated Circuits and Systems*, vol. 9, no. 6, pp. 629–641, 1990.

[11] N. Saxena and E. McCluskey, "Control-flow checking using watchdog assists and extended precision checksums," *IEEE Transactions on Computers*, vol. 39, pp. 554–559, April 1990.

[12] A. Mahmood and E. McCluskey, "Watchdog processors: Error coverage and overhead," *Dig. 15th Annual Int. Symp. Fault Tolerant Computing*, pp. 214–219, 1985.

Author Index